Undergraduate Lecture N

For further volumes:
http://www.springer.com/series/8917

Undergraduate Lecture Notes in Physics (ULNP) publishes authoritative texts covering topics throughout pure and applied physics. Each title in the series is suitable as a basis for undergraduate instruction, typically containing practice problems, worked examples, chapter summaries, and suggestions for further reading.

ULNP titles must provide at least one of the following:

- An exceptionally clear and concise treatment of a standard undergraduate subject.
- A solid undergraduate-level introduction to a graduate, advanced, or non-standard subject.
- A novel perspective or an unusual approach to teaching a subject.

ULNP especially encourages new, original, and idiosyncratic approaches to physics teaching at the undergraduate level.

The purpose of ULNP is to provide intriguing, absorbing books that will continue to be the reader's preferred reference throughout their academic career.

Series Editors

Neil Ashby
Professor, Professor Emeritus, University of Colorado Boulder, CO, USA

Professor, William Brantley
Furman University, Greenville, SC, USA

Michael Fowler
Professor, University of Virginia, Charlottesville, VA, USA

Michael Inglis
Associate Professor, SUNY Suffolk County Community College, Selden, NY, USA

Elena Sassi
Professor, University of Naples Federico II, Naples, Italy

Helmy Sherif
Professor, University of Alberta, Edmonton, AB, Canada

Douglas W. MacDougal

Newton's Gravity

An Introductory Guide to the Mechanics
of the Universe

Springer

Douglas W. MacDougal
Portland, Oregon, USA

ISSN 2192-4791 ISSN 2192-4805 (electronic)
ISBN 978-1-4614-5443-4 ISBN 978-1-4614-5444-1 (eBook)
DOI 10.1007/978-1-4614-5444-1
Springer New York Heidelberg Dordrecht London

Library of Congress Control Number: 2012951423

© Springer Science+Business Media New York 2012
This work is subject to copyright. All rights are reserved by the Publisher, whether the whole or part of the material is concerned, specifically the rights of translation, reprinting, reuse of illustrations, recitation, broadcasting, reproduction on microfilms or in any other physical way, and transmission or information storage and retrieval, electronic adaptation, computer software, or by similar or dissimilar methodology now known or hereafter developed. Exempted from this legal reservation are brief excerpts in connection with reviews or scholarly analysis or material supplied specifically for the purpose of being entered and executed on a computer system, for exclusive use by the purchaser of the work. Duplication of this publication or parts thereof is permitted only under the provisions of the Copyright Law of the Publisher's location, in its current version, and permission for use must always be obtained from Springer. Permissions for use may be obtained through RightsLink at the Copyright Clearance Center. Violations are liable to prosecution under the respective Copyright Law.
The use of general descriptive names, registered names, trademarks, service marks, etc. in this publication does not imply, even in the absence of a specific statement, that such names are exempt from the relevant protective laws and regulations and therefore free for general use.
While the advice and information in this book are believed to be true and accurate at the date of publication, neither the authors nor the editors nor the publisher can accept any legal responsibility for any errors or omissions that may be made. The publisher makes no warranty, express or implied, with respect to the material contained herein.

Springer is part of Springer Science+Business Media (www.springer.com)

*To my Mom and Dad and my family
Kathy, Colin and Emily, Aidan and Ella,
and Ben and Sam*

Frontispiece photo of Comet Lovejoy C/2011 W3, December 24, 2011, courtesy of James Tse from Christchurch, New Zealand.

Preface

My father showed me a comet through his binoculars after dinner one day when I was 6. I saw it suspended in space, hung among the stars of the Big Dipper, a white, ghostly wisp from somewhere distant. I knew the planets moved around the Sun. I had heard about comets. But to actually *see* one was breathtaking, different. It was not a picture. It was *there*. It was silent and mysterious, seemingly from another time. Years later, when I was a middle-school student in Florida, our school librarian displayed a copy of Newton's *Principia* prominently on a stand in the library. It was laid open to some pages of intriguing, complex-looking geometrical drawings, including a dramatic illustration of a comet. I had no real comprehension of the strange book's contents, but was drawn to flipping through its pages every time I passed by it. In there, I had been told, was the first real explanation of how things in the sky moved. I later learned how the riches in that book, first published in the late 1680s, really did open the door to understanding celestial phenomena. It was a revelation to discover that things I saw in the sky could be known in a completely new way, through the language of mathematics.

The universe we inhabit can be known on many levels even to those who have never seen a country-dark sky. No words can adequately convey the ethereal majesty of comet Hale-Bopp's twin tails seen from the pitch-dark skies of Haleakala, in Maui, but the mathematics of its motion are also fascinating and beautiful, and last well beyond the actual experience. The motions of the bodies in our solar system are a phenomenon of a more abstract kind, most accessible through the figures and operations of mathematics. One can try to describe the movements of the celestial bodies in language, and many have written wonderfully poetic descriptions of heavenly phenomena. But to begin to understand how it all *works*, one needs to penetrate into the core of things, into masses, forces, and accelerations, which describe quantitatively how bodies affect each other even over staggeringly huge distances. It is not widely appreciated that with modest familiarity with the tools of high-school mathematics, one may gain surprisingly clear insight into the mysteries of celestial motions. It is a deep and rich world of sublime, subtle relationships. Some of them, like Kepler's Harmonic Law, are especially beautiful, echoing the harmonies in the physical world. With the language of mathematics, one can more deeply appreciate the workings of creation in a way unknown to those who do not take the trouble to learn it. After a while, the relatively few key equations become as familiar as old friends. And one does not have to take anyone else's word for how it all fits together: it *does*. One can see for oneself how gravity works and bodies move by doing the manipulations and the calculations. It is the next best thing to being there!

It is one purpose of this book to convey the power of simple mathematics to tell fundamental things about nature. Many people, for example, know the tides are caused by the pull of the Moon and to a lesser extent the Sun. But very few can explain exactly how and why that happens. Fewer still can calculate the actual pulls of the Moon and Sun on the oceans. The book attempts to show this with simple tools. The book endeavors to cross disciplines to provide context – history, astronomy, physics and mathematics – and effort has been made to explain things frequently passed over or taken for granted in other books. It does not purport to be a comprehensive textbook or tome on every aspect of classical celestial mechanics. Rather, it samples key areas of interest and invites further inquiry. The emphasis is on intuitive appreciation rather than rigor. The book hopefully will lead readers to investigate the fundamentals of mass and motion on their own, and to puzzle through the problems that Newton and others faced in trying to make sense of why things move as they do. It is also hoped that the book will encourage a sense of wonder at the beauty of the physical world and an appreciation of the brilliant minds who struggled to comprehend and express it in almost equally beautiful mathematics.

The focus of the book is Newton's, rather than Einstein's, gravity. In other words, it deals with classical mechanics, which originated in the seventeenth century and which remains the basis for the core problems of celestial mechanics today. It therefore does not treat the curved spacetime of Einstein's General Theory of Relativity. There is nothing in the book about the behavior of masses at relativistic speeds, black hole physics, or other aspects of Einstein's geometrical view of gravity.

The book has three main characteristics that define it:

- First, it concentrates strongly on the historical development of the mathematics and science of orbital motion, beginning with Galileo, Huygens, Kepler, and Newton, each of whom is prominently represented. Quotes and problems from Galileo's *Dialogs Concerning Two New Sciences*, Huygens's *The Pendulum Clock*, and particularly Newton's *Principia* should help the reader get a little bit inside the minds of those thinkers and see the problems as they saw them, and experience their concise and typically eloquent writing.
- Second, it is problem based: it uses concrete, hopefully interesting problems and case studies to teach and illustrate. This method is critical for a hands-on understanding of this topic. Many of the problems use actual historical data, and results are compared with those obtained using modern data and methods. To underscore the relevancy of the original thinking on these issues, modern problems dealing with near Earth asteroids, NASA missions, and newly

discovered dwarf planets are set next to historical problems that deal with the same mathematical or physical principles. Emphasis is on problems with dramatic interest and power of illustration.

Third, its mathematics is the simplest possible. The math is generally at the high-school or early college level (algebra and the most basic geometry and a little trigonometry), with detailed explanations of the methods needed to solve the problems and understand the concepts. Calculus is the standard method for presenting this subject in most textbooks, and it produces quick, concise results. But it does not necessarily follow that calculus is needed for those results, or that a calculus-based presentation is the most intuitive vehicle for a beginner learning the fundamentals. I use methods and derivations that have appeared to be the most intuitively comprehensible, with stress on practical application. The surprise is how deeply one can dive with only the most basic mathematics and an intuitive grasp of the physics. This having been said, certain fundamental pre-calculus concepts involving limits introduced by Newton in his *Principia* are dealt with in this book, and should help one who has not had calculus get a foothold on the subject.

I wish to acknowledge two people who were the guiding lights for me in working on this book. The first is my father, who, when he was alive, always encouraged my interest in science, even to the point of driving me on my birthdays when I was a young boy to any observatory of my choice in the West. The other is my wife, Kathy, whose has given total support for this project even as it emerged from preparing casual collections of class notes to the far greater and more taxing commitment of a book. I am also very grateful for her editorial good judgment on the random questions that I would pose to her.

I also wish to thank the editors at Springer for their excellent editorial suggestions that have helped add clarity in many places in the book.

Portland, Oregon, USA   Douglas W. MacDougal

About the Author

Douglas W. MacDougal has a degree in mathematics, with a minor in physics, from the University of Vermont, and is an adjunct professor at Portland State University teaching celestial mechanics and (previously) astronomy. For many years he has also taught courses in astronomy and mathematics in Portland's Saturday Academy, whose classes typically include gifted middle-school and high-school students. MacDougal, a lifelong amateur astronomer, teaches every summer at the Oregon Museum of Science and Industry (OMSI) Astronomy Camp in the Oregon desert.

Contents

1 **Introduction: The Twin Mysteries of Mass**............... 1
   The Quality of Inertia.. 1
      Straight-Line Motion..................................... 2
      Forces in Combination................................... 3
      The Independence of the Effects of Perpendicular
      Forces on Straight-Line Motion......................... 4
      Inertial Frames of Reference and Relative Motion........ 5
   The Attraction of Masses.................................... 7
      The Force of Gravity: Falling........................... 7
      The Attractive Force of Gravity......................... 8
      Gravity and Distance.................................... 9
      Newton's Universal Law of Gravitation................... 9
      Gravitational Acceleration.............................. 10
      Gravitation and the Laws of Motion..................... 11
   Gravity, Inertia, and Curvilinear Motion in Space............ 13
   Reference.. 15

2 **Galileo's Great Discovery: How Things Fall**.............. 17
   The Distance a Thing Falls.................................. 17
   The Meaning of Constant Acceleration........................ 20
      Graphing Velocity and Time in Uniformly
      Accelerated Motion..................................... 23
      If Galileo Only Knew Calculus: A Quick Look at
      Instantaneous Velocity................................. 27
      Using Limits to Find the Acceleration
      of a Freely-Falling Body............................... 31
   A Summary of Galilean Equations............................ 33
      Notes on Using the Law of Conservation
      of Mechanical Energy to Derive the Galilean
      Distance Equation...................................... 33
   Reference.. 36

3 Christian Huygens' Remarkable Pendulum 37
Of Planes and Pendulums: A Historical Sketch 37
How Gravitational Acceleration Was Measured
with a Pendulum 42
   Determining How Far a Body Falls in One Second
   from the Length of a Pendulum 43
Deriving an Equation for the (Approximate)
Period of a Pendulum 46
The Pendulum's Satisfying Coincidence
with Circular Motion 53
References 58

4 The Geometry of the Solar System: Kepler's Laws
of Planetary Motion 59
The Basic Geometry of the Ellipse 64
   Determining the Eccentricity of the Moon's Orbit 67
   Deriving Kepler's Law of Areas 70
   Determining the Semi-major Axis
   of the Asteroid Ceres 73
   How Conic Sections May Be Generated
   by One Equation 74
   Graphing the Orbit of Comet Schwassmann-Wachmann
   3 Using Rectangular Coordinates 78
References 82

5 How the Moon Falls Toward the Earth
(but Keeps Missing It) 83
The Facts of Inertia: Newton's First Law 84
What Is Centripetal Acceleration? 87
Comparing Centripetal Accelerations in Different Orbits 90
   Proving Principia's Proposition IV: The Proportionality
   of Centripetal Forces in Circular Orbits 92
Determining if Newton's Centripetal Acceleration
for Circular Orbits Is Consistent with the Galileo's
Distance-Time Squared Rule for Falling Bodies 95
Deriving Galileo's Equation Geometrically from Newton's
Equation for Centripetal Acceleration 97
Reflections Upon Centripetal Acceleration 99
Inertial ("Centrifugal") Force 103
A Useful Notation for Circular Motion 103
References 106

6 Newton's Moon Test 107
Summary of Newton's Moon Test 109
The First Part of the Test 109
The Second Part of the Test 110

Contents

  The Third Part of the Test.................................... 110
  The Fourth Part of the Test................................... 110
   Newton's Demonstration that "The Moon
   Gravitates Towards the Earth".............................. 111
   A Simple Confirmation of the Inverse Square Law
   Between Earth and Moon Using Modern Data................. 116
  A Geometric Approximation of the Moon's Period
  (and Other Diversions).. 118
  A Pendulum in Space... 122
  References.. 125

**7 Newton Demonstrates How an Inverse Square Law Could
Explain Planetary Motions**.. 127
  Clearing the Mathematical Path to the Inverse
  Square Law.. 127
   And Conversely... 132
   Summary of Above Corollaries to Proposition IV............... 133
   A Small Lesson... 133
   Of Circles and Ellipses..................................... 133
  Evidence from Distant Worlds: The Moons of Jupiter
  and Saturn Obey Kepler's Third Law, and Therefore
  the Inverse Square Law....................................... 134
   The Motion of Jupiter's Satellites as a Test
   for the Universality of the Inverse Square Law................. 135
  Finding the Keplerian Proportionality Constant
  in the Jovian Satellite System................................. 138
   Using a Modern Jovian Proportionality Constant to Find
   the Periodic Time of Jupiter's Inner Satellite Amalthea........... 140
   Applying the Proportionality Constant in Kepler's Third
   Law for the Earth–Moon System to Find the Distance
   to a Geosynchronous Satellite............................... 142
  References.. 146

8 Newton's Master Stroke: The Universal Law of Gravitation....... 147
  Working with Forces: Deriving the Inverse Square Law
  from Newton's Second Law.................................... 149
  Constructing Newton's Gravity Equation......................... 150
  Reflections upon the Equilibrium of Gravitational and Inertial
  (Centrifugal) Forces of a Mass in Orbit......................... 155
  Kepler's Third Law as Modified by Newton...................... 158
  Why, After Newton, It Became Evident that Kepler's Proportions
  Are Not Strictly Satisfied for Large Secondary Masses............. 161
   The Strict Proportionality of Kepler's Third Law................ 161
   The Deviation from Strict Keplerian Proportionality............. 162
   Comparing Periods with and Without the Secondary Mass........ 163

	Example: The Significant Consequences of Disregarding the Mass of Jupiter	164
	Reference	165
9	**Gravity on This and Other Worlds**	167
	How Planetary Radius and Density Affect g	167
	All the Mass of a Body May Be Regarded as Concentrated in a Single Point	169
	The Earth Spins	169
	Observed g and True g	170
	Newton's Analysis of the Effect of the Earth's Spin on Its Gravitational Pull	171
	The Earth Is Not a Perfect Sphere	175
	Finding the True Value of g at the Equator	177
	Using g to Derive Velocities in Circular Orbits	179
	Using g to Derive Periods in Circular Orbits	181
	Comparing Velocities and Periods at Increasing Distances from Earth	183
	Calculating the Lunar g: Gravitational Acceleration on the Surface of the Moon	185
	Calculating the Solar g: Gravitational Acceleration of the Sun	187
	References	191
10	**A Binary System Close to Home: How the Moon and Earth Orbit Each Other**	193
	Locating the Center of Mass	196
	Finding the Earth–Moon Barycenter	197
	Deriving Equations for Velocities Around the Center of Mass in Circular Binary Orbits	199
	Calculating the Orbital Velocities of the Earth and Moon Around Their Center of Mass	204
	The Relations Between Masses, Accelerations and Period in Circular Binary Orbits	205
	Reference	211
11	**Using Kepler's Third Law to Find the Masses of Stars and Planets**	213
	The Once Unknown Mass of Pluto	213
	Using Kepler's Third Law to Find Mass	215
	Ratios of Ratios, All Is Ratios: The Historical Use of Kepler's Third Law to Determine Mass by Proportions	216
	How Newton Estimated Planetary Masses	218
	Newton's Calculation of the Mass of Jupiter	220

"Weighing" Jupiter the Modern Way 223
  Calculating the Combined Masses of Quaoar
  and Its Satellite Weywot 225
Manipulating Units to Simplify Equations 227
  Reducing Kepler's Third Law to Its Simplest Form 228
  Finding the Combined Masses of Sirius and Its Companion 230
  Determining the Individual Masses in the Double
  Star System Alpha Centauri 231
What Is the Mass of Our Galaxy? (And Observations
on Dark Matter) 234
References 238

12 Motion in Elliptical Orbits 239
Velocity Along the Apsides of Elliptical Orbits 239
The Orbital Velocities of Earth Around the Sun 242
Comparing Circular and Elliptical Orbits 242
Angular Velocity in Circular and Elliptical Orbits 244
Gravitation and Elliptical Orbits 245
The Orbital Velocities of Mars 246
Kepler Revisited 248
Finding the Velocity of an Artificial Satellite
in Earth Orbit When Just Its Perigee
and Apogee Are Known 249
Deriving Apsidal Velocities in an Elliptical Orbit
from the Laws of Conservation of Energy
and Momentum 251
The Gaussian Constant in Celestial Mechanics 258
  Using the Gaussian Constant to Find Heliocentric Periods 259
  Applying the Gaussian Constant to Find Heliocentric
  Orbital Velocities 260
  Simple Computation of Apsidal Velocities of Objects
  in Heliocentric Orbits 262
  Exploring Sedna's Orbit 262

13 The Energy and Geometry of Orbits 267
The Total Energy in a Circular Orbit 268
The Total Energy in an Elliptical Orbit 269
Velocity Anywhere Along the Elliptical Orbit 271
The Very Particular Parabolic Orbit
and the Velocity of Escape 272
Hyperbolic Trajectories 274
Summary of Orbital Energy Relationships 274
  Determining the Velocity of a Near Earth
  Asteroid Which Passed by Earth 278
  Newton, Halley and the Great Comet of 1680 281
References 288

14 Introduction to Spaceflight 289
Using a Hohmann Transfer to Achieve
a Geostationary Orbit 290
Designing an Orbit for a Lunar Mission 293
Planning a Mission to Mars 299
Calculating the Velocity Needed for the Trip to Mars 300
Using the Energy Equations to Define
the Orbit to Mars 303
    Summary of Some Key Energy-Derived Equations 304
Notes on the Apollo 11 Moon Mission 306
Reference 312

15 Getting Oriented: The Sun, the Earth
and the Ecliptic Plane 313
Visualizing the Heliocentric Elements
of an Elliptical Orbit 313
Schematic of Selected Elements of the Orbits of Mars,
Jupiter and Saturn 317
The Heliocentric Longitude of a Body 319
Summary of the Elements of the Heliocentric Orbit 319
Method for Determining Earth's Heliocentric Longitude 321
    Finding the Earth's Heliocentric Longitude
    at Close Encounter with Asteroid 2010 TD54 322
Calculating the Mean Anomaly and Heliocentric
Longitude of the Near Earth Asteroid 2010 TD54 324
Graphing the Inner Orbit of Comet West C/1975 V1-A 327
References 335

16 An Introduction to Kepler's Problem: Finding the True
Anomaly of an Orbiting Body 337
The True Anomaly Is the Body's Actual Position in Orbit 337
The Limitations of Using Mean Daily Motion
to Find Position in an Elliptical Orbit 339
Kepler's Auxiliary Circle: Determining the True Anomaly
from the Eccentric Anomaly 339
Kepler's Equation: Finding the Eccentric Anomaly
from the Mean Anomaly 341
    Determining the Eccentric Anomaly of Near
    Earth Asteroid 2010 TD54 at the Time of Its Closest
    Approach to Earth 341
    Finding the True Anomaly of Near Earth Asteroid 2010
    TD54 at the Time of Its Closest Approach to Earth 343
    Asteroid 2007 WD5's Passed by Mars: How Close
    Did It Come? 347
References 354

17	**What Causes the Tides?**	355
	Calculating the Differential Gravitational Forces Exerted by the Moon on the Earth	358
	Deriving a General Equation for Determining Tidal Forces	360
	Visualizing Tidal Pull by Comparing Theoretical and Actual Orbital Velocities at Opposite Points on Earth	365
	Estimating the Lifting Force of the Moon	367
	References	375
18	**Moons, Rings, and the Ripping Force of Tides**	377
	Deriving an Equation for the Roche Limit and Applying It to the Earth–Moon System	378
	Finding the Roche Limit for Saturn and Its Innermost Satellite Pan	384
	Comet Shoemaker-Levy 9's Fatal Encounter with Jupiter	388
	How Comet Lovejoy C/2011 W3 Barely Survived the Solar Furnace	391
	References	395
19	**Hovering in Space: Those Mysterious Lagrangian Points**	397
	Deriving Equations for Finding the L1 and L2 Points in the Earth–Sun System	399
	Determining the Accelerations on a Satellite at the Sun–Earth L1 Point	404
	Calculating the Heliocentric Orbital Velocities of a Spacecraft at the Sun–Earth L1 and L2 Points	408
	Where the Fictional "Counter-Earth" Would Be: The L3 Point in the Earth–Sun System	410
	Space Station Parking Spot for Lunar Exploration: Determining the L1 Point in the Earth–Moon System	411
	The Equilateral L4 and L5 Lagrangian Points in the Earth–Moon System	413
	References	418
Appendix: Solutions to Problems		419
Index		427

Chapter 1
Introduction: The Twin Mysteries of Mass

The Quality of Inertia

Imagine a mass of any size, such as a bowling ball, bike, or boat, and a place where there are no obstacles or inducements to its movement. In this perfectly level place there is no friction to slow it, and no wind to propel it. Under such ideal conditions, any material thing will resist movement solely in proportion to its mass. Even in our non-ideal world where friction exists, we perceive this relationship. We cannot flick a bowling ball into motion with our fingers. To start someone on a swing takes an initial hard push, but then it is easier. We cannot even slide a large and slippery block of ice across the floor without some effort to get it going. Its heftiness alone makes it hard to move. As experience shows, the larger the object, the greater is the required effort. This resistance to movement has the name *inertia*. Likewise when a thing is moving it can be hard to stop, and the faster it moves the tougher it is to brake it. We feel the hard work needed to slow a gently rolling car, and the impact on our hands of catching a fast-pitched ball can be painful. The lighter the object, the easier it is to slow it down. It is easier to slow a drifting canoe than a large boat. These too are the effects of inertia. To move or stop anything, then, takes effort, or *force*. It takes force to overcome inertia to get something to move. But once moving, it takes force to stop it. Any *change of motion* requires force to make it happen: force in the direction of motion or braking. To go from stop; to stop from go: change in motion is called *acceleration*. Acceleration means change of speed. To drive from 10 mph to 20 requires acceleration. Once I am at that speed, I am no longer accelerating. When I brake to slow back to 10 mph, or stop, I am *decelerating*. Or I could say I have negative acceleration. So it is right to say that I encounter inertia when I accelerate or decelerate. The engine has to power up, to apply force to the wheels, to get me going, to accelerate me. The brake exerts a strong force of resistance on the wheels and car to slow me down, more so on a heavy car than a bike.

The most perfect expressions of the classical concepts of inertia are found in the famous three laws of motion of Isaac Newton (1642–1727), which we will discuss in Chap. 5. If a thing is moving, it will tend to keep moving; if it is stopped, it tends to stay stopped (the gist of his First Law). And the amount of force it takes to move (or stop) a thing – to accelerate or decelerate it where friction is ignored, depends *only* on its mass (the essence of his Second law). In fact, if one pushes on a mass, it so resists the push of my hand as to push my hand back with equal force (his Third Law). These are the inherent, inertial qualities of *mass*, so far as we know, everywhere in the universe.

Mathematically, the Second Law is expressed like this: $F = ma$, where force equals mass times acceleration. In words, it takes force to accelerate a mass. To achieve a certain acceleration, the quantity of force needed to overcome inertia depends only on the amount of mass being accelerated. Again, to go anywhere from rest (that is, to accelerate), a force must be applied in the direction of motion. Increasing the force in a given direction increases the velocity – accelerates it – in that direction until the force is no longer applied, whereupon (in a place void of resistance) it will continue at the velocity it had when the force ceased. A constant force in one direction will yield a constant acceleration in that direction: indeed, in that circumstance, the velocity will continue to increase – it will keep gaining speed – in direct proportion to the time. Similarly, force in the opposite direction will cause acceleration in the opposite direction, such as the case of the Apollo lunar landers decelerating before touchdown on the Moon.

Straight-Line Motion

Once inertia is overcome and a mass is induced by some force to move, the mass will continue to move *forever* in a straight line unless something changes its state. We are used to air resistance slowing things down. But in space, a mass in motion will continue in motion in a straight line at the same velocity it started with unless something – another mass or masses (large, small or particulate), slows it down or speeds it up. This again is the nature of inertia. The Voyager and Pioneer spacecraft, launched decades ago, are still sailing silently through space. They glide without power, and long ago left our solar system. They will go forever in a straight line, unless eons from now impacts or the attractions of other bodies deflect that course. The concept of eternal straight-line motion is completely different from Aristotle's idea of motion. He believed that a thing continues to move until it loses its "impetus." A projectile would therefore fall abruptly to Earth when it suddenly lost its impetus during its flight. One can see his influence in oddly asymmetric diagrams of cannonball trajectories in books on military artillery written two millennia later, in the early 1600s.

Aristotle of course had no concept of outer space or vacuums. But through experiment and insightful analysis of the motion of rolling balls, Galileo Galilei was close to discovering Newton's First Law. He knew that rolling balls starting from equal heights and falling along smooth planes *at any angle* would reach the bottom of the slope with the same velocity.

Ball rolling down and up planes of identical height

No matter the inclination of the plane, a ball released from the same height would attain the same velocity at the bottom. Moreover, under these theoretical friction-free conditions, he deduced that the ball would rise to the same place at the top of another plane set at the same angle. Galileo was reinforced in these conclusions by his observation of pendulums, which he noticed swung to the same height on the other side of the swing no matter how far the pendulum is pulled back, save only a small portion it might lose from friction in passing through air.

The next step is to imagine the second plane slanted more toward the horizontal on each try. The ball, still tending to attain its original height (in a theoretical friction-free environment), will roll and roll to whatever distance necessary until it does. Finally, make the second ramp horizontal, and the ball will roll forever.

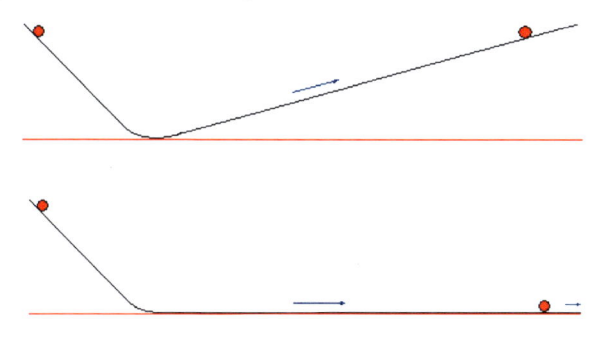

Ball without friction continuing forever

Galileo did not know about the laws of conservation of energy and momentum that would be formulated in the future, but his intuition was keen. It was left to Newton to realize that the inertial tendency of a body in motion to remain in motion was what kept the moons and planets in their orbits.

Forces in Combination

Several forces may be instantaneously impressed on a single body, and the *resultant* force, which is the vector sum of all the forces combined, will still result in straight-line motion. A sailboat courses in a direction that is the resultant path of wind in one direction and the resistance of the keel in another. Newton's Corollary I to his Laws states: "A body, acted on by two forces simultaneously, will describe the diagonal of a parallelogram in the same time as it would describe the sides by those forces separately." Newton showed this principle using the diagram below, which is here reproduced just as Newton drew it, except for the arrows on the vectors. Assume a body is at *A*. Suppose in a given time, a force M acting alone would carry it from *A* to *B*, and force N acting alone would carry it from *A* to *C*. Acting together, the resultant of the two forces would in the same time carry it from *A* to *D*.

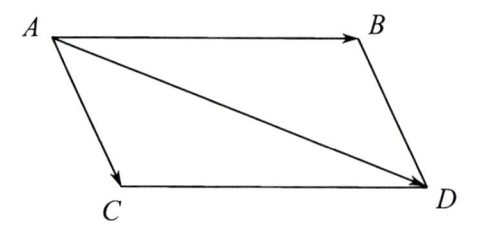

Combination of forces along AB and AC
Resultant force is along AD

Newton reasoned that, since motion arises *only* in the direction of the applied force, the force N in the direction of *AC* will not alter the velocity generated by the other force M in the direction of *AB*. So the body moved by force M along *AB* will arrive somewhere along the line *BD* at the same time regardless of whether force N is impressed or not. Likewise the body moved by force N will arrive somewhere along *CD* at the same time regardless of whether force M is impressed or not. Therefore the body will be found at *D* where the lines meet. "But it will move in a right line from *A* to *D*, by Law I."

It follows from this argument that a force O along the diagonal *AD* can be *resolved* into forces along *AB* and *AC*, or along any perpendicular axes.

The Independence of the Effects of Perpendicular Forces on Straight-Line Motion

Suppose you are on a cliff with two rocks in your hands. You manage to drop a rock off the cliff at the same time as you throw the other rock horizontally, such that they leave your hands from exactly the same height from the ground below.

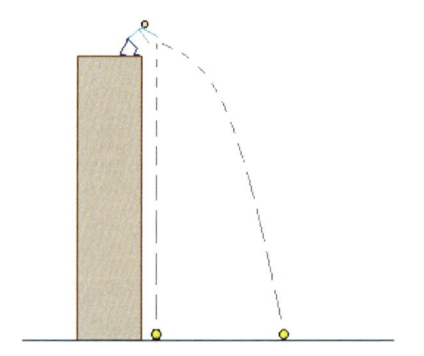

The independence of horizontal and vertical motions

One rock goes straight down, the other goes out. Assuming no friction, which will hit the ground first? As students in every physics class are taught, they hit at the same time. This is because the motion of the rock in the vertical direction is independent of the motion of the rock (and the throwing force that induced that motion) in the horizontal direction.

We can generalize this principle. Imagine a bowling ball travelling in uniform (unaccelerated) motion through space, where there is no friction of any kind to slow it down. By Newton's First Law, if it encounters no other forces on its way, it will travel forever in a straight line. Now suppose you are somehow able to arrive in your space suit and give it a shove *at right angles* to its motion. Because you pushed it perpendicularly, the ball is not speeded up or slowed down. By this we say that no component of your pushing force can be resolved along its axis of motion to accelerate or retard it. But the perpendicular shove does deflect it off its course. Consistent with Newton's Second Law, the ball will respond to the shove by going in a straight line only in the direction of the applied force. *In that perpendicular direction, the ball was originally at rest*, even though in the other direction, along the axis of motion, it was not. Your push did not alter its motion in the original direction of travel. The shoving force only affected the ball *in the direction of the shove*. The resulting direction of travel for the ball will be the composite of the two motions. If the shove made the ball go perpendicularly at the same velocity as the original ball was travelling, the resultant motion will be along the 45° diagonal between them. *Forces and resulting motions along an axis in three-dimensional space are completely independent of forces and resulting motions along any other axis.* So long as no component of the force is along the original axis, the original velocity will be unchanged. This independence of forces with respect to each axis in three-dimensional space was a principle insisted upon by Galileo to his skeptical friends.

Inertial Frames of Reference and Relative Motion

Again suppose the bowling ball is moving past you through space in its original state of constant velocity. Now you are somehow able to hop on the bowling ball and travel with it. Assume there is nothing around you to tell if you are moving or not. From your perspective, the bowling ball will now appear to be at rest. Without acceleration, you don't know if you are sitting still or riding along on it at constant speed in some direction. It seems like it is at rest. Your friend now gives you and the ball a shove perpendicularly to the right. From your perspective, his push will appear to be an initial acceleration from rest to a certain velocity off to the right. You will feel that acceleration (as you feel acceleration when an elevator goes up or down) but once he stops pushing, you will again perceive you are at rest. Unless there is some acceleration (or external reference point) to inform you if you are being shoved to the right or pushed to the left, or kicked from behind or slowed down, everything seems to be at rest. This is called an *inertial frame of reference*. It only exists in situations of constant velocity (or rest) when there is no acceleration in any direction acting upon you. This helps explain why forces act completely independently along each of the three-dimensional axes of space. Any mass, moving or not, that is not accelerated may be deemed in motion or at rest in some frame of reference, and along some axis. Any applied force on that mass will move it from that rest frame of reference to another inertial frame of reference which will persist when the force ceases. The application of Newton's Laws to that mass in any inertial frame of reference will yield results consistent with the application of those laws in any other inertial frame of reference.

Textbooks often state that "acceleration is the change of velocity or direction of an object." But in fact they are the same, because *any change of direction of motion of an object is a change in velocity* along one or more axes in your chosen frame of reference. If a particle is moving with uniform velocity along an x axis, and a small force displaces it perpendicularly along the y axis, it will then have motions along both the x and y axes: the velocity along the x axis will be unaffected (due to the independence of axial motions, by Newton's Second Law), but the velocity along the y axis will increase from zero to its new velocity in the y direction. This change in velocity is of course acceleration, but along one axis only. Any change of velocity may be seen as a change along one or more axes of motion, independent of the motion along any axis where no force has been applied, just as in the case above of two rocks falling and thrown outward from a cliff.

Consider these principles with respect to any given mass particle or object, where the mass is at rest or in uniform, rectilinear motion (that is, not being accelerated):

- Any three-dimensional coordinate system can be constructed in any orientation with the object its origin. That coordinate system will be the frame of reference for that object.
- Any object in uniform motion can be made to appear at rest by adopting the three-dimensional coordinate system as the frame of reference for such mass, with the mass at its origin.
- Any axis of a three-dimensional coordinate system can be aligned along the path of uniform motion of a mass. In such case, the mass will be at rest with respect to the other two axes.

Where the velocity of the mass changes (accelerates), we can add these principles:

- Any change of direction of a mass is a change in velocity along one or more axes of a three-dimensional coordinate system. Any such change in velocity of the mass is called acceleration in that reference system.
- Any change, with respect to a mass, of the velocity of the three-dimensional coordinate system itself, such that the mass appears to be changing velocity along one or more axes of a three-dimensional coordinate system, is also called acceleration of the mass with respect to that reference system.
- A frame of reference that accelerates in tune with physical objects in a given location that are subject to gravity and no other forces is called a *free-falling frame of reference*.

Full exploration of the implications of these principles is beyond the scope of this book. Suffice to say, it was through the deep pondering of these ideas of inertial and accelerated frames of reference that Einstein derived his theories of relativity.[1] Keeping these principles in mind will assist in understanding many aspects of motion and forces.

The Attraction of Masses

The Force of Gravity: Falling

Another observable characteristic of masses is that they have *weight*. No bodies in our terrestrial experience have no weight. They want to gravitate toward the Earth. If dropped, they *fall*. The felt weight is a force pulling down which, as with inertia, becomes greater as the mass increases. Boulders have more mass than marbles, so they weigh more. Rockets weigh more than birds, and take more force to get them off the ground. The weights of objects at equal distances from the center of the Earth in fact depend only on the quantity of mass. Why don't heavier objects, with more mass, fall faster? It is because the inertias of the objects resisting their downward movement also depend on their quantities of mass – heavier objects being also harder to move. The greater the mass, the greater is the gravitational force making it fall, but greater too is its inertia resisting its movement. Any advantage of weight for a swifter fall is exactly offset by a corresponding disadvantage of inertia resisting the fall. The result is that, assuming no air resistance, all objects fall at the same rate, as Galileo observed. Another way to look at it is to consider that the gravitational force on each particle of mass is resisted by the inertia inherent in each particle of mass. Given this rule, it makes no difference how big or small the mass is, everything falls at the same rate. This is true whether you consider a single particle, a cluster of particles or a conglomeration of millions of particles. This may be made clearer if you imagine a collection of particles of equal mass falling next to each other. They arrive at the ground at the same time. Now imagine that they are tied to each other with thin threads, or even so close that they may be adhered to each other, and form an aggregate mass of all the single particles. They still fall at the same time, regardless of the horizontal component of being adhered to each other in a solid mass. If you take away the resistance of air, and they fall from the same height, a marble will hit the ground at the same time as a boulder, a rocket or a bird.

The rule may be simply stated: the forces pulling bodies down are proportional to their masses; but because their inertias are also proportional to their masses, the accelerations each mass experiences will be the same. This is true on Earth, the Moon and every other planet. This applies when the resistance of the atmosphere is neglected, which otherwise slows things unequally, depending on their shape.

---

[1] See, for example, Peter G. Bergmann [1] for a clear explication of these concepts.

The Attractive Force of Gravity

It is the phenomenon of gravity that each particle of mass attracts every other particle of mass. The force of attraction of two bodies on one another is the product of their masses. The Earth pulls on a 2 kg rock with twice the force that it pulls on a 1 kg rock. The whole gravitational pull of the Earth is essentially doubled up on the rock weighing twice as much as the other rock, and weighs twice as much on a scale. A 1,000 kg boulder experiences a 1000 times more force than does a 1 kg rock. We could increase the rock's size without limit and the Earth's pull on it would also increase without limit. The gravitational attraction of mass is cumulative. It grows and grows as more and more matter is gathered.

This can be made clearer if we imagine two equal particles of the smallest size you may wish to consider, each attracting each other. We might represent their mutual attraction like this:

Two particles of mass mutually attracting

We could think of the line as a *line of force* between the two particles, pulling on each mass equally. Now let's picture three particles of mass on the right and four particles on the left, tightly clustered on each side so they touch (but expanded here for the sake of clarity of illustration) so their internal attractions to each other are irrelevant. Now diagram all the attractions of each side to each.

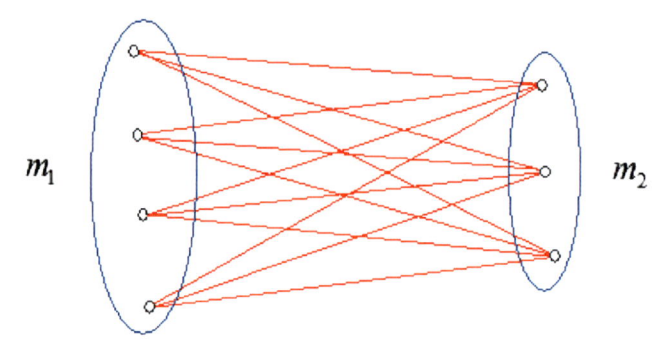

The mutual pull of gravity on each particle of mass

Count the lines of force. If each of the four particles attracts each of the three particles, we can assume that the total forces are four times three, or twelve times the force of a single pair of particles alone. The diagram shows 12 connecting lines of force among the particles. If the cluster of four particles on the left are called one mass and labeled m_1, and the cluster of three particles on the right are called m_2, their mutual gravitational attraction would be proportional to the *product* of the masses, or $m_1 m_2$. We can expand this notion to include hundreds, thousands millions, billions or more particles in the left-hand mass, and no particle in the right-hand mass would be diminished or diluted by having to "connect" with more and more on the left. We would not say "The lines of force for that particle have been all used up." As long as their distance from each group of masses remains the same, there would be no dilution of any line of force, no matter how many lines we "burden" each particle with. We could add particles upon particles of mass on either side, or both, and there would be no impairment of any individual line of force. In other words, the capability for attraction of each particle to any number of others appears, quite mysteriously to our intuition, to be unlimited.

One could also think of the above example in a three by four matrix, with each little force line represented by the symbol $f_{x,y}$:

$m2$	3	$f_{1,3}$	$f_{2,3}$	$f_{3,3}$	$f_{4,3}$
	2	$f_{1,2}$	$f_{2,2}$	$f_{3,2}$	$f_{4,2}$
	1	$f_{1,1}$	$f_{2,1}$	$f_{3,1}$	$f_{4,1}$
		1	2	3	4

$m1$

The total of all the forces is the total number of elements in the matrix, which again is the product of column by row; that is, the total of all the individual masses. Note too that the attractions of the particles to each other on each side, whether or not they are in contact, do not diminish the gravitational attraction represented by the lines of force to m_2. The mutual attraction of each particle of the 5.97×10^{24} kg comprising the Earth's mass in no way detracts from the attraction of those particles to other masses. Similarly, each grain of interstellar dust and molecule of gas coalescing from a cloud to become a star is capable of attracting every other grain and molecule, and themselves together attracting other clouds of gas and dust.

Gravity and Distance

The strength of the gravitational force between two particles is dependent on the particles' distance apart. If particles of mass are moved farther away from each other, their mutual gravitational attraction will diminish, not linearly, but more so. It diminishes in a special way: twice the distance, one-fourth the force; three times the distance, one ninth the force. This is an *inverse square* relationship, and so far as we know, it holds to the ends of the universe.

Newton's Universal Law of Gravitation

To summarize the ideas of attraction as the product of the masses, and strength of that attraction varying as the inverse square of their distance apart, we can use a concise mathematical symbolism:

$$F \propto \frac{m_1 m_2}{r^2}$$

The equation tells us that the force F between two objects is proportional to the product of their masses m_1 and m_2 and inversely as the distance r between them squared. To actually calculate the force, in newtons, the units of distance and mass must be given in meters and kilograms (if we use the mks system), and a constant must be employed to make the units come out right. Physicists use G, the *gravitational constant*, which has been found (in the mks system) to be 6.674×10^{-11}. Using this constant, the full equation of Newton's universal law of gravitation becomes,

$$F = \frac{Gm_1m_2}{r^2}$$

Gravitational Acceleration

Suppose for the sake of simplicity we have four particles in m_1 and only one particle in m_2. The lines of force in our example now look like the diagram below. The diagram represents the gravitational force exerted on a single "unit mass" by the particles of mass in m_1. It is no longer 12 lines of force but 4. The secondary mass has been stripped down to one, and we have isolated the "grabbing potential" of the primary mass m_1 alone. This is a convenient way of acknowledging the amount of gravitational acceleration induced on any object (of any mass) by the mass m_1.

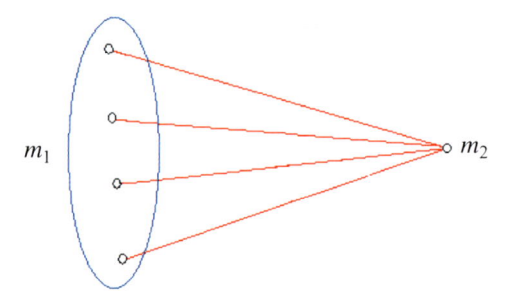

Gravitational force on a unit mass
(gravitational acceleration)

Since m_2 is unity, we have done the mathematical equivalent of dividing each side of the above equation by m_2. We could therefore write the above equation this way:

$$\frac{F}{m_2} = \frac{Gm_1}{r^2}$$

But since by Newton's Second Law ($F = mf$) acceleration (here denoted by f) is force per unit mass ($f = F/m$), we make this substitution on the left side of the equation and it becomes,

$$f = \frac{Gm_1}{r^2}$$

Because there is no secondary mass m_2 in the equation, it should be apparent from this equation that the gravitational acceleration acting upon *any* mass falling to Earth from the same height will be the same, be it an elephant or egg. Hence (again) they will hit the ground at the same time.

Gravitation and the Laws of Motion

The physical interpretation of the above result involves equating the gravitational and inertial forces on a given mass. Hold a rock m_2 in your hand. By Newton's Third Law, the gravitational force of Earth, of large mass m_1, creates an equal resisting force on the mass in the opposite direction. The resisting force of inertia on this second mass, whose state of rest wants not to change, can be expressed by Newton's Second Law. The two forces are therefore,

$$F = \frac{Gm_1 m_2}{r^2} \qquad\qquad F = m_2 f$$

Force of gravity pulling the mass down (and the Earth up) Force of inertia resisting any change of velocity

Observe that the mass m_2 appears in each equation. If it has a large value in one equation, it is just as large in the other. However massive and heavy the rock m_2 may be, the gravitational pull on it in the left-hand equation will be exactly offset by the force of inertia resisting that pull, in the right-hand equation. Again, that is why (in spaces devoid of friction) objects of any mass in the same gravitational field fall at the same rate. The quality of mass that is both resistive to acceleration (inertia) and a source of gravitational attraction is called the *principle of equivalence*.

Setting the forces equal to each other as Newton's Third Law requires, the m_2 terms cancel, and we again find the acceleration induced on any mass by mass m_1 at distance r:

$$f = \frac{Gm_1}{r^2}$$

In summary, the Earth's gravity will set up an *acceleration* in any mass whatsoever, whose change of motion is always resisted by the equivalent inertial capacity of that mass, be it a feather, a stone, the Moon, a distant planet or the sea. This acceleration is thus independent of the mass pulled; it depends *only* upon the Earth's mass and the distance of the thing from the center of the Earth. It will be a *continuous* acceleration. This follows from the continuous application of the constant pull of Earth's gravity on it at every moment. Drop a stone from any height and it will accelerate all the way to the ground, going initially from a state of rest to a state of motion downward, and increasing its velocity at every instant.

This acceleration, while continuous, is not the same at every distance. The acceleration equation tells us that. Because acceleration is force per unit mass, gravitational acceleration naturally diminishes with distance just the way gravitational force diminishes. The acceleration of the Earth's gravity at the surface of the Earth, at "one g" (where r is one Earth radii), calculated from the above equation, is about 9.8 m/s, every second: a falling object gains 9.8 m/s of velocity in every second of its fall. As we go up and away from the Earth, the gravitational acceleration diminishes rapidly by the inverse square rule. At two Earth radii, the gravitational pull of the Earth is one fourth g. At the Moon, which is distant from us by about 60 Earth radii, the inverse square weakening of the Earth's gravitational acceleration has taken its toll. The gain in velocity every second is only one sixtieth this value squared, $1/60^2$, or about 2.7 *mm*, which is what our Moon experiences.

An object achieving *escape velocity* will journey fast enough to forever leave the gravity of the body it left behind. It "leaves" gravity by achieving such distance from the mother planet (or satellite, or star) that the holding power of its gravitational force becomes ineffectual to stop it. Mathematically, the rapidly increasing denominator r squared makes the value of acceleration f ultimately shrink to relative insignificance. A baseball batted off a small asteroid could possibly achieve that velocity; if so, it would fly forever in a straight line. But if it does not achieve the necessary escape velocity, it would be bent around by gravity and return, ultimately orbiting or colliding with the body it sought to leave. The simple equation telling how acceleration relates to escape velocity will be discussed later.

Given the idea of gravitational acceleration, which will be the same on all masses large or small at the same distance from the center of mass of Earth, what then is weight? In the context of Newtonian gravitation, it is the *force* of gravity pulling on the mass. And, from Newton's Second Law, the gravitational force needed to accelerate a mass is equal to the mass acted upon times that acceleration:

$$f = \frac{Gm_1}{r^2} \Big\} \text{ Gravitational acceleration induced by Earth, } m_1$$

$$F = m_2 f \} \text{ Force resisting any such acceleration on mass } m_2$$

The acceleration in the first equation is now substituted into the second equation:

$$F = m_2 \left(\frac{Gm_1}{r^2}\right) \Big\} \text{ Weight of } m_2$$

And the latter expression is of course Newton's gravitation equation:

$$F = \frac{Gm_1m_2}{r^2}$$

It is not always easy to understand that the acceleration due to gravity, which acts upon the rock held at rest in your hand, and is perceived as weight, is the same kind of acceleration needed to change the velocity of an object, from rest or otherwise. Both involve force. And if it weren't for the chemical energy in your muscles resisting the fall of the rock, its velocity would indeed change.

Gravity, Inertia, and Curvilinear Motion in Space

A concise description of the balance of forces that must exist to keep the planets from falling into the Sun was given in the preface to Newton's *The System of the World* (or *Treatise on the System of the World*), published in 1728.[2] The likely author of the preface was Newton's translator, Andrew Motte. He wrote:

> Since then the Planets by radii drawn to the Sun describe areas proportional to the times, it follows that the compound force which keeps them in their regular motions according to that law, is compounded of two forces; one which impels them according to the tangents of their orbits, and the other impels them toward the Sun; the first of which we may call the Projectile, the other the Centripetal Force.

According to Motte, all orbital motion, in simple two-body Keplerian mechanics, may be understood as the composition of those two forces: the tangential, "projectile" force, and the centripetal, inward-pulling force. In reality, however, the first "force" is nothing more than the simple resistance (inertia) to being compelled away from its Newtonian straight-line motion. The resistance arises from the fact that moving objects tend to move in a straight line, forever, at constant velocity unless acted upon by a force (such as friction) which changes that. The idea of "projectile" motion, however, accurately conveys the *rectilinear* nature of that motion. Were there nothing to slow down or change the path of a cannonball shot from a high mountain, it would proceed forever in a straight path at the same velocity.

The force in space that changes that projectile motion is the gravitational force from another body, pulling the object into a *curvilinear* path. That force, as Newton first showed, is directed toward one focus of an elliptical orbit, and to the center of a circular orbit. The effect of the inward pulling force on a rotating object is again

---

[2] The work is believed to be an earlier and more accessible draft of what eventually became Book III the *Mathematical Principles of Natural Philosophy* or *Principia*. Sir Isaac Newton, *Mathematical Principles of Natural Philosophy*, p. 395 (Translation by Andrew Motte 1729, revised by F. Cajori, Berkeley: University of California Press 1949). All excerpts from the *Principia* discussed in this book are drawn from this translation.

a resistance to going in a curve. It is sometimes popularly called inertial force, or centrifugal force, familiar to us when we go around a sharp curve in a fast moving car, and lunge toward the left side of the car. Centrifugal force again is not an actual force pulling an object outward, but is the product of *inertia* – the "want" of an object to go in a straight line.[3] Newton's Second Law ($F = mf$) tells us that the more massive the object is, the harder it is to pull it off its straight, projectile motion. It is said to have more inertia.

Forces can combine upon a mass to create a motion that looks like a continuous curve. But as Newton showed, that path may be viewed as a composite of instantaneous straight-line motions. The Moon both falls toward Earth and goes around it. It falls toward the Earth's center directly like a dropped stone, at 2.7 mm/s per second acceleration. But by inertia it also tends to continue in a straight line tangent to its orbit – at right angles to the straight line of its fall. The combination of these two motions traces a path that is the curve of its orbit. Were it not for its centripetal motion, toward Earth, it would continue its projectile motion forever in a straight line, at the speed it had from its origins. But it is accelerated toward the Earth – it is falling.

Let us imagine the gravitational force of Earth as occurring in pulses, repeated in the smallest instants of time you can conceive. At every instant, Earth's gravity tugs the Moon away from its straight line, rectilinear motion. At that instant, it moves along a new tangent line. And in the next instant, with the next pulse of gravity, it is tugged again to yet another straight line, which has no "memory" of where it was before – the new straight line motion could continue for a fraction of a second or a million years; but in fact the pull of gravity tugs it to a new straight line again and again and again. Each pulse is a fresh tug, causing a fresh fall away from the straight line, into another straight line. But at the lunar distance, the tug is not so great as to pull it into the Earth, nor is the Moon's velocity so great as to allow it to escape. All is in balance. Now imagine this process repeated in almost infinitesimally small time intervals, with the straight lines becoming shorter and shorter, so that the little, infinitely small, saw tooth patterns of this motion smooth out to create a path lasting over the whole course of the Moon's orbit.

All the mathematics for two-body problems is a simple combination of the description of the gravitational forces and inertial, rectilinear motion; the combination of the gravitational force of a mass pulling an object inward toward it, and the would-rather-keep-going-in-a-straight-line inertial tendency of a mass to resist

---

[3] The real forces acting on an orbiting object are the inward, gravitational force and the inertial tendency to move in a straight-line direction tangent to the orbit. The resistance to the deflection off the straight line course is due to the inertia of the moving mass, not some new force pulling it outward. It happens whenever a reference frame is rotating, causing thereby a circular acceleration. For this reason, centrifugal force is often called a "fictitious force."

the inward pull. These attributes of mass and motion are expressed in the two families of equations which when intermixed produce Kepler's laws as one of their many progeny. And it was Newton who, relying upon the great minds before him, showed this comprehensively in his *Principia*, mostly with the aid of geometry.

Reference

1. Bergmann PG (1968) The riddle of gravitation. Charles Scribner's Sons, New York

Chapter 2
Galileo's Great Discovery: How Things Fall

Galileo Galilei (1564–1642), the famous Italian mathematician at the leading edge of the scientific revolution that was to sweep Europe, was curious about motion. He was an experimentalist who for the first time had the insight and talent to link theory with experiment. He rolled balls down an inclined plane in order to see how things fell toward the Earth. He discovered in this way that objects of any weight fell toward the Earth at the same rate – that they had a uniform acceleration. He surmised that if they fell in a vacuum, where there was no air resistance to slow some objects more than others, even a feather and a cannon ball would descend at the same rate and reach the ground at the same time. He also explored the motion of pendulums and other phenomena. He is perhaps most famous for his 1610 telescopic discoveries of the moving moons of Jupiter, the phases of Venus, and the craters of the moon, all of which convinced him, against the ages-old wisdom of Aristotle and of the Catholic Church, of the rightness of the Copernican heliocentric view of the solar system.

In his investigations of motion, Galileo was the first clearly to understand that the forces acting upon objects could be broken into *independent components*; that a thrown stone had a force pulling it down as well as the force throwing it horizontally outward. These insights would be of great use to Isaac Newton, born the year Galileo died, in devising the calculus and his universal laws of gravity and motion.

The Distance a Thing Falls

Galileo was interested in understanding how things moved and fell. What laws of motion governed them? Determining physical laws from experiment was completely new in early Seventeenth Century Italy, but Galileo was intellectually adventurous enough to try. One of Galileo's most famous experiments, in 1604, was his inclined plane experiment, where he measured the distance a ball rolled down a ramp in each unit of time. Since forces in different directions act independently, he could time the descent of the ball and learn how forces act on the ball as if it were only moving in the down direction. From that he could deduce how a freely falling object would move.

Why did he use a ramp to measure fall? It is because with the limited technical means then available to Galileo, he could not possibly have timed, with any reasonable accuracy, the rapid descent of a vertically dropped ball. Using an inclined plane allowed him to dilute the force of gravity and slow the ball down so he could time it with a water clock, where he could then compare the weight of water that poured out before and after each event. This told him the time intervals elapsed during the ball's fall down the ramp, which were not necessarily the equivalent of the seconds of a modern clock.[1] Galileo repeated this experiment many times to help remove some of the subjectivity in his measurements and thereby gain greater accuracy of the result.

Galileo's idea of using an inclined plane to accurately measure free fall as noted took advantage of his insight that forces act independently in each dimension. As we saw in Chap. 1, the downward, vertical force on the rolling ball (the gravitational force) can be analyzed separately from the horizontal force that moves it laterally along the plane. The movement down the plane is the result of the *combination* of the downward gravitational force on the ball and the force of the plane on the ball, acting perpendicular or normal to the plane, resisting it and deflecting it horizontally. Recall our example of a sailboat where the resultant path is the combination of the forces of wind on the sail and resisting water on the keel. Studying the downward *component* of the forces acting on the rolling ball, then, is equivalent (if we disregard the frictional forces acting on the ball) to studying the force drawing down a freely falling object, where all the forces act downward, in

---

[1] It is quite possible that Galileo initially timed the rolling ball by means of musical beats and later confirmed their accuracy by means of the water clock. Galileo was musically inclined, accomplished on the lute, and his father and brother were musicians. Setting adjustable gut "frets" on the inclined plane would enable the ball to make an audible bumping sound as it passed over the frets. Adjusting the spacing of these frets so the bumps occurred at exactly even intervals, according to his internal sense of rhythm, could easily have been done. Indeed, that method was likely far more accurate than any clocks of the day, which could not measure times shorter than a second. This idea was advanced by the late Stillman Drake, Canadian historian of science and Galileo expert. See Stillman Drake, "The Role of Music in Galileo's Experiments", *Scientific American*, June, 1975. The important thing is that the time intervals be deemed to be equal, whatever those intervals may be:

> The phrase "measure time" makes us think at once of some standard unit such as the astronomical second. Galileo could not measure time with that kind of accuracy. His mathematical physics was based entirely on ratios, not on standard units as such. In order to compare ratios of times it is necessary only to divide time equally; it is not necessary to name the units, let alone measure them in seconds. The conductor of an orchestra, moving his baton, divides time evenly with great precision over long periods without thinking of seconds or any other standard unit. He maintains a certain even beat according to an internal rhythm, and he can divide that beat in half again with an accuracy rivaling that of any mechanical instrument. Ibid., 98.

the y direction. This was a fundamental intuition, and helped lay the conceptual foundation for Newton's work on the action of forces, and the concept of vectors.

Using the ramp, however, came with a price. Any object falling through the atmosphere will experience friction, whose effects will vary with the weight and shape of the object. Using an inclined plane introduces a whole new element of friction into the experiment. Galileo tried to minimize it by covering the plane with parchment. Air resistance on a heavy, slow-moving ball would probably have been negligible. Energy went into rotating the mass, though, and with the remaining friction the ball would descend at some fraction of the speed with which it would roll freely on a frictionless surface. But in this experiment Galileo was seeking only proportions: the relationship between time and distance of fall. Even if the ball was retarded by some unknown amount of friction, the relationship between distance and time after it got going should be affected more or less equally in each interval.

Galileo's famous *Dialog Concerning Two New Sciences*[2] of 1638 is his rich and delightful inquiry into fundamental physical questions about the strength of materials and motion. It is highly readable, even after more than three and a half centuries. Here is Galileo's description of his inclined plane experiment, stating his findings regarding the relation between distance and the time of fall:

> A piece of wooden moulding or scantling, about 12 cubits long, half a cubit wide, and three finger-breadths thick, was taken; on its edge was cut a channel a little more than one finger in breadth; having made this groove very straight, smooth, and polished, and having lined it with parchment, also as smooth and polished as possible, we rolled along it a hard, smooth, and very round bronze ball. Having placed this board in a sloping position, by lifting one end some one or two cubits above the other, we rolled the ball, as I was just saying, along the channel, noting, in a manner presently to be described, the time required to make the descent. We ... now rolled the ball only one-quarter the length of the channel; and having measured the time of its descent, we found it precisely one-half of the former. Next we tried other distances, comparing the time for the whole length with that for the half, or with that for two-thirds, or three-fourths, or indeed for any fraction; in such experiments, repeated a full hundred times, *we always found that the spaces traversed were to each other as the squares of the times, and this was true for all inclinations of the plane*, i.e., of the channel, along which we rolled the ball.[3]

We can confirm this mathematical relationship found by Galileo between the distance the ball went and its time of descent. Below are some of Galileo's measured times and distances for the rolling ball as it progressed down the inclined plane. Galileo used a measuring unit called points, each of which was about 29/30 mm. As shown on the chart, the ball rolled about 2104 × 29/30 mm or about 2,034 mm (2.34 m), less than 7 ft, in eight intervals (again, not necessarily seconds):

---

[2] *Dialog Concerning Two New Sciences* (translated by Henry Crew and Alfonso de Salvio, Macmillan 1914). This classic translation is also available online: http://galileoandeinstein.physics.virginia.edu/tns_draft/index.html. See also Hawking [1] (Short title) *Dialogs*, which uses the same translation.

[3] *Dialogs*, 136–37 (my italics).

Time (equal intervals)	Distance (in points)	Distance divided by 33
1	33	1.00
2	130	3.94
3	298	9.03
4	526	15.94
5	824	24.97
6	1,192	36.12
7	1,620	49.09
8	2,104	63.76

The last column we added to show the multiples of the initial distance. The idea is to try to intuit a pattern in the numbers, by dividing the initial distances by the first distance number (33) and seeing if the other numbers are multiples of it. It is apparent that, after dividing each distance (which we will call s) by 33, the last column increases approximately as the square of the time. Indeed it does not take a great deal of rounding of this experimental data to make the relationship just that, as Galileo saw. The exact relationship is confirmed by many repeated observations. It can be expressed by a simple equation:

$$t^2 = \frac{s}{k}$$

Where k is 33, hence,

$$s = kt^2$$

We have not explained what the k means in this equation. It is a "proportionality constant" whose value determines the distance in the equation, in units of our choice. As we shall see, the constant is a constant of acceleration, the gradual incremental addition to velocity in the presence of a force inducing it.

The Meaning of Constant Acceleration

To understand uniformly accelerated motion, we shall return to the man who articulated it in simple terms. In his *Dialogs*, Galileo reasoned that constant acceleration implied steady, incremental additions of velocity evenly in proportion to time:

> When, therefore, I observe a stone initially at rest falling from an elevated position and continually acquiring new increments of speed, why should I not believe that such increases take place in a manner which is exceedingly simple and rather obvious to everybody? If now we examine the matter carefully we find no addition or increment more simple than that which repeats itself always in the same manner. This we readily understand when we consider the intimate relationship between time and motion; for just as uniformity of motion is defined by and conceived through equal times and equal spaces (thus we call a

motion uniform when equal distances are traversed during equal time-intervals), so also we may, in a similar manner, through equal time-intervals, conceive additions of speed as taking place without complication; *thus we may picture to our mind a motion as uniformly and continuously accelerated when, during any equal intervals of time whatever, equal increments of speed are given to it.* Thus if any equal intervals of time whatever have elapsed, counting from the time at which the moving body left its position of rest and began to descend, the amount of speed acquired during the first two time-intervals will be double that acquired during the first time-interval alone; so the amount added during three of these time-intervals will be treble; and that in four, quadruple that of the first time interval. To put the matter more clearly, if a body were to continue its motion with the same speed which it had acquired during the first time-interval and were to retain this same uniform speed, then its motion would be twice as slow as that which it would have if its velocity had been acquired during two time intervals.[4]

In other words, in *uniformly* accelerated motion, the velocity will be proportional to time,

$$v \propto t$$

The proportionality constant to convert the proportion $v \propto t$ into an equation must be the amount by which increases in velocity occur steadily with time. This is the definition of constant acceleration. In a freely falling object, acceleration in falling due to the Earth's gravitational field is known from experiment (and theory) to be 9.8 m/s per second.[5] Hence for such cases the equation may be written $v = 9.8\ t$.

To make this clearer, suppose a rocket is boosting a probe into deep space at a constant acceleration of 10 m/s, every second. Constant acceleration means constant increase in velocity – in heaps of 10 m/s velocity in every new second.

Time in seconds	Increments of velocity (m/s)	Total velocity (m/s)
1	10	10
2	10 + 10	20
3	20 + 10	30
4	30 + 10	40
5	40 + 10	50

Every second begins with the velocity the object had at the end of the previous second. The incremental addition to velocity in each second – the constant by which the time is multiplied – is 10, which is the probe's constant acceleration. Where acceleration (again symbolized by a) is constant, the equation for velocity is,

$$v = at$$

---

[4] *Dialogs*, 123 (my italics).

[5] The acceleration of a falling body near the surface of the Earth due to gravity is 9.8 m/s per second, written usually as 9.8 m/s^2. This means it gains 9.8 m/s in velocity each second of its fall. As noted in Chap. 1, this value for acceleration, or "g" as it is called, will diminish as we move farther away from the Earth's surface (diminishing, in fact, with the square of the increasing distance from the center of the Earth).

The *distance* the spacecraft will go in a given time at this rate of velocity increase is also some function of time. Let's see what it is. If an object moves at *constant velocity* (zero acceleration), the distance it travels is,

$$s = vt$$

where s is our symbol for distance. But the velocity in our case keeps increasing every second, at a constant rate. For example, it goes 10 m/s in the first second and 20 m/s in the next second. So the overall velocity during these 2 s will not be 10 or 20 but the *average* of them, or 15 m/s. In 2 s, it will actually traverse 30 m. Given an average velocity $\bar{v}$, the distance equation becomes,

$$s = \bar{v}t$$

The distance the spacecraft travels will equal its average velocity times time. But is there a simpler way of computing average velocity when we know acceleration? To compute the average of first two velocities, as we did above, we mentally did this:

$$\bar{v} = \frac{v_2 - v_1}{2}$$

To compute the average velocity over a span of time, we need to know how many seconds elapsed, from the time the motion started, t_s to the time it ended, t_e. Since velocity will in each case be equal to acceleration times time, the above equation becomes,

$$\bar{v} = \frac{at_e - at_s}{2}$$

Or, to simplify, assuming the start time is zero seconds, letting t just be elapsed time,

$$\bar{v} = \frac{1}{2}at$$

So in our example the average velocity in 5 s will be 25 m/s. How far will it go? To find out, substitute the above equation for average velocity into the distance equation, $s = \bar{v}t$:

$$s = \frac{1}{2}at^2$$

In 5 s the spacecraft in our example with constant acceleration of 10 m/s, every second (expressed as 10 m/s^2) will therefore travel (½ × 10 × 5^2 =) 125 m.

This is the relationship between distance and uniformly accelerated motion which we saw was discovered by Galileo from his experiments with rolling balls. The k in the relation $s = kt^2$ must therefore equal half the acceleration: $k = ½a$, where the constant acceleration g (of 9.8 m/s^2) was furnished by the Earth's gravity. Rewriting the equation specific to Earth's gravitational acceleration we have,

$$s = \frac{1}{2}gt^2$$

Notice that there is no horizontal component to this equation. The vertical distance is, again, independent of any additional thrust, push, shove or shot in a direction perpendicular to line of descent. This equation applies to the little man we encountered on the cliff in Chap. 1 dropping and throwing the rocks. If it takes 3 s for the rocks to reach the ground, the balls were dropped from about 44 m (almost 145 ft):

$$s = \frac{1}{2}(9.8)(3)^2$$

$$s = 44.1 \text{ meters}$$

Graphing Velocity and Time in Uniformly Accelerated Motion

Problem Show *graphically* that the distance traveled by a *freely falling* object under uniform gravitational acceleration equals half the velocity times the time of the fall. In doing so, use the value of the Earth's gravitational acceleration as the uniform acceleration constant.

Given

$v = at$	Velocity in uniformly accelerated motion
$a = 9.8 \text{ m/s}^2$	The acceleration due to Earth's gravity at its surface

Assumptions We will assume that the object moves without resistance, and disregard any slowing effects of friction, as if it were falling in a vacuum.

Method By inspection of the equation $v = 9.8t$, one can see that it must be a straight line, with a slope (or rate of change of s/t) of 9.8 and an s intercept (on the y axis) of 0. This tells us that the velocity increases at a constant rate with time. The change of velocity with time is by definition acceleration. Velocity is changing at each instant, so to capture the true $s = vt$ relationship for every changing increment of time, we need to find the total area under the graph.

Calculations The relationship between time (in seconds) and the ever-increasing velocity during fall is an upwardly sloping straight line, apparent on the graph below.

It is evident that a falling object speeds up with a *constant* or uniform acceleration, with a *velocity proportional to time of the object's descent*. That the slope is 9.8 to 1 means that the velocity increases 9.8 m/s every second, in the progression,

9.8, 19.6, 29.4, 39.2... That is the "the amount of speed acquired" in each interval of time, where the intervals are 1 s each.

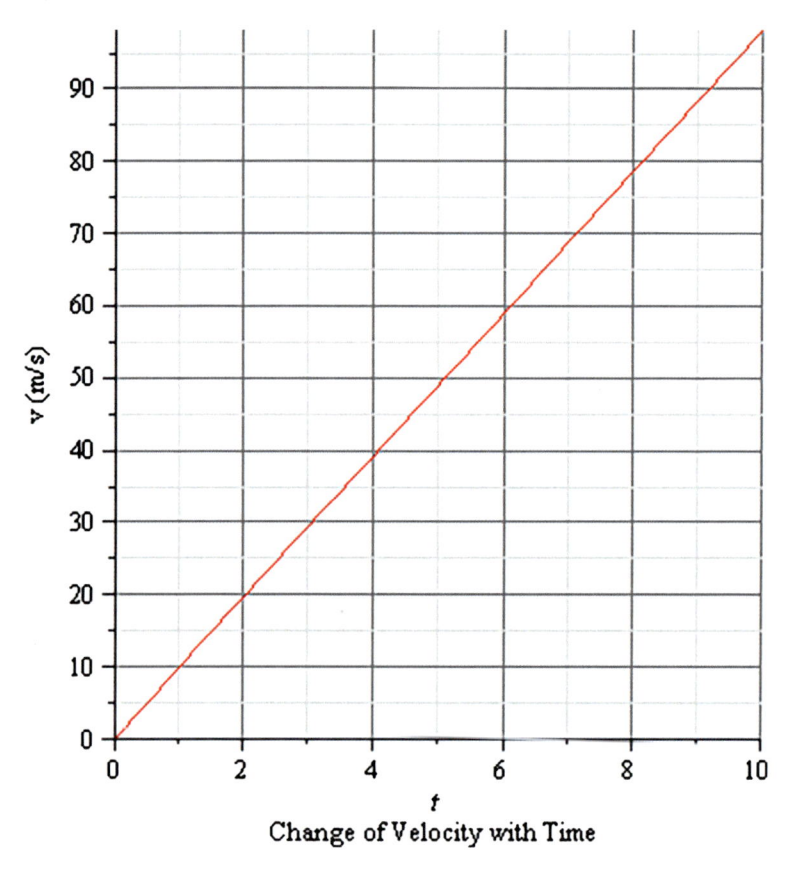

Change of Velocity with Time

Observations

1. How can we find the distance when the time and velocity at each point are known? It may be tempting to conclude that the total distance traveled by the object is found merely by multiplying v times t in the chart, under the assumption that $v = st$. But here velocity is not constant, it is changing. The object is falling at 98 m/s after 10 s – but the total distance it fell is not 980 m. For the first half of the trip down (zero to 5 s), it was falling at *less* than 50 m/s, and for the second half (after 5 s) it was falling at *more* than 50 m/s. So we must multiply the *average* velocity of the voyage by time.[6] Average velocity is just $v_{avg} = (0 + 9.8)/2$, or 49 m/s. This multiplied by the full time of fall of 10 s is 490 m, the

---

[6] Galileo's first theorem in the section of his *Dialogs* titled "Naturally Accelerated Motion" states: "The time in which any space is traversed by a body starting from rest and uniformly accelerated is equal to the time in which that same space would be traversed by the same body moving at a uniform speed whose value is the mean of the highest speed and the speed just before acceleration began." *Dialogs*, 132.

The Meaning of Constant Acceleration

total distance of fall. The equation for finding distance, when the elapsed time and uniformly changing velocity from rest are known, is $s = \frac{1}{2}vt$. This procedure is equivalent to finding the area of the triangle under the sloped line. The area of a triangle is half the base times the height, so

$$s = \frac{1}{2}vt.$$

2. In his investigations of accelerated motion, Galileo utilized both experimental and theoretical methods. Since measuring time precisely and easily was not possible in his day (the water clock as we mentioned above was probably used by him to confirm, rather than discover, these laws), he sought to understand his experimental results by mathematical reasoning, using geometry to illustrate *proportional relationships*.[7] The rolling ball experiment was his attempt to find the proportional relationship between distance traveled and time. This yielded the law that distance was proportional to time squared.[8] Since he had concluded that in uniformly accelerated motion an object's velocity increased in a manner that was proportional to time, he could intuit the correctness of the time-squared law for distance mathematically. Here is an intuitive, non-geometrical shorthand version of his reasoning: The distance covered by a non-accelerating object is proportional to velocity times time. But in a uniformly accelerating body, that velocity is *itself* proportional to time. Hence the distance is proportional to times squared. Expressed symbolically:

$$s \propto vt \text{ in uniform motion; but}$$

$$v \propto t \text{ in a uniformly accelerating body, so}$$

$$s \propto (t)\, t \text{ in a uniformly accelerating body, or}$$

$$s \propto t^2 \quad \leftarrow$$

This result corresponds to the relation Galileo obtained in his rolling ball experiment. Hence he could feel confidence in the truth of his theorem: "The spaces described by a body falling from rest with a uniformly accelerated motion are to each other as the squares of the time-intervals employed in traversing these distances."[9]

3. Galileo's own reasoning is a good example of geometrical thinking in an era where quantifiable results were difficult to come by. Below is his line of argument and diagram proving the same problem we worked through above[10]:

---

[7] The method of finding the limit of a function to quantitatively derive velocity or acceleration was unknown, and lay three-quarters of a century in the future.

[8] *Dialogs*, 133: Theorem II, Proposition II in "Naturally Accelerated Motion."

[9] Ibid.

[10] Galileo is here proving his Theorem I, Proposition I: "The time in which any space is traversed by a body starting from rest and uniformly accelerated is equal to the time in which that same space would be traversed by the same body moving at a uniform speed whose value is the mean of the highest speed and the speed just before acceleration began." *Dialogs*, 132.

Let us represent by the line AB the time in which the space CD is traversed by a body which starts from rest at C and is uniformly accelerated; let the final and highest value of the speed gained during the interval AB be represented by the line EB, drawn at right angles to AB; draw the line AE, then all lines drawn from equidistant points on AB and parallel to BE will represent the increasing values of the speed, beginning with the instant A. Let the point F bisect the line EB; draw FG parallel to BA, and GA parallel to FB, thus forming a parallelogram AGFB which will be equal in area to the triangle AEB, since the side GF bisects the side AE at the point I; for if the parallel lines in the triangle AEB are extended to GI, then the sum of all the parallels contained in the quadrilateral is equal to the sum of those contained in the triangle AEB; for those in the triangle IEF are equal to those contained in the triangle GIA, while those included in the trapezium AIFB are common. Since each and every instant of time in the time-interval AB has its corresponding point on the line AB, from which points parallels drawn in and limited by the triangle AEB represent the increasing values of the growing velocity, and since parallels contained within the rectangle represent the values of a speed which is not increasing, but constant, it appears, in like manner, that the momenta assumed by the moving body may also be represented, in the case of the accelerated motion, by the increasing parallels of the triangle AEB, and, in the case of the uniform motion, by the parallels of the rectangle GB. For, what the momenta may lack in the first part of the accelerated motion (the deficiency of the momenta being represented by the parallels of the triangle AGI) is made up by the momenta represented by the parallels of the triangle IEF. Hence it is clear that equal spaces will be traversed in equal times by two bodies, one of which, starting from rest, moves with a uniform acceleration, while the momentum of the other, moving with uniform speed, is one-half its maximum momentum under accelerated motion. Q.E.D.[11]

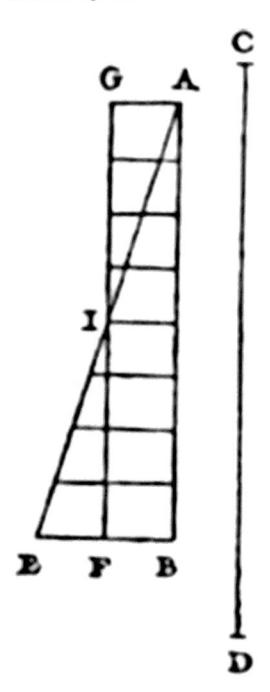

---

[11] Ibid. 132–33.

If Galileo Only Knew Calculus: A Quick Look at Instantaneous Velocity

We noted above that Galileo realized that the forces acting upon objects could be broken into independent components, which we now typically represent as vectors. An object is pulled directly downward by gravity with uniform acceleration whether or not it is flung outward or just dropped. The idea of separating forces and motions into components and analyzing them separately is a method of analysis now commonplace in the study of dynamics. The path of a falling object can be analyzed using a graph in terms of change along the y axis and change along the x axis. The ratio of those quantities, called the slope, tells about the rate of change of the curve at the point in question. Let us look again at the simple case of objects falling – or rolling – freely in a gravitational field. It will help us understand how the ever-changing velocity entailed in accelerated motion ultimately required Newton to explore what happens when segments under investigation are broken down into smaller and smaller parts, which was the heart of his remarkable discovery of the calculus. The principles are best conveyed through the use of a few simple problems.

Problem The actual initial distance a dropped object falls to Earth in the first second is about 4.9 m (16 ft). Using only the distance-time-squared relationship discovered by Galileo, expressed in the equation $s = kt^2$, where the constant $k = 4.9$ (since at $t = 1$, $k = s = 4.9$), and the distance and time data plotted below, derive an equation for the instantaneous velocity of a freely-falling boulder, during any moment of its descent.

Given Below is the distance in meters the freely-falling boulder covers in 10 s.

Time (s)	Time squared	Distance (m) $s = kt^2$
1	1	4.9
2	4	19.6
3	9	44.1
4	16	78.4
5	25	122.5
6	36	176.4
7	49	240.1
8	64	313.6
9	81	396.9
10	100	490

Assumptions We will as before ignore the role air friction plays, and assume it falls in a vacuum toward the Earth.

Method Examining the plot of data below, we note that the rate of descent appears non-uniform. Earlier parts of the curve are shallow, gradually becoming steeper and steeper with increasing time. Steeper means faster, since there is increasingly more change along the y axis – distance – in a unit of time than where the curve is more horizontal. We can see from both charts, too, that the numbers grow faster with each second. This conforms to our intuition, since things do fall faster and faster as they drop, as one can see by looking at a waterfall, where the water descends with increasing rapidity as it nears the bottom.

The *rate* of this descent – the velocity of the object – will differ at each point along the curved graph. How can we find the velocity of our rock at any particular point? We must pick a small place on the curve beginning with that point, illustrated below by small boxes, and find the *slope* of the line there. The slope is the difference in the boxes between the two s points on the y axis (how distance has changed), which is Δs, divided by the difference between the two t points on the x axis (how time has changed), which is Δt. This relationship $\Delta s/\Delta t$, increment of distance per increment of time, is the mean velocity within the box. We can see how this looks on a graph of the 10 s of data:

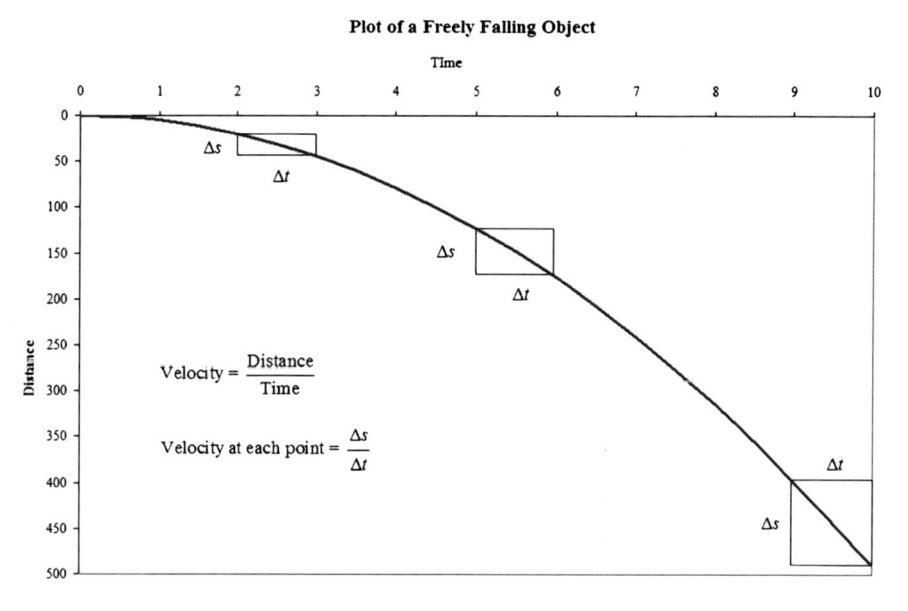

Plot of a Freely Falling Object

$$\text{Velocity} = \frac{\text{Distance}}{\text{Time}}$$

$$\text{Velocity at each point} = \frac{\Delta s}{\Delta t}$$

Little rectangles are drawn at arbitrary times (2, 5 and 9 s) to illustrate that that the slope gets increasingly steep with time. The first rectangle is elongated horizontally, and the second less so, and the last is beginning to be elongated vertically. Since velocity is the ratio of distance to time, we can see the velocity of the rock (and the slope, $\Delta s/\Delta t$) is increasing as time goes on. If we call these boxes Box A, B and C respectively, and compute $\Delta s/\Delta t$ for each, we get the following results:

Box	Time t when entering box	Distance s when entering box	Distance s' when leaving box	Difference in s values: Δs	Δt	Average velocity: $\Delta s/\Delta t$
A	2	19.6	44.1	24.5	1	24.5
B	5	122.5	176.4	53.9	1	53.9
C	9	396.9	490	93.1	1	93.1

The Meaning of Constant Acceleration

The velocity increases dramatically in each box. But what do those represent? They are, again, the *average* velocities of the object within each box. The rock took 1 s to drop through each box, and the length of that drop was the Δs value applicable to each. If Δt is as big as it is, then we must be content with this approximation of velocity at the location of each box. Within the boxes there is an approximation: the curve changes, albeit slightly inside each box, so the slope in each is really not the same throughout the whole box. We still don't know the actual velocity at any particular point. If one were trying to determine exactly the force on a bungee cord, or the speed for a soft landing on the Moon, however, one would need to know the exact, not merely average, velocity. How can we do this? We can try by making Δt smaller. Since Galileo's equation for this curve is known, we can see what happens to s when Δt is added to the equation; specifically, how the equation may change when we reduce the value of that Δt to near zero. This is called finding the *limit* of a function, as Δt *approaches* (but does not actually reach) zero. It is equivalent to reducing the time intervals to ever smaller units, as if the boxes were made smaller and smaller, and seeing how the distance changes in those smaller units of time. If the boxes are made almost *infinitely* small, such that they become virtually *points*, the velocity at those points essentially becomes the *instantaneous* velocity we seek instead of the average velocity.

Calculations The average velocity in any box is given by this relation:

$$\bar{v} = \frac{\Delta s}{\Delta t}$$

We eventually want to find what the instantaneous velocity is as Δt approaches a limit, in this case zero, expressed mathematically by this notation:

$$\lim_{\Delta t \to 0} v = \frac{\Delta s}{\Delta t}$$

Calling the distance at entry into the box s and the distance at exit s', and the small increment of time Δt, the velocity equation before we limit the size of Δt is,

$$v = \frac{s' - s}{\Delta t}$$

Now let us modify Galileo's equation by adding the Δt to the time increment:

$$s = kt^2$$

$$s' = k(t + \Delta t)^2$$

and the equation for velocity in the box becomes,

$$v = \frac{k(t + \Delta t)^2 - kt^2}{\Delta t}$$

This can be expanded, then simplified,

$$v = \frac{k(t^2 + 2t\Delta t + \Delta t^2) - kt^2}{\Delta t}$$

$$v = \frac{k\Delta t(2t + \Delta t)}{\Delta t}$$

$$v = k(2t + \Delta t)$$

From this equation the velocity through any size box can be computed. All we do is adjust the size of Δt and make the box as small as we choose. Now we can see what happens when we make Δt approach zero:

$$\lim_{\Delta t \to 0} v = k(2t + \Delta t)$$

$$v = 2kt$$

which, under our initial assumption that an object falls toward Earth a distance of 4.9 m in the first second (that is, $k = 4.9$) reduces to,

$$v = 9.8t$$

Observations

1. This of course is the familiar $v = at$ equation discussed earlier, where a is the value of acceleration due to gravity near the Earth's surface of 9.8 m/s^2.
2. Inserting the values of 1 at 2, 5 and 9 s, respectively, for Δt in the equation $v = k(2t + \Delta t)$ yields 24.5, 53.9 and 93.1 m/s. These are the same values for average velocity given by the numerical method in the chart above. When we reduce Δx to zero, to find instantaneous velocity, the equation $v = 9.8\,t$ applies. The *instantaneous* velocities at those times are 19.6, 49 and 88.2 m/s, respectively.
3. To see how the initial values of velocity *converge* on these final values as Δt gets smaller, take the middle box as an example, where the object has fallen for 5 s and traveled 122.5 m. Beginning with the equation $v = k(2t + \Delta t)$ with Δt at 1, we reduce the size of Δt by increments before finally bringing it to zero. The average velocity of the object through the box between 5 and 6 s becomes the instantaneous velocity of 49 m/s at exactly 5 s:

Δt	Velocity at 5 s (m/s)
1	53.9
.5	51.45
.25	50.225
.1	49.49
.01	49.049

(continued)

Δt	Velocity at 5 s (m/s)
.001	49.0049
.0001	49.00049
0	49

4. The method of finding the limit of the equation as we shrink Δt is at the heart of the differential calculus, invented independently at about the same time by Englishman Newton and the German Gottfried Wilhelm Leibniz (1646–1716). It sparked a revolution in mathematics and was Newton's prime tool for his seminal analysis of the motion of moving bodies, published in 1687 as *Philosophiae Naturalis Principia Mathematica*, or *The Mathematical Principles of Natural Philosophy*, (often referred to as the *Principia*). The increment Δt (or any such small increment of a value, be it time, velocity, distance, etc.) was called by Newton a "fluxion" and a "differential" by Leibniz. Leibniz's terminology and symbols have survived.

Using Limits to Find the Acceleration of a Freely-Falling Body

We have seen that the velocity of a falling object near the Earth increases at a constant 9.8 m/s^2. That is, the rate of change of velocity (the acceleration) is 9.8 m/s^2. This result we *infer* from the equation for velocity, $v = 9.8\,t$. Let us now use some of the same limit concepts to find it directly.

Problem Derive the acceleration of a freely falling object using the same techniques as were used in the above discussion.

Given

$v = 9.8t$	The equation for velocity of a freely falling object toward Earth
$\lim\limits_{\Delta t \to 0} a = \Delta v / \Delta t$	Acceleration (change of velocity with time) as Δt approaches zero

Assumptions We will again ignore the role air friction plays, and assume the object falls in a vacuum toward the Earth.

Method To assist our analysis, we consider the object's velocity at entry into any arbitrarily small box and the velocity at exit, after a small increment of time has passed. The difference may then be found between the velocity after the small time increment and the velocity before it, as the numerator in the above limit. We then calculate the expression when Δt goes to zero. Graphically, the result is the same as before: Recalling the straight-line graph of velocity (on the y axis) and time (on the x axis) we can again visualize smaller and smaller increments of time, and correspondingly smaller and smaller increments of velocity as a function of time.

Calculations If we let the velocity be v at entry into any arbitrarily small box and the velocity at exit be v', then $\Delta v = v' - v$. The small increment of time we again represent as Δt, so

$$\lim_{\Delta t \to 0} a = \frac{\Delta v}{\Delta t}$$

Since $\Delta v = v' - v$, the acceleration equation before we limit the size of Δt is,

$$a = \frac{v' - v}{\Delta t}$$

Now let us modify the velocity equation by adding the Δt to the time increment:

$$v = 9.8t$$

$$v' = 9.8(t + \Delta t)$$

Substituting these values into the equation for acceleration, the expression becomes,

$$a = \frac{9.8(t + \Delta t) - 9.8t}{\Delta t}$$

$$a = \frac{9.8t + 9.8\Delta t - 9.8t}{\Delta t}$$

Simplifying and dividing by Δt, we find,

$$a = 9.8 \, \text{m/s}^2$$

Observations

1. The approach is called differentiation, and with it we arrive at the same result as we found before: that the acceleration is indeed a constant. The important point is the method, unknown to Galileo, by which Newton was able to unlock the secrets of lunar and planetary motions.
2. In summary, we used the method of differentials on two equations above. It was applied to the distance-time-squared equation $y = kt^2$ and then the other was $y = 9.8\,t$. The method yielded $2kt$ and 9.8, respectively. The first produced the rate of change of distance with respect to time, or velocity. The second yielded the rate of change of velocity with time, or acceleration. It looks as if the same operation was applied to each equation. Can we generalize the method? Indeed, it can be shown that, in most cases, all one needs to do to find the rate of change of such a function is first to multiply the function by its exponent and then reduce

the power of the exponent by one.[12] Suppose we try it on a cubic equation, say $y = Zx^3$, where Z is any constant, returning to the conventional x, y notation for the abscissa and ordinate. If the general rule is correct, then the rate of change of y with respect to x, written as $\Delta y/\Delta x$, should simply be $3Zx^2$. Using the above methods, try this for yourself to see if you agree.

A Summary of Galilean Equations

It is worthwhile thoroughly to know the basic equations of motion. Below is a summary of the relationships among distance, velocity, acceleration and time, which we will sometimes call the "Galilean equations" even though Galileo himself never actually expressed these relationships in this algebraic form:

Summary of Galilean equations

Condition	Proportion	Proportionality constant	Equation
Distance at time t assuming continuous, uniform motion (i.e., where velocity has been constant)	$s \propto vt$	1	$s = vt$
Distance at time t after uniform acceleration from rest (i.e., where initial velocity was zero)	$s \propto vt$	½	$s = \frac{1}{2}vt$
Velocity at time t after uniform acceleration from rest	$v \propto t$	a	$v = at$
Distance at time t after uniform acceleration from rest	$s \propto t^2$	½a	$s = \frac{1}{2}at^2$

Notes on Using the Law of Conservation of Mechanical Energy to Derive the Galilean Distance Equation

We may use energy concepts to derive the Galilean distance-time-squared relation. The law of conservation of mechanical energy was not well-understood until the nineteenth Century. The subject is treated extensively in most physics textbooks; we will give only a brief summary here of the key concepts relevant to our discussion. Energy is the ability to do *work*, which is the ability of a force to move an object through a distance. *Kinetic energy* is the energy of motion. It has the ability to do work upon impact with another object. It is expressed by the relation: $KE = \frac{1}{2}mv^2$. That is, the kinetic energy of a mass m varies as half the square of its velocity. The kinetic energy per unit mass m thus equals $v^2/2$. *Potential energy* is a kind of stored energy. When a spring is wound, it can do work when unwound. When an object is raised above the ground it acquires potential energy, the greater

---

[12] A calculus course is advised to properly examine the many nuances and variations of this idea to more complex functions, and its inapplicability to certain classes of functions not considered here.

with height, which is converted to kinetic energy as it falls and gains speed. It is expressed by the following relation near the surface of the Earth: $PE = mgh$. That is, the kinetic energy of a mass m increases as it is raised a distance h against the accelerative force of gravity g.[13] The potential energy per unit mass m thus equals gh. It is a fundamental law of physics that total mechanical energy in a system can neither be created nor destroyed. So in any "closed" mechanical system, where we assume no gains or losses of energy from external sources (for example, in the form of heat due to friction) then $KE + PE =$ constant. The total energy in the system is $E_{total} = KE + PE$.

Now with these ideas in mind, imagine a rock high up on a cliff, whose height is h. You are about to push it off. Before the rock is pushed, its velocity is zero, so $v = 0$, and its total energy is mgh, which is entirely potential. After you shove it off, it loses height but gains velocity: the potential energy is "exchanged" for kinetic energy as it falls. As it hits the ground, where $h = 0$, potential energy has diminished to zero,[14] and its total energy, now all kinetic, is ½ mv^2. The total energy from top to bottom, however, has always remained constant:

$$E_{total} = \frac{mv^2}{2} + mgh$$

The kinetic energy at the bottom of the cliff thus equals E_{total} and that is the same value as the potential energy at the top. Equating them, cancelling the mass terms, relabeling acceleration a and the height s, and isolating the velocity term, we have

$$v^2 = 2as$$

Given that in the case of uniform acceleration, the velocity at time t is equal to acceleration times time elapsed, or $v = at$, we can square that expression (to become $v^2 = a^2 t^2$) and substitute the right-hand side for v^2 in the previous equation:

$$a^2 t^2 = 2as$$

Solving for distance yields,

$$s = \frac{1}{2} at^2$$

We have derived the Galilean relationship from purely energy concepts unknown in Galileo's time. Had he known, how much time it would have saved him!

---

[13] As one ventures far from Earth the value of g changes with the inverse square of the object from the center of the Earth, so in those cases the equation must be modified to take that into account, and the equation is $PE = -GMm/r$.

[14] We here adopt the surface of the Earth as the arbitrary zero reference point for potential energy, since the rock can descend no lower.

Exercises: Playing Around with Gravity Galileo undertook his experiments with early nineteenth Century materials and concepts. We now have modern mathematics and far greater knowledge of the motion of bodies both on this and other worlds. We also have experience in doing thought experiments using mathematical models to study and predict outcomes. The problems below invite you to model simple dynamic situations in various contexts to help you gain greater familiarity with the basic Galilean concepts.

Problems

1. An amateur rocket test-fired from rest travels horizontally with constant acceleration. After 30 s of flight it has achieved a velocity of 900 m/s. What is its horizontal acceleration and distance travelled?
2. Ignoring friction, at what height must the horizontally-aimed rocket in the previous problem be launched in order for it not to hit the ground before 30 s of flight?
3. A rocket has been fired vertically from the ground with steady acceleration. Its velocity at height h is 300 m/s. What is height h?
4. A small balloon ascends to an altitude of 1 km above the surface of the earth, whereupon a 1 kg ball is dropped. What potential energy has the ball gained at the top of its ascent? Ignoring friction, (a) what would be the ball's kinetic energy and velocity be when it hits the earth; and (b) assuming a gravitational acceleration of 9.8 m/s per second, how long would the fall take?
5. Mars has a surface gravity that is 3.71 m/s^2. If Galileo's rolling ball experiment were set up on Mars, estimate the numerical results that would be obtained. Ignore friction. Explain your reasoning.
6. Neptune has surface gravity that is about 1.138 times that of the Earth. If Galileo's rolling ball experiment were set up on Neptune, estimate the numerical results that would be obtained on that planet. Ignore friction. Explain your reasoning.
7. Galileo's rolling ball experiment is primarily an illustration of: (a) how fast things fall; (b) how friction affects the speed of moving objects; (c) how objects in motion tend to stay in motion; (d) how vectors work.
8. Suppose you drop a tennis ball from a height of 1.5 m while your friend holds a gun and, at the moment you drop the ball, shoots it perfectly horizontally from the same height. Ignoring friction, which should hit the ground first: the ball or the bullet? Explain your answer.
9. Galileo discovered that a dropped object will fall (a) equal distances in equal times; (b) twice as fast in the second second as in the first second; (c) as far proportionally as the square of the time; (d) half the acceleration times the time.
10. Imagine a 1 kg ball is rolled down a frictionless ramp 10 m long whose high end is 1 m off the ground. The ramp merges into a perfectly flat pathway at ground level. Describe and compute the forces, accelerations and velocities of the ball at each second of its motion down the ramp and along the level plane.

Reference

1. Hawking S (ed) (2002) On the shoulders of giants: dialogs concerning two new sciences. Running Press, Philadelphia

Chapter 3
Christian Huygens' Remarkable Pendulum

The pendulum is curiously related to circular motion in a gravitational field, and its investigation helped lay the foundations of celestial mechanics. With it, the brilliant polymath Christiaan Huygens (1629–1695) in the mid-1600s was able deduce the vertical fall of an object in a given time and determine the pull of Earth's gravity with remarkable accuracy. His results, developed in the course of his invention of the pendulum clock, were used by Newton and, as will be discussed in Chap. 6, gave critical, empirical support to his universal theory of gravitation.

Of Planes and Pendulums: A Historical Sketch

It is not easy, even nowadays with a reliable stopwatch, to time a falling stone with absolute exactness. How much greater was the challenge in the 1600s. Yet if an object's fall in just 1 s could be known, not just approximately, but with precision, it would open the door to a more important and sensitive measure, then unknown: the constant of acceleration due to gravity, Earth's "g". But the crude measuring devices of the day would not allow it. Counting pulse beats or weighing water in a water clock was too cumbersome and ultimately subjective to measure anything fast. The inventive Galileo chose to run his ball down an inclined plane so he could slow its motion down, and discern the crucial proportional relationships between distance and time. But as to actually quantifying the speed, to know its velocity exactly, this was still out of the question. It was again the question of accurate timekeeping. Perhaps there would be another way to find out.

The steady, silent rhythm of the swinging pendulum has captivated humans for ages. Galileo noticed that no matter how far a pendulum of a given length is pulled back, it appeared to have the same period.[1] He also keenly deduced that the period of a pendulum is proportional to the square root of its length.[2] Would measuring the *period* of a pendulum of a known length be somehow sufficient to tell the distance of fall in a short period of time? Huygens, building on Galileo's work, unlocked the key to pendulum motion in his 1673 book, *The Pendulum Clock or Geometrical Demonstrations Concerning the Motion of Pendula as Applied to Clocks*.[3]

Why should the fall of an object in a gravitational field be determinable by a pendulum? Recall the independence of forces emphasized by Galileo in his experiments with ramps. A pendulum bob, held to one side and released, will experience a downward force of gravity, and a transverse pull from the string holding it. The downward force of gravity is independent of the transverse pull of the cord drawing it to the center of its swing. If we can determine the downward fall, the vector component that acts just vertically—if we isolate that in our thinking—the fall downward should be no different than any dropped ball. It should cover its descent distance in the same amount of time as if it were dropped free of the pendulum cord. Hence, the Galilean equation for distance (proportional to time squared) should apply to a falling pendulum bob the way it applies to any other falling object. Christiaan Huygens demonstrated this in simple propositions. He began by re-proving Galileo's findings about objects on inclined planes. He then imagined a series of connected planes and extrapolated this to motion along the arc of a pendulum. His Proposition IV draws directly from Galileo's discussion of balls moving on two sloping planes, connected at the bottom, which we discussed in Chap. 1:

PROPOSITION IV

If a heavy body begins to move upwards with the same velocity acquired at the end of a descent, then in equal parts of time it will cross the same distances upwards as it did downwards, and it will rise to the same height from which it descended. Also in equal parts of time it will lose equal amounts of velocity[4]

---

[1] Hawking [1] (Short title) *Dialogs*, 65: "But observe this: having pulled aside the pendulum of lead, say through an arc of fifty degrees, and set it free, it swings beyond the perpendicular almost fifty degrees, thus describing an arc of nearly one hundred degrees; on the return swing it describes a little smaller arc; and after a large number of such vibrations it finally comes to rest. Each vibration, whether of ninety, fifty, twenty, ten or four degrees occupies the same time: accordingly the speed of the moving body keeps on diminishing since in equal intervals of time, it traverses arcs which grow smaller and smaller." This is known as *isochronism*, and has since been found to be true for small angles of swing, but only to a fair approximation for larger angles. See the discussion about the period of the pendulum in the text.

[2] *Dialogs*, 73: "As to the times of vibration of bodies suspended by threads of different lengths, they bear to each other the same proportion as the square roots of the lengths of the threads." Compare this result with the equation for the period of the pendulum given below in the text.

[3] Huygens [2] (Short title) *Pendulum Clock*. The *Pendulum Clock* was first published in 1673.

[4] *Pendulum Clock*, 38: Proposition IV. Along these lines, Galileo had earlier stated: "[It] is very likely that a heavy body falling from a height will, on reaching the ground, have acquired just as much momentum as was necessary to carry it to that height; as may be clearly seen in the case of a rather heavy pendulum which, when pulled aside fifty or sixty degrees from the vertical, will acquire precisely that speed and force which are sufficient to carry it to an equal elevation save only that small portion which it loses through friction on the air." *Dialogs*, 72.

Thus the distances and velocities at any point on one side will equal the distances and velocities at the same points on the other side. From this, Huygens, echoing Galileo, tells us that the velocities acquired by objects "falling through variably inclined planes are equal if the elevations of the planes are equal."[5] And (by Proposition IX) any such object "will rise to the same height from which it came no matter how many contiguous plane surfaces it may have crossed...or what their inclinations are."[6]

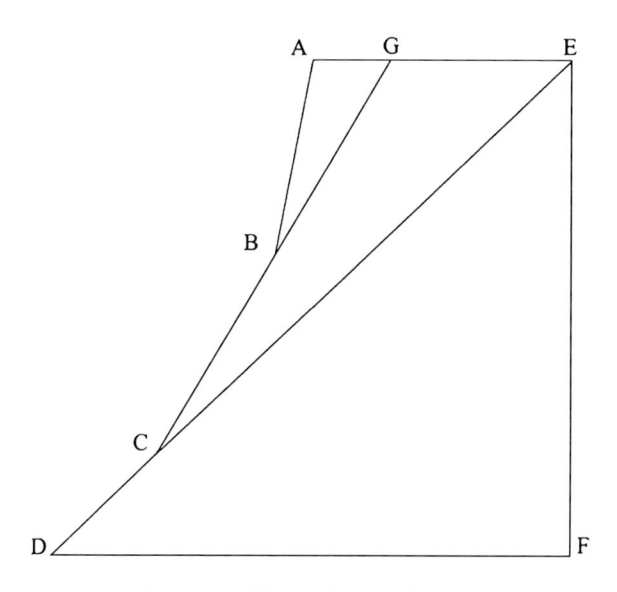

Huygens' Illustration of Continuous
Planes Forming a Curve

---

[5] *Pendulum Clock*, 43: Proposition VI. Again, from Galileo: "[W]e may logically infer that a body which descends along any inclined plane and continues its motion along a plane inclined upwards will, on account of the momentum acquired, ascend to an equal height above the horizontal...and this is true whether the inclinations of the planes are the same or different ... But by a previous postulate the speeds acquired by fall along variously inclined planes having the same vertical height are the same" *Dialogs*, 172. That is, the speeds at the bottom of the plane will be the same if the fall is from the same height; the angle of the plane, be it sloped at a low angle or almost vertical, or absolutely vertical, will make no difference to this conclusion. This result follows from considerations of potential and kinetic energy discussed in Chap. 2.
[6] *Pendulum Clock*, 46: Proposition IX.

Huygens has conceived here of any number of plane surfaces made up of smaller and smaller linked segments, so that it approaches a smooth curve.[7] This analysis anticipates Newton's division of arcs into small linear increments, merging ultimately (as the segments become infinitesimally small and almost infinitely more numerous) to a smooth curve:

> [A] body descending through the circumference of a circle or through any curved line (for here we can consider any curve to be composed of an infinite number of straight lines) will always acquire a velocity equal to what it would acquire by descending from an equal height. This velocity will be equal to the velocity acquired in a perpendicular fall from the same height.[8]

This was the beginnings of calculus, and it is clearly present in Huygens. In short, the fall of a pendulum bob from a given height, measured vertically, will equal the fall of the bob as if dropped freely.

Nowadays we would confirm the truth of this proposition from the principles of conservation of mechanical energy. Suppose a pendulum is 5 m long and is pulled to the right by 3 m (about .64 rad or almost 37°), such that the bob when released would fall 1 m, as in the figure.

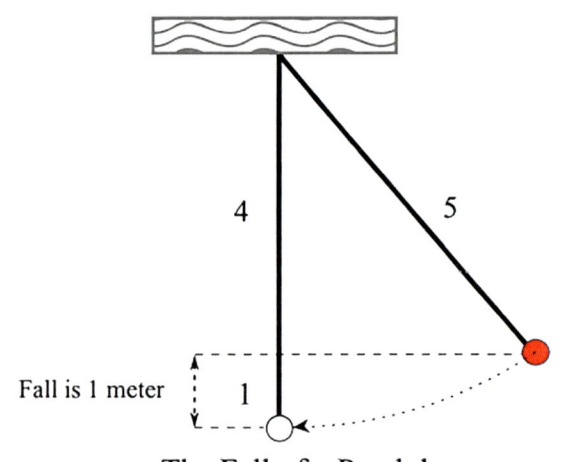

The Fall of a Pendulum

---

[7] Huygens' proof of proposition VIII may be excerpted as follows, with reference to the above illustration copied from *Pendulum Clock*, 45, Fig. 13:

> Let AB, BC, and CD be contiguous planes whose terminus A has a height above the horizontal line DF.... And let a body descend through these planes from A to D. Now I say that at D it will have the same velocity which it would have at F by falling from E. For when CB is extended it cuts AE at G, and likewise DC, when extended, cuts AE at E. Now a body descending through AB will acquire the same velocity at B as a body descending through GB [Proposition 6]. Hence it is clear that, if change or direction at B has no effect, the body will have the same velocity when it arrives at C as if it had descended through the plane GC, which is the same as it would have by descending through EC. Hence, in the same way, it will cross the remaining plane CD as if it had passed through EC. And thus at D it will have the same velocity as if it had descended through the plane ED, that is, the same velocity acquired by a fall through the perpendicular EF. Q.E.D.

[8] Ibid.

What is the speed of the bob at the lowest point, and how does this compare with the speed of the bob if dropped vertically? It is apparent that the sum of kinetic and potential energies at the top and bottom of the pendulum's swing must be the same, just as in the case of the rock dropped from the cliff in Chap. 2. The velocity squared at the end of the fall is therefore, $v^2 = 2as$, where a is the steady acceleration and s is the fall distance. The acceleration is due to gravity, so we can write the equation for velocity this way:

$$v = \sqrt{2gs}$$

Inserting for g the Earth's gravitational acceleration of 9.8 m/s^2, the fall distance of 1 m, and neglecting the effects of air friction, the speed of the bob at the bottom of its swing is,

$$v \simeq 4.4 \text{ m/s}$$

From the equation alone it is apparent that this result is the same as if the bob, freed from the pendulum cord, were simply dropped 1 m to the ground: the final speeds attained from the same height are always the same. Of course if we pull the pendulum back farther, the bob will swing through a steeper angle, the initial height at release will be greater, and the fall distance, and hence the maximum velocity of the pendulum, will be greater. Pull the bob up to 45° from vertical ($\pi/4$ rad) and the fall distance increases to about 1.5 m; the maximal velocity (at the bottom of the swing) increases to about 5.4 m/s.[9] A lesser angle will produce a lesser velocity. Deflect it 9° ($\pi/20$ rad) and the fall distance becomes .06 m, and it swings past vertical at only 1.1 m/s. Except for the means of accounting for friction, these results are no different than would be obtained by experimenting with variably inclined planes and observing the end velocity's dependence upon initial height. Notice too that the speed is independent of the mass of the bob; it does not appear in the above velocity equation.[10] This is consistent with Galileo's early observations that, disregarding friction, objects of unequal mass falling from the same height will still fall to the ground at the same speed—the principle of equivalence discussed in Chap. 1.

We said that Galileo noticed constancy in the period of a pendulum: no matter how far he pulled it back, the oscillations seemed to take the same amount of time. Why should this be? Let's continue the thought experiment in which we pulled a pendulum back far, at a steep angle, then less so. When the bob is pulled back far, its end velocity is greater but it travels through a longer arc than when deflected only slightly. A pendulum swung through a small angle similarly has, as we noted, a slower velocity, but less distance to travel than when deflected more. In all, shouldn't the swing times, the *periods* of the pendulums therefore all be the same? The answer is, to a fair approximation, yes. The arc distance to midpoint of the 9° pendulum is about .785 m, and the pendulum's mean velocity is about .55 m/s.

---

[9] If θ is the pull-back angle of the pendulum from the vertical, and l the length of the pendulum, the drop distance s can be calculated from the equation $s = l(1-\cos\theta)$.

[10] The mass terms cancel out in the energy equations: $KE_{top} + PE_{top} = KE_{bottom} + PE_{bottom}$

For the 45° pull-back angle, the pendulum's arc length is about 3.93 m and its mean velocity is about 2.75 m/s. Divide the arc length by the velocities, and the time to travel these arc lengths in each case is the same: about 1.4 s. When Galileo observed this remarkable property of pendulums, called *isochronism*, he had the idea that it could be used to tell time far more accurately than ever possible, from which seed of inspiration Huygens' pendulum clock was conceived half a century later.

How Gravitational Acceleration Was Measured with a Pendulum

Before Huygens, the constant of acceleration due to gravity was unknown. It could have been determined by the Galilean equation $s = \frac{1}{2} gt^2$, where g is the acceleration due to gravity: if one knows the distance s a thing falls in a second, one can solve the equation for g. But to do so in the 1600s would as we noted have required a timekeeping device accurate to within at least hundredths of a second, and some automatic means of recording the start and end of fall, which as we mentioned did not exist. To return to our original question, the key issue was how to know exactly the distance an object of any mass would fall in a given time: from this the acceleration of gravity could be found. Galileo gave us proportions: the distance fallen is proportional to the square of the time. But he could not reliably quantify how far an object would far in the space of, say, 1 s. This quantification, as we shall see in Chap. 6, would be vital to Newton's calculations of how the Moon's motion conformed to his evolving theory of gravitation.

Huygens' addressed this challenge in the last proposition of his book: "How to define the space which is crossed by a heavy body falling perpendicularly in a given time."[11] He acknowledges the limitations of experiments, due to the speed of the fall:

> Those who have previously studied this measure [of the fall of a body] agree that it is necessary to consult experiments. But the experiments which have been conducted so far do not easily give an exact determination of this measure, because of the speed which the falling body has acquired at the end of its motion. But if one uses our Proposition ... in The Falling of Heavy Bodies ..., and if one knows the length of a pendulum which marks off seconds, then he can explain this matter as a certain derivation without experiment.[12]

In other words, the perpendicular fall can be measured by the period of a pendulum whose length is adjusted just so it completes one oscillation—one arc sweep in one direction (half a full period – just the tic part of the pendulum clock)—in 1 s. The period of such a pendulum is 2 s (the tic and the tock of the clock). Such a pendulum is called a seconds pendulum and is a fruitful way of simplifying calculations. This was an excellent development since it avoided the rather clumsy inclined planes, with all their friction and imprecision of measurement.

---

[11] *Pendulum Clock*, 170: Proposition XXIV.

[12] Ibid., 170–71: Proposition XXVI.

Determining How Far a Body Falls in One Second from the Length of a Pendulum

Huygens succeeded in finding the relation between the period of a pendulum and the fall of an object. Isaac Newton concisely described Huygens' result in these words: "[T]he space which a heavy body describes by falling in one second of time is to half the length of this [seconds] pendulum [of Christiaan Huygens] as the square of the ratio of a circle to its diameter (as Mr. *Huygens* has also shown)..."[13] The next problem will show how this is so.

Problem Given the information below, derive Huygens' result, as expressed by Newton, to show that the fall distance s of an object equals half the length of a seconds pendulum times the "square of the ratio of a circle to its diameter."

Given

$s = \frac{1}{2} at^2$	The Galilean equation for the distance an object moves in time t under the influence of a constant acceleration a. Where the constant acceleration is induced by gravity, g, the equation becomes $s = \frac{1}{2} gt^2$
$P = 2\pi \sqrt{\frac{l}{g}}$	Equation approximating the period P of a pendulum of length l

Assumptions The pendulum formula is a useful approximation for small angles of swing. We will assume that the swing angle of the seconds pendulum is small, say a few degrees. We will also disregard the fact that the value of g varies slightly at different locations on the Earth, and at different altitudes, a fact unknown by Huygens at the time.

Method Since the Galilean equation relates fall distance to acceleration, it may be rearranged to isolate acceleration. This we can substitute for acceleration in the pendulum equation. Doing this should yield a value of s in terms of l, the length of the pendulum, with the only other variable being t. Since the time we are considering is 1 s, we can set $t = 1$ and simplify the result. Finally, we must set the period equal to two for a seconds pendulum.

Calculations Starting with the Galilean distance-time-squared equation, $s = \frac{1}{2} gt^2$, we can isolate the acceleration term:

$$g = \frac{2s}{t^2}$$

---

[13] Sir Isaac Newton [3] Book III, Proposition IV. Theorem IV.

Now we can substitute the right-hand side of the above equation into the pendulum equation given above:

$$P = 2\pi\sqrt{\frac{l}{g}}$$

$$P = 2\pi\sqrt{\frac{l}{\left(\frac{2s}{t^2}\right)}}$$

Let us now set $t = 1$ and square each side:

$$P^2 = \frac{2\pi^2 l}{s}$$

Since the period of a seconds pendulum is 2 s, we set $P = 2$, and solve for s:

$$s = \frac{l}{2}\pi^2$$

This is the result we sought, telling how far, s, a body will fall in 1 s, a relation discovered by Huygens and (as we shall see later) used by Newton, where the "square of the ratio of a circle to its diameter" is of course π^2.

Observations

1. This equation means that a seconds pendulum of length l will complete one swing, say right to left (but not back) in the time a free-falling object at that location will fall through distance s. This was a valuable equation. With it, neither Huygens, nor Newton, nor anyone else in the 1670s or 1680s would need a stopwatch or electronic timer to try to time the fall of an object. Longer durations of time could be measured accurately, over many swings of the pendulum, and the result averaged to get a fairly accurate time for the period of each swing. Then, carefully calibrating the pendulum length so the swing intervals would come out to be exactly seconds, one could then calculate the 1-s free fall distance.[14]
2. How can we find an equation for the length of a seconds pendulum? Since, by the pendulum equation, $P^2 = 4\pi^2 l/g$, we can set period equal to 2 s again, and solve this for l. The result is,

$$l = \frac{g}{\pi^2}$$

---

[14] See, e.g., *Pendulum Clock*, 170–71: Proposition XXVI, and the discussion by H.J.M. Bos in the introduction to the *Pendulum Clock*, xvi–xvii.

Try inserting values for g into that equation and see what lengths you find for the seconds pendulum under the assumed values you chose.

3. Assume that we are Huygens, and we don't know the value of g, but have painstakingly, by trial and error, experimentally found the right length of a seconds pendulum. Let us say the magic length is found to be .993 m. Given this, let's determine g. Since from the above discussion, $l = g/\pi^2$, then

$$g = l\pi^2$$

$$g = .993(3.14..)^2$$

$$g = 9.8 \, \text{m/s}^2$$

Christiaan Huygens did not know the value of gravitational acceleration, g, but he was the first to determine it, and he did so with amazing accuracy.[15] He went on to propose that the length of a seconds pendulum be adopted as a universal standard for the meter. His efforts were supported by many, but ultimately did not win the day. At the time it was not known that the value of g was not exactly the same all over the Earth.

4. Let us take the last two derived equations, eliminate the length term, and see what emerges when we solve for g:

$$s = \underbrace{\frac{\pi^2}{2}l}_{\text{Distance of fall related to length of pendulum}} \qquad l = \underbrace{\frac{g}{\pi^2}}_{\text{Length of pendulum related to } g}$$

We can take the expression for l on the right and substitute it into the left-hand distance equation, and we obtain this result,

$$g = 2s$$

This should not be a completely surprising conclusion, since we get the same thing by setting time equal to 1 s in the equation $s = \frac{1}{2} gt^2$.

5. If the length of the seconds pendulum is about .993 m, what is the fall distance s of an object in 1 s?

$$s = \frac{\pi^2}{2}l$$

---

[15] *Pendulum Clock*, 171: Proposition XXVI. Actually, Huygens found the value of 2s which, from the equation $s = \frac{1}{2} gt^2$ (where time is 1 s) is readily seen to be equivalent to the value of g, as discussed in the following text. See Chap. 6 for the discussion of how Newton used the value found by Huygens for his famous "Moon test."

$$s = \frac{(3.14..)^2}{2}(.993)$$

$$s = 4.9 \text{ meters}$$

Since $g = 2s$, acceleration due to gravity should be 9.8 m/s^2, which it is.

Deriving an Equation for the (Approximate) Period of a Pendulum

We gave the equation for approximating the period of a pendulum whose swing was small. Now we can derive that equation. In doing so, we can review a few fundamentals of circular geometry that will help us later on.

Problem Derive an equation for the approximate period of a pendulum, swinging in small arcs.

Given A pendulum of any length, depicted in the figure below.
These basic concepts of Euclidean geometry:

1. A central angle is measured by its intercepted arc. Hence $\angle AOB = \overset{\frown}{AB}$ where the notation $\overset{\frown}{AB}$ means the length measured along the arc.
2. An inscribed angle (the angle formed where two chords of a circle meet at one point) is measured by one half its intercepted arc. Hence $\angle APB = \frac{1}{2}\overset{\frown}{AB}$, and $\angle APB = \frac{1}{2} \angle AOB$. Likewise $\angle APB = \angle CBA$

Assumptions We will ignore the effects of air friction on the bob and string, and assume no forces other than gravity are present to disturb its motion.

Method Understanding the geometry of the pendulum will help in the understanding of orbits and Newton's breakthroughs discussed in the next few chapters. We first examine some basics of circular geometry, circular motion and radian measure. Suppose an object is moving at constant speed around a circle. It takes a certain amount of time to complete one revolution, and that time is the period P. Now allow that the object moves only part way around the circle, through an angle of θ in time t. We don't need to know what θ is yet since we are at the moment just exploring angular relationships. The circumference of the circle is π times the diameter, or

Deriving an Equation for the (Approximate) Period of a Pendulum

2π times the radius r. Taking the radius as one unit for convenience, we can derive this proportion:

$$\frac{t}{P} = \frac{\theta}{2\pi}$$

That is, the ratio of the time of a partial rotation θ is to the period (the time of full rotation) as the angle described in that time is to the whole circle. The time elapsed is therefore,

$$t = P\frac{\theta}{2\pi}$$

We will keep this equation in mind for use in solving the problem. Next, the distance along the circumference in a given time is important to know, and can be conveniently expressed as units of radii, or *radians*. We have to use angular units of radians, rather than degrees, to do so.[16] The distance along any arc is the angle in radians times the radius: $r\theta$:

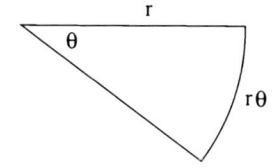

Arc length from angle
in radians and radius

For instance, in circle whose radius is 5 m, the arc distance encompassed within 15°, which is .2618 rad, would be 5 × .2618 = 1.31 m. This is a wonderfully useful mathematical tool! Now refer to the accompanying diagram.

---

[16] An angle in radians can be thought of as a fractional part of a circular arc. For example, 1 rad is one whole radius laid out along the circumference of a circle of radius r. Half a radian is the projection or trace of half the length of a radius on the arc. An angle where $\theta =. 25$ rad (14.324°) describes a quarter of a radius along the arc, and so on. There are 2π radians in a whole circle. To convert degrees to radians, multiply the angle in degrees by $\pi/180$. 1 rad is about 57.3°.

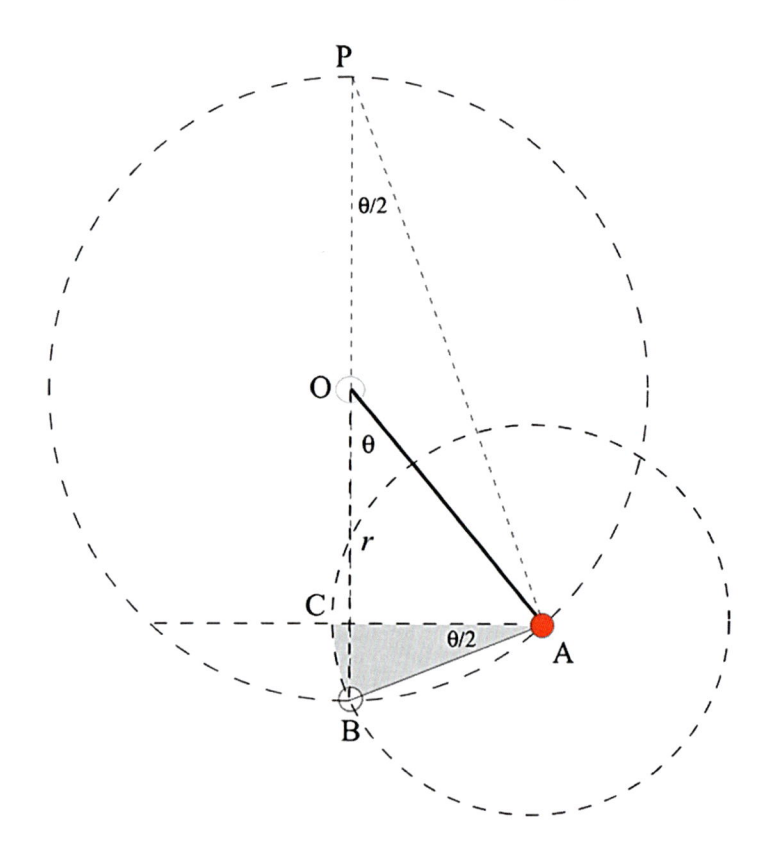

Pendulum Geometry

In reviewing the figure, it is helpful to recall some principles you likely learned in geometry: a central angle is measured by its intercepted arc, and an inscribed angle is measured by one half its intercepted arc. Our overall task will be to relate the geometry of the circle in the figure to parameters of pendulum length, gravitational acceleration, and period. To find the period of a pendulum we should first investigate the equations we have already discussed. The Galilean equation $s = \frac{1}{2} gt^2$ may be useful, for if we know the fall distance s we can relate it to gravitational acceleration g. The fall distance s in the equation may be approximated by BC. If we can find BC, we can begin to merge the geometry with equations of motion. How do we find the fall distance? In the triangles above consider the radius r of the larger circle, equal to OA and OB. If we multiply this radius by the angle θ, the result will be the arc length from A to B: $r\theta$ will give the arc $\widehat{AB}$. If we use this arc length as an *approximation* of the chord AB (which is truer and truer as an approximation as θ gets smaller and smaller) then that chord becomes the radius of the *smaller* circle. Thus AB × $\theta/2$ should give us (roughly) BC. We've used one radius times its angle (θ) to leap frog to another radius, which, times its angle ($\theta/2$),

Deriving an Equation for the (Approximate) Period of a Pendulum

gets us what we want. Once we have the fall distance, what do we do with it? We will then need to find a way of putting it in terms of period, hopefully getting rid of the angle terms. Since we know the value of t from the equation $t = P\theta/2\pi$ we found above, that should open the door to the solution. That will let us put s in terms of period and angle. Let's see where all this takes us.

Calculations The length along the arc generated by the angle θ is $\widehat{AB}$, so that,

$$\widehat{AB} = r\theta$$

Note that this arc length is not the same length as the chord AB (written commonly with a bar over it to distinguish it from arc AB) but as the angle θ becomes small, the approximation is better. We will consider that, as θ approaches a very small size, then,

$$\overline{AB} \simeq \widehat{AB}$$

So that,

$$\overline{AB} \simeq r\theta$$

Now chord $\overline{AB}$ is the radius of the smaller circle, and its equivalent $r\theta$ times the angle $\theta/2$ will give the approximate length of arc $\widehat{BC}$:

$$\widehat{BC} = r\theta\left(\frac{\theta}{2}\right)$$

$$\widehat{BC} = \frac{r\theta^2}{2}$$

Since we are assuming that θ approaches very small size, then,

$$\widehat{BC} \simeq \overline{AC} \simeq s \simeq \frac{r\theta^2}{2}$$

Now we can write the equation $s = \frac{1}{2}gt^2$ this way:

$$\frac{r\theta^2}{2} \simeq \frac{1}{2}gt^2$$

$$r\theta^2 \simeq gt^2$$

Earlier we found that $t = P\theta/2\pi$, so we are closer to our goal since we can make the substitution for time and get this result,

$$r\theta^2 \simeq g\left(\frac{P\theta}{2\pi}\right)^2$$

Doing the simplifying algebra, canceling the θ terms and letting the length of the pendulum l be the radius, we arrive at this satisfying conclusion, most accurate for small angles:

$$P \simeq 2\pi\sqrt{\frac{l}{g}}$$

Observations

1. Inspection of the equation shows that the period is independent of the mass of the bob. Compare that conclusion with Galileo's discovery that, in the absence of the effects of friction, objects of unequal mass fall to the ground at the same time.
2. The mesmerizing swings of a pendulum or the oscillations of a spring are examples of what is called *simple harmonic motion*. Let us take a brief look at the forces acting on the pendulum to understand what simple harmonic motion is and how it works.

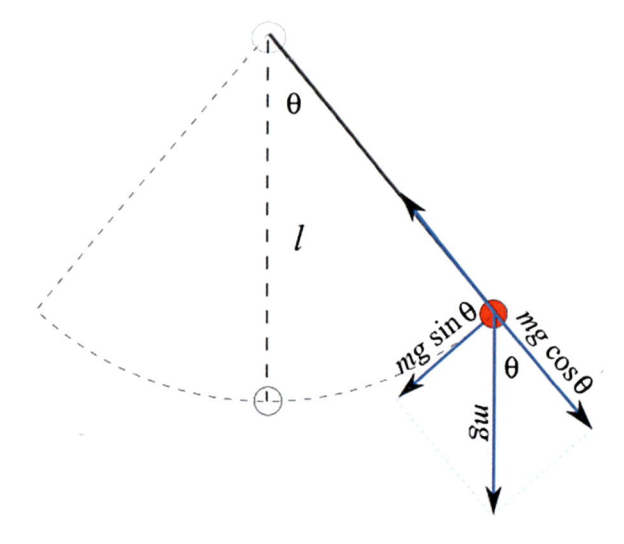

Forces on a Pendulum

The pendulum's motion begins at rest, and upon release, it experiences the downward pull of gravity and slantward pull of the string. The forces are not opposing, but offset, depending on the angle of the swing, as seen in the figure. Resolving the forces into rectangular vector components, the sideways force tugging on the bob is $mg \sin \theta$ where θ is the angle of the swing, m is the mass of the bob, and g is the gravitational acceleration. At the center of the swing, the pull on the string is aligned with the pull of gravity, and the sideways force is zero. At that point, if the string were cut, the sideways motion of the bob would be tangent to the arc, and were it not for gravity and the restraint of the string, would continue in a straight, horizontal line, impelled by inertia, just like the rolling ball after it reached the end of its inclined plane. But the pendulum is pulled back by the cord, and up, and the restraining force on its upward swing is $F = -mg \sin \theta$, growing larger as the angle θ increases. (We use the negative sign to in indicate the reverse direction of the swing.) For very small angles, $\sin \theta \approx \theta$, so we can say that $F \approx \pm \theta$, where the restraining force is proportional to the displacement θ. This is an important indicator of simple harmonic motion: in a physical system where the force is proportional to displacement, things behave like springs and other harmonic oscillators, and can be similarly described mathematically.

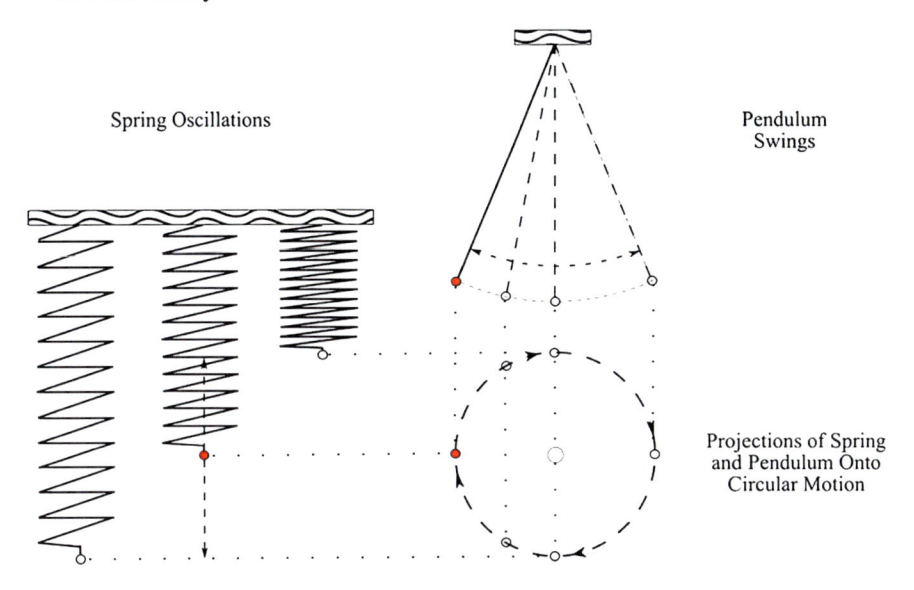

Spring Oscillations

Pendulum Swings

Projections of Spring and Pendulum Onto Circular Motion

Examples of Simple Harmonic Motion

An edgewise perspective of an object in uniform circular motion creates a pattern which mimics those of harmonic oscillators, such as springs and pendulums. In other words, the sideways projection of circular motion moves with the same sinusoidal motion of a spring or pendulum. On a moving graph, the pendulum, spring and circular moving object all create sine waves. It is for this reason that when you line up a pendulum and circle whose respective length and radius match, and give them the same period, they will move in synch.

2. One can derive the equation for the period of a pendulum from the principles of simple harmonic motion. This is classically illustrated by the motion of springs. A stretched spring will obey Hooke's law (after Robert Hooke who discovered it in 1678), and require a force to stretch it that is *proportional to the distance it is stretched*. After it is stretched, a restoring force pulls it back to its equilibrium position. The force works both in the stretch and compression of a spring, and its quantity for a given distance of stretch or compression will vary with the nature of the spring. The force is given by the equation $F = -kx$, where $-x$ is the amount of displacement of the spring—how far it is stretched or compressed from its rest position—where k is the *spring constant*. (The negative sign indicates that the restoring force is in the opposite direction of the spring's stretch.) A large k means a stiff spring. This linear relationship is the basis for spring scales. The total *work* done in stretching a spring to distance x is the sum of the work done in each little displacement of the spring along the way, and turns out to be $W = \frac{1}{2}kx^2$ (where the work units are joules). When stretched or compressed, the spring is storing energy: its potential energy. When released, the spring's potential energy is initially converted to the kinetic energy of motion, then back to potential energy at the end of the oscillation. Its acceleration is always changing as it elastically extends and compresses. A spring hung vertically, affixed with a mass, and given a downward tug will thus oscillate with simple harmonic motion. Likewise a pendulum swinging through small angles is an example of simple harmonic motion. If we pull the pendulum aside and let it go, its potential energy is converted into kinetic energy of motion, which is at its maximum at the middle of the oscillation, just as in the case of a spring. Since the total $KE + PE$, as before, is always constant, then

$$\frac{1}{2}mv^2 + \frac{1}{2}kx^2 = E_{Total}$$

When x (the spring displacement) is at its maximum, velocity will be zero; when the velocity is greatest, at the center of oscillation, the potential energy will be zero. We can thus equate them and solve for velocity:

$$v = x\sqrt{\frac{k}{m}}$$

Since the projection of uniform circular motion onto a straight line reveals the pattern of simple harmonic motion, the displacement of the spring, x, can be considered equivalent to the projected radius of a circle. By analogy to circular motion, the period to complete of one cycle of motion is,

$$P = \frac{2\pi x}{v}$$

Making the substitution for velocity from the equation just before, and cancelling the x terms yields this equation, which is starting to look like the pendulum equation, but is not quite there:

$$P = 2\pi\sqrt{\frac{m}{k}}$$

We need to find out what k is. From the discussion above, we know that force is proportional to the angle θ for small angles (i.e., proportional to displacement of the pendulum, the indicator of simple harmonic motion). If the angle is in radians, then the distance d along the arc of the swing will be the angle times the length l of the pendulum, $\theta = d/l$. Remembering that, for small angles, $F = mg\theta$, then,

$$F = \left(\frac{mg}{l}\right)d$$

Now step back and look at this equation. Does it resemble the form of a harmonic oscillator, where the force to move a mass through a distance is directly proportional to the displacement, $F = kx$? Indeed it does, and since d is the displacement, the other constants in the parenthetical must be k:

$$k = \frac{mg}{l}$$

Taking this value for k and referring again to the period equation, we can insert this ratio for k, and the period becomes,

$$P \simeq 2\pi\sqrt{\frac{l}{g}}$$

This again is the period of a pendulum, derived through its connection with the harmonic oscillator.

3. In comparing the periods of pendulums and springs, one notices that for a given pendulum length, P is proportional to $1/\sqrt{g}$; for a given mass stretching a spring to a certain length, P is proportional to $1/\sqrt{k}$. One may regard g and k as "stiffness constants" whose value for those conditions determines the rate of oscillation. The greater the value of those constants, the shorter are the periods of each.

The Pendulum's Satisfying Coincidence with Circular Motion

Problem Take a pendulum and a circularly revolving object where the length l of the pendulum is equal to the radius r of the circle. From the above period equation for a pendulum, find the ratio of the period of the revolving object to the period of the pendulum; then make the period of each the same and find an equation for the velocity of the revolving object.

Given

$P \simeq 2\pi \sqrt{\frac{l}{g}}$	The approximate period of a pendulum, derived above
$v = \frac{2\pi r}{P}$	Velocity of an object in uniform circular motion

Assumptions We will again ignore the effects of friction, and assume the pendulum deflection is rather small. For the moment, we also offer no particular hypothesis (such as gravitational attraction) why the object should be moving in uniform circular motion.

Method The second equation can be rearranged into an equation for period, after which the ratios of the two periods may be found. Making the periods equal will then give us the solution to the second part of the problem. Use of subscripts will keep us reminded of which variables belong to which entity.

Calculations First we find the period of the circular revolution:

$$P_{circle} = \frac{2\pi r}{v_{circle}}$$

Now write the ratios of the two periods:

$$\frac{P_{circle}}{P_{pendulum}} = \frac{\left(\frac{2\pi r}{v_{circle}}\right)}{2\pi \sqrt{\frac{l}{g}}}$$

We can cancel the 2π terms and, since the radius of the circle equals the length of the pendulum, simplify the ratio to this:

$$\frac{P_{circle}}{P_{pendulum}} = \frac{\sqrt{gr}}{v_{circle}}$$

Because the problem states that the periods are the same, the left-hand side is unity and the circular velocity is,

$$v_{circle} = \sqrt{gr}$$

Thus if an object in uniform circular motion has period and radius respectively the same as the period and length of a pendulum, it will have a velocity equal to the square root of g times the radius.

Observations

1. We will see later that this is the exactly the equation for circular orbital motion around a moon or planet in a gravitational field, where g is the value of gravitational acceleration at distance r from the primary object (e.g., not at the Earth's surface). This equation also hints at an equation for acceleration in circular motion, discussed in the next chapter, by solving for g. For now, we posited no reason *why* the object was revolving with uniform circular motion. We can see it merely as a geometric fact that when the periods and lengths (of radius and pendulum) are the same, the object will revolve with a velocity that is proportional to the square root of the radius times some "stiffness constant" which we are calling g. The higher the constant, the more rapid is the circular velocity, and the shorter the period. If we imagined the projection of this circle against an oscillating spring, with the same constants, the result would be the same, another consequence of simple harmonic motion.
2. The velocity equation for circular motion can be derived from the pendulum period equation, and that same velocity equation, working the other way, yields the period equation applicable to circular as well as pendulum motion, where the same constants are used. For uniform circular motion,

$$P = \frac{2\pi r}{v}$$

replacing the velocity term with circular velocity found above,

$$P = \frac{2\pi r}{\sqrt{gr}}$$

again yields the period equation for circular motion,

$$P = 2\pi \sqrt{\frac{r}{g}}$$

As we will see later, this too is the equation for the period of circular orbital motion around a moon or planet in a gravitational field, where g is the value of gravitational acceleration at distance r from the primary object.

Exercises: Pendulum Puzzles and Diversions Pendulums are fun to watch. There is a fascinating, almost mesmerizing, quality about them. As tools in the hands of able thinkers like Galileo and Huygens, they began to open the door to the mystery of gravity. The foundations laid by these gentlemen helped Newton to pull together vital pieces of his evolving theory of gravitation. The questions, problems, and diversions below are intended to test and enhance your own understanding of the simple, silent, mysterious pendulum and the principles that underlie its motion.

Questions

- Explain what a seconds pendulum is.
- Given two pendulums, what is the ratio of their periods to their lengths?
- Explain simple harmonic motion.
- Explain Hooke's Law
- Compared to its period on the surface of the Earth, do you think a pendulum of the same length would swing faster, slower, or at the same rate as on the Moon? Why?
- In Proposition VI of *The Pendulum Clock*, Huygens states: "The velocities acquired by bodies falling through variably inclined planes are equal if the elevations of the planes are equal." Explain the truth of this proposition using energy concepts.
- Explain the exchanges between potential and kinetic energy as the pendulum bob goes through its swing.
- Imagine an enormous pendulum suspended over the plane of the solar system, such that its bob, when resting in the vertical position, hangs over the Sun. That is, the plane of the solar system is perpendicular to the hanging pendulum. The pendulum's length is very great: about five or so astronomical units long (about as far away vertically from the Sun as the orbit of Jupiter is from the Sun along its plane). Would Hooke's law apply if we pulled the bob to the orbit of Mercury, then Venus?
- If the solar system were arranged on a principle of simple harmonic motion, how would the Sun's gravity change with increasing distance away from the Sun? How does it actually change? Explain any differences. Do you think that would be a viable arrangement for a solar system?

Things to Do

- Make a seconds pendulum and measure the value of g as accurately as you can.
- Test Huygens' Proposition VI ("The velocities acquired by bodies falling through variably inclined planes are equal if the elevations of the planes are equal") by measuring the terminal velocities of rolling balls on planes of different inclinations, but with the same starting elevation. If velocity measuring equipment is unavailable, measure the time two balls released simultaneously on two differently inclined planes of the same elevation take to reach the terminus of their slopes and compute average velocity.
- Test Huygens' Proposition VII ("The times of descent on variably inclined planes whose elevations are equal are related to each other as the lengths of the planes") by measuring the times of descent of rolling balls on planes of different lengths, but with the same starting elevation.
- Determine the actual change in acceleration on a rolling ball descending down a sloping plane which about halfway down completes its descent on a connecting plane of a lesser slope.

Problems

1. Suppose the value of g is 9.782 m/s² in the Canal Zone and 9.818 m/s² in Stockholm. What is the difference in length of a seconds pendulum in those two places?
2. Assuming you could place a pendulum on Jupiter, and that its surface gravity at that location is 2.53 times Earth's mean surface gravity, how long would a seconds pendulum be?
3. Mars has a surface gravity that is 3.71 m/s². What is the length of a seconds pendulum there? Use the period equation to confirm its period.
4. Using the result of the previous problem, confirm how far a rock would fall in 1 s on Mars?
5. You are an astronaut who has landed on the asteroid Pallas. You brought with you a pendulum whose cord is 1 m long. After a series of trials, you determine its period to be about 14.32 s. What is the approximate surface g of Pallas?
6. Using the result of the previous problem, calculate the length of a seconds pendulum on Pallas.
7. Two planets have the same diameter. On each is a pendulum of equal length, but the period of the first is one third the period of the second. What is the ratio of the masses of the planets? What is the ratio of their densities?
8. Explain why a pendulum's period should be independent of the mass of the bob.
9. Recall from Chap. 1 the equation for gravitational acceleration induced by Earth, which is the terrestrial g. Acceleration is determined there by mass m, the gravitational constant G, and the radius of the Earth r. Write the equation for the period of a pendulum expressed in terms of those values, instead of g. If the length of the pendulum cord were the same as the radius of the Earth, what would the period equation look like? Recast that equation using k for all the constants in the equation except r. Interpret your answer. (Keep this answer in mind as you read through the next chapter).
10. Assume a pendulum could somehow be anchored far out in space, beyond the orbit of the Moon, and that its length happens to be 384,400 km, which is just the mean distance from Earth to the Moon. By this configuration, the pendulum's string falls toward the Earth, and the bob is at the location of where the Moon would be in its orbit. Now assume it is pulled back a reasonably small angle in relation to its length, so that its period approximates that of the pendulum equation given above. Find the period of the pendulum in days. Compare this with the period of the Moon in days. Explain your result. [*Hint*: What is the value of Earth's g at lunar distance (refer to Chap. 1)?]

References

1. Hawking S (ed) (2002) On the shoulders of giants: dialogs concerning two new sciences. Running Press, Philadelphia
2. Huygens C (1986) The pendulum clock or geometrical demonstrations concerning the motion of pendula as applied to clocks. University of Iowa Press, Ames
3. Sir Isaac Newton (1949) Mathematical principles of natural philosophy (translation by Andrew Motte, 1729, revised by F. Cajori). University of California Press, Berkeley, p 408

Chapter 4
The Geometry of the Solar System: Kepler's Laws of Planetary Motion

The German mathematician Johannes Kepler (1571–1630) worked tirelessly for years trying to make sense the data of his employer Tycho Brahe, a Danish nobleman-astronomer. Tycho was a meticulous observer. From his self-financed and constructed observatory in Denmark, Tycho in the late 1500s had gathered an enormous set of positional data on the movements of the planets, including Mars. Later (in 1600) Tycho hired Kepler to be his assistant in Prague, in the court of Rudolph II, ruler of the Holy Roman Empire. Though Kepler was Tycho's able assistant, Tycho was not generous with sharing his data with Kepler. This frustration ended upon Tycho's death, Kepler succeeded to his position as imperial mathematician, and inherited all of Tycho's precious notes. Tycho's data on Mars in particular was thorough and precise, far more accurate than any observations in the history of astronomy.

The orbit of that planet, however, had defied understanding. Since the time of ancient Greeks it had been assumed that all the planets moved in perfect circles. But the observed varying velocities of Mars in its orbit did not conform to that of a body in circular motion. Since Kepler's observation platform, the Earth, was also moving in orbit, it was necessary for him to understand its motion as well; he could then make corrections for his own movement relative to Mars and therby deduce the true motion of Mars. He struggled with the computations for years, recording all of his mis-steps with excruciating candor, but his persistence paid off. From analysis of the varying velocities of Earth in its orbit, and reduction of the precise positions of Mars, Kepler concluded that the ancient idea of the uniform, circular motions of the planets in their orbits had to be discarded. The orbits were not circles. The planets did not move at uniform rates. Those ancient deeply rooted ideas whose lineage could be traced as far back as Plato had to be thrown out. It was revolutionary, but it was only the first step. For in concluding the planets did not travel at uniform speed, he searched for the geometric form which would account for their motion. He had in mind an oval, an egg-shaped orbit for Mars pointed at its perihelion. But the data still did not match. During these efforts, he began to intuit the physical nature of planetary orbits; this was not just a geometrical exercise – there must be a *cause* for the motions of the planets. But Newton had not yet been born, and the universal law of gravitation was unknown. Yet Kepler perceived that the Sun must be the "seat of *virtue* that impelled the planets in their orbits." Whether the cause was some form of magnetic induction, or swirling vortices, or some kernel of the idea of gravitation between masses, he assumed that whatever the effect was, it would decrease with distance from the Sun and increase with proximity. He tested the Earth's velocities at each of its apsides (closest and farthest points from the Sun) and found that those velocities were indeed proportional to distance from the Sun. This was important, but this relationship was not true everywhere else in the orbit.

As he was twirling these ideas around in his mind, Kepler made an imaginative leap. He speculated that the Earth swept out equal areas of its orbit in equal times, tested it with his data, and found it to be true – and thus was born a fundamental law of planetary motion. Though it was discovered first, it later became known as his second law of planetary motion. Still perplexing was the unanswered question of the exact shape of the Martian orbit. In puzzling over the shape of the Martian orbit, Kepler had faced the problem of determining why Tycho's observations showed that Mars' orbit was not perfectly circular, but yet deviated from the perfect circle only to a very small degree: about 429 parts in 100,000. In other words, if a circumscribed circle is one unit in radius, then the Martian orbit left a small gap of about .00429 radii, as shown in the following diagram, which has for illustration greatly exaggerated the elliptical shape of the orbit of Mars.

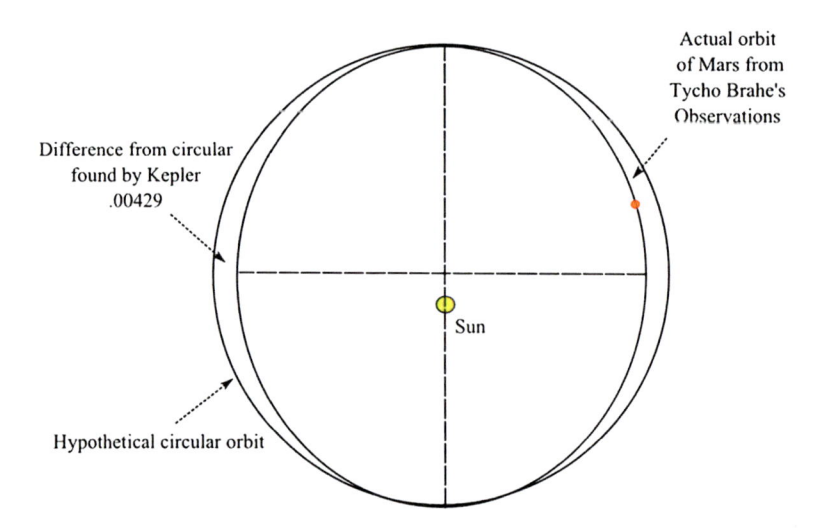

The Problem Faced by Kepler in Determining the
True Shape of Mars' Orbit and its Deviation from
a Circluar Orbit

After some false starts, and pages upon pages of calculations, he at last "awoke as from sleep," in his words, and tried an ellipse. He re-did his calculations, and behold, Tycho's observational data fit! This became the basis for his first law, that all the planets travel in elliptical orbits, with the Sun at one focus of the ellipse.

Kepler's breakthough involved deciding at last that the shape of the orbit must be an ellipse with the Sun at one focus of the ellipse. If the outer circle is one unit in radius, the Sun is offset from the center of that circle by the distance e which is in fact the *eccentricity* of the ellipse – a measure of its flatness. Kepler discovered that the width of the little lacuna on each side was actually the ratio of the radius of the circle over the short axis. What was exciting was that this ratio is geometrically related to the eccentricity of an ellipse, just the way it should be if the planet's orbit were in fact an ellipse. In other words, the empirical data of Tycho Brahe matched the geometry of an ellipse. In modern parlance, the width of each little lacuna is a function of the ratio of the semi-major axis a over the semi-minor axis b, and the length of b is a function of a and the eccentricity e.

To illustrate one reason why the determination of the ellipticity of Mars' orbit was difficult for Kepler, and the considerable accuracy of Tycho's observations, we need only look at modern data of that planet's eccentricity. A perfect circle has 0 eccentricity. Mars' eccentricity is, most accurately, 0.0933941, which is very close to circular.[1] In terms of the width of the lacuna by modern measures, we need to calculate the ratio of major axis (or semi-major axis) over minor axis (or semi-minor axis). The major axis of the Martian ellipse is 1.52371 AU.[2] The minor axis of the Martian ellipse is 1.51705 AU. The ratio of these is not far from the .00429 Kepler was working with. Incidentally, the width of the lacuna on the scale of the Martian orbit is a little more than 80% of the diameter of our Moon's orbit around the Earth – a miniscule displacement on the scale of our solar system. To detect such a variation from a true circle in the late Sixteenth and early Seventeenth century required observations of unprecedented precision, and Tycho Brahe's data had revealed that Kepler's earlier orbits were incorrect by only the tiniest fractions of a degree.

That Kepler did not rest until he eliminated this error, even if it meant discarding the sacred baggage of ancient assumptions, was a watershed event in the history of science. Kepler's travails are documented in his 1609 work, *Astronomia Nova seu Commtaria de Motibus Stellae Martis*, usually abbreviated as *Astronomia Nova*.[3] It is one of the great works in the history of science. Kepler's third law, relating the squares of the periods of the orbits to the cube of their distances from the Sun, came a decade later. Here together are Kepler's famous three laws of planetary motion.

KEPLER'S LAWS OF PLANETARY MOTION
  I. The planets move in elliptical orbits with the Sun at one focus.
  II. A line from the Sun to a planet sweeps out equal areas in equal times.
  III. The square of the period of an orbit (measured in years) is equal to the cube of the semi-major axis of the orbit (measured in AU).

---

[1] This data can be found at the NASA Jet Propulsion Laboratory website: http://ssd.jpl.nasa.gov/txt/p_elem_t1.txt .

[2] An astronomical unit (AU) is the mean Earth-Sun distance, the astronomers' standard yardstick.

[3] For a lively account of Kepler's progress and setbacks, see A. Koestler [1]. A clear and concise account can be found in "Kepler as an Astonomer" by W. Carl Rufus [2].

These laws were of course completely contrary to the ancient, geocentric view of the solar system (and universe), and helped further vindicate the Copernican heliocentric view. We shall briefly explain their meaning here.

Law I *The planets move in elliptical orbits with the Sun at one focus.* According to Kepler's First Law, all the planets orbit the Sun in elliptical orbits. Ellipses differ from circles by virtue of their eccentricity. Circular orbits are actually a special case of elliptical ones, where the eccentricity (degree of flattening) is zero.

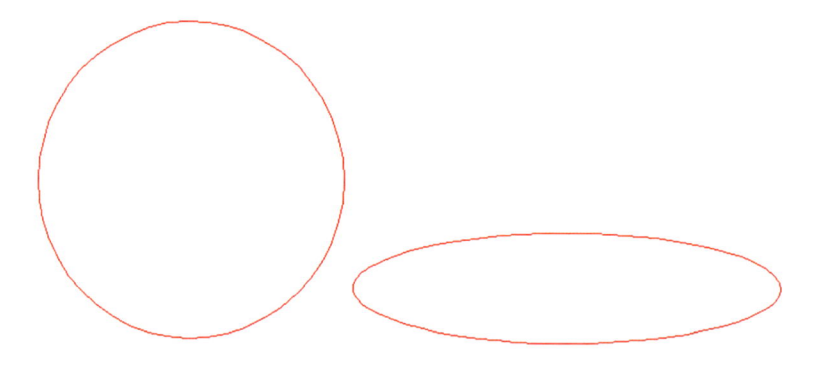

Earth's orbital eccentricity: .017   Comet Halley's orbital eccentricity: .967

Here are the eccentricities of the orbits of the (traditional) planets of the solar system:

Planet	Eccentricity of orbit
Mercury	.206
Venus	.007
Earth	.017
Mars	.093
Jupiter	.048
Saturn	.054
Uranus	.047
Neptune	.009
Pluto	.249

It is evident that the planetary orbits are not entirely circular, with Venus and Neptune being the closest to perfect circles.

Law II *A line from the Sun to a planet sweeps out equal areas in equal times.* This suggests that the planets' orbital velocities increase as they are nearer the Sun (being fastest at perihelion), and are at their slowest when farthest away (aphelion), which is the case. This law is derived below.

Law III *The square of the period of an orbit (measured in years) is equal to the semi-major axis of the orbit (measured in AU)*. Let's test this on the solar system[4]:

Planets and Pluto	Semi-major axis distance (a) in AU	Cube of distance (a^3)	Period in years (P)	Square of period (P^2)
Mercury	.387	.058	.241	.058
Venus	.7233	.378	.615	.378
Earth	1.00	1	1.00	1
Mars	1.524	3.54	1.881	3.54
Jupiter	5.203..	140.8	11.86..	140.7
Saturn	9.537..	867.3	29.447..	867.1
Uranus	19.189..	7.066	84.017..	7.059
Neptune	30.070..	27.189	164.79..	27.156
Pluto	39.48..	61.546	247.92..	61.464

Simply stated, the law says that for any set of planets of respective periods P_n and distances a_n from the Sun, then

$$P_1^2 : a_1^3 :: P_2^2 : a_2^3$$

Looking at the table, one can see how the third and fifth columns match. The lack of exactitude for some of the outer planets in the matches, however, reveals something about Kepler's law. The overwhelming primary mass of the Sun in relation to the other planets means that, for the smaller planets in particular, the orbits obey this ideal proportion, or at least that deviations from it are not noticeable at reasonable levels of accuracy. But as we will see later, the great masses of Jupiter and Saturn, and to a lesser extent Uranus and Neptune, do prevent the periods squared and distances cubed of those planets from being perfectly proportional without some correction for their masses, though this was not detectable in Kepler's time. Given this, Kepler's Third Law is actually more of an approximation than a law. Newton later showed how to take these secondary masses into account and render exact the period squared, distance cubed relationship in an ideal two-body system (that is, a system that ignores the effects of other planets). Even with Newton's modification of the law, however, there are the unaccounted-for influences of one planet on another. If we had even more refined numbers we might discern perturbations or gravitational interference of the large planets in particular on the others, for example Jupiter on Saturn when they are near each other. These perturbations tend to force the perturbed planet out of a purely elliptical orbit. It was just the lack of exact conformity to Kepler's Third Law, even after accounting for the masses and gravitational pull of Uranus and the other gas giants, which led to Leverrier's and Couch's prediction of the existence of Neptune.

---

[4] The data on masses and sidereal periods were taken from the Jet Propulsion Laboratory website http://ssd.jpl.nasa.gov/?planet_phys_par and the distances from the Sun were derived from http://ssd.jpl.nasa.gov/txt/p_elem_t1.txt. The ellipsis after some numbers indicates truncation (not rounding).

The Basic Geometry of the Ellipse

Let's look a little more at the ellipse, which, as Kepler showed is a form of fundamental importance in orbital mechanics. We begin with the basic terminology. The definitions below apply to the accompanying ellipse figure:

The Ellipse
(Showing Sun and planet with eccentricity greatly exaggerated)

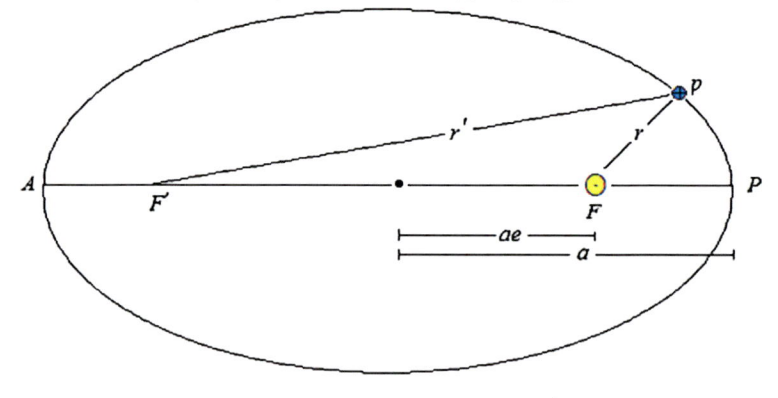

A is aphelion   *a* is the semi-major axis

P is perihelion   *e* is the eccentricity

p	is the point (such as a planet), located at distance r from the *focus F*
r	is the *radius vector*, which is the line between *F* and *p* (here, *p* is for planet), always in the direction of *p* with a magnitude equal to the distance between *F* and *p*, and which varies as *p* sweeps counterclockwise around in its orbit. In an ellipse, by definition $r + r'$ is always constant. At one point (on the semi-major axis) $r + r' = 2a$, so we know it is equal to $2a$ everywhere else on the ellipse
☉	is the mass orbited, here shown as the symbol for Sun. It is at *F* the focus of the ellipse
F	is at one *focus* of the ellipse and is the location of the center of mass of the system, which in the solar system is the Sun[a]
F'	is the *empty focus* of the ellipse, with no physical significance
a	is the *semi-major axis* of the ellipse. For a circle, $a = r$. The line *AP*, whose distance is $2a$, is the *major axis* of the ellipse
P	is the location of the *perigee*, which is the point of shortest distance to *F* from the orbit. If *F* is the Sun, another star or our Moon, *P* would be called the *perihelion*, *periastron* or *perilune*, respectively.[5] It is often generically referred to as the *periapsis*

(continued)

---

[5] Perihelion is nowadays often denoted by the letter *q*.

A	is the location of the *apogee*, which is the point of farthest distance to F from the orbit. If F is the Sun, another star or our Moon, A would be called the *aphelion*, *apastron* or *apalune*, respectively.[6] It is often generically referred to as the *apoapsis*
ae	is the displacement from the center of each focus, where e is the *eccentricity* of the ellipse. The eccentricity is the distance between the two foci, F and F', divided by the major axis, $2a$

[a]As we will see later, the center of mass of the solar system is near, but not at the center of the Sun, and in fact depends upon on the alignment of the planets, particularly Jupiter and Saturn

An ellipse may be drawn with a length of string attached to the two focus points, and a pencil moving around the outermost perimeter permitted by the string. In that case, $r + r'$ will always be constant, being the total length of the string. One of the most important things to learn about an ellipse is the concept of eccentricity. Varying the distance between focus points will vary the eccentricity. It is a measure of the flatness of the ellipse. A circle has an eccentricity of 0 (where the distance ae is zero). The flattest ellipse has an eccentricity of just less than 1.

The long *major* axis is $2a$, and half this, the *semi-major* axis is a. The body orbited, whether it is the Sun or the Earth or any other planet, would in this diagram be at F. As a circle is stretched to become an ellipse, the distance between F and F' becomes greater. Pinch them back together and the ellipse becomes a circle. The amount of flattening, again, is determined by e, the eccentricity. The eccentricity is a number between zero (a circle) and just less than one (an almost flat loop), one step away from being "broken" into a parabola. So an ae of zero means there is no distance between the two foci and the shape is, again, a circle. As ae approaches one, the distance between the center and each focus approaches a.

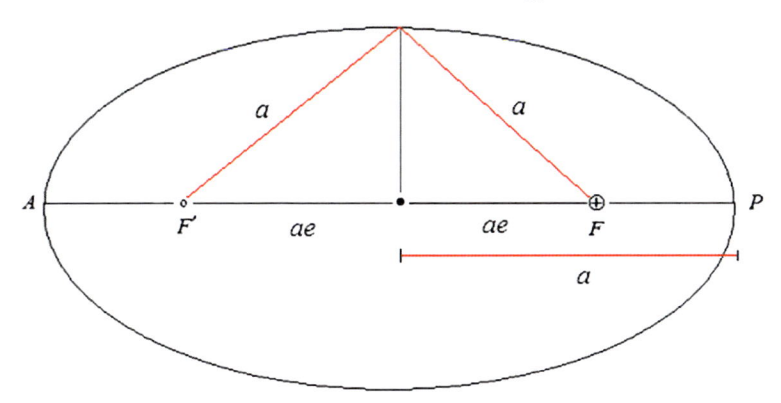

a is the semi-major axis
e is the eccentricity

---

[6] Aphelion is often denoted by the letter Q, which we will use later to avoid confusing it with period P of a body in orbit.

Note how the semi-major axis also forms an isosceles triangle above (or below) the center of the ellipse, as shown.

Now notice r, which is the distance from the focus F and the perimeter. It can move about the focus, as the angle θ sweeps around. If the Sun were at F and the Earth were at p, the line r would be the distance from the Earth to the Sun at every moment of its orbit. As noted, the point P is the place in the orbit nearest the Sun, the perihelion, and point A is the aphelion. These are derivations from the Greek language meaning near and far from the Sun, respectively. In any ellipse, the sum of the distances from p to F and to F' is always constant. Now we can put the value of r in elliptical terms, at the perihelion and aphelion points:

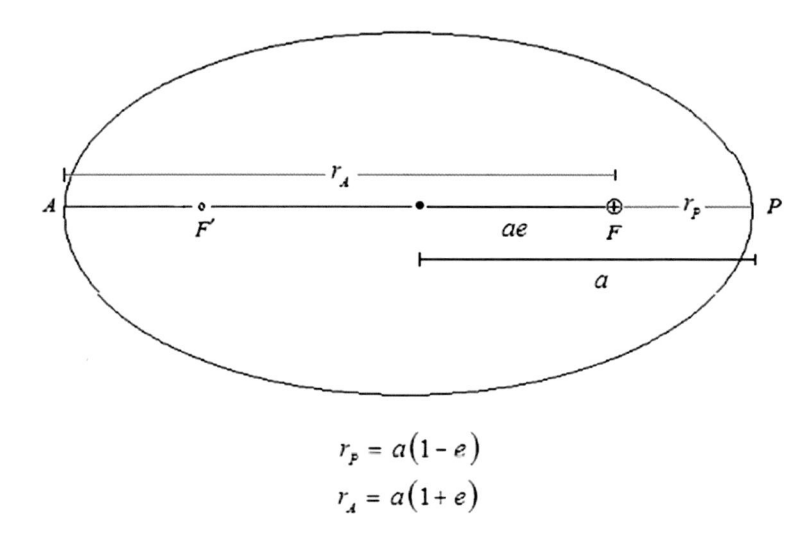

$$r_p = a(1-e)$$
$$r_A = a(1+e)$$

At both perihelion and perihelion, the r line lie lies flat along the major axis, the line of *apsides*. At perihelion the value will be $a - ae$ and at aphelion the value will be $a + ae$. So the values of r at these points are:

$$r_P = a(1-e)$$

$$r_A = a(1+e)$$

The Basic Geometry of the Ellipse

We can see another feature of the ellipse, the *semi-minor axis, b*:

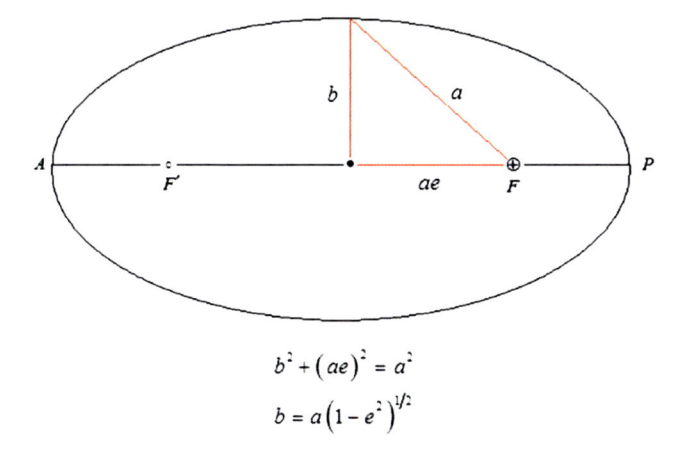

$$b^2 + (ae)^2 = a^2$$
$$b = a(1-e^2)^{1/2}$$

Using the Pythagorean theorem, the value of the semi-minor axis can easily be calculated:

$$b = a\sqrt{1-e^2}$$

As the eccentricity approaches zero, then the semi-major and semi-minor axes become equal, as in a circle. At the point of equality, they both represent the radius. Compare the area of a circle with that of an ellipse:

$$A_e = \pi ab \} \text{ Area of an ellipse}$$

$$A_c = \pi r^2 \} \text{ Area of a circle}$$

In the case where the two axes of the ellipse are equal, the equations become identical.

Determining the Eccentricity of the Moon's Orbit

Problem The Moon's perigee (minimum distance from the Earth's center) is 363,104 km, and its apogee (maximum distance) is 405,696 km. Derive an equation for finding the eccentricity of an ellipse, where only perigee and apogee are known; and, using that formula and the information above, find the eccentricity of the Moon's orbit.

Given The accompanying diagram of the ellipse.

Assumptions We will use this idealized ellipse to represent the Moon's orbit around the Earth, though the Moon's actual orbit is not so elongated. The Earth will be assumed to be at the focus F.

Method We notice that the only place e appears in the diagram is in the length ae. To solve the problem, we will therefore need to find an equation relating ae to perigee P and apogee A. There are several ways to do this. One way is to add or subtract ae from the semi-major axis value a. This will yield the equation.

Calculations

For perigee:
$$A = a + ae$$
$$A = a(1 + e)$$

For perigee:
$$P = a - ae$$
$$P = a(1 - e)$$

We can eliminate the a term by dividing one expression into the other:

$$\frac{A}{P} = \frac{a(1+e)}{a(1-e)}$$

$$\frac{A}{P} = \frac{(1+e)}{(1-e)}$$

$$A(1-e) = P(1+e)$$

Next we solve for e,

$$A - Ae = P + Pe$$
$$A - P = e(A + P)$$
$$e = \frac{A - P}{A + P}$$

Since we have found an easy expression for determining eccentricity, we can calculate it from the data given. Since perigee is 363,104 km and apogee is 405,696 km, we can compute eccentricity:

$$e = \frac{404696 - 363104}{405696 + 363104}$$

$$e = .0554$$

Observations

1. This value is therefore just off from fairly circular. The "eccentric" ae for the Moon is 21,296 km, or about 3⅓ Earth radii off from a circular orbit.
2. If we know P and A, we can of course find the semi-major axis a since $A + P = 2a$. Hence $a = (A + P)/2$, which we can interpret by saying that a is the average or *mean distance* of object p from F during the course of its orbit.

The Basic Geometry of the Ellipse

3. Likewise a can be derived from either A or P. The useful resulting expressions are:

$$a = \frac{A}{1+e} \quad a = \frac{P}{1-e}$$

4. It is instructive to see how the apsidal distances relate to the semi-minor axis, depending on eccentricity. We refer again to this diagram of the ellipse:

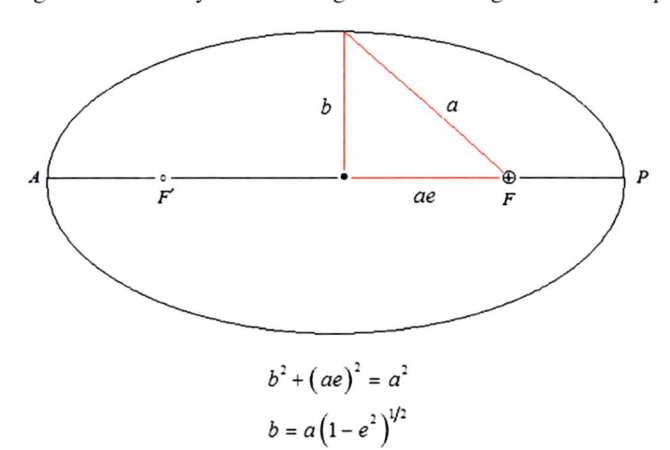

$$b^2 + (ae)^2 = a^2$$
$$b = a(1-e^2)^{1/2}$$

Mars, for example has a semi-major axis of 1.5237 AU. Its eccentricity is .093394. Its semi-minor axis is therefore,

$$b = a\sqrt{1-e^2}$$

$$b = 1.5237\sqrt{1 - .093394^2}$$

$$b = 1.517 \text{ AU}$$

So we can see that in an orbit like Mars', where the eccentricity is fairly small and the orbit not far from circular, the semi-minor axis differs only slightly from the semi-major axis, as noted in above in the discussion of Kepler's struggle to find the true shape of its orbit.

5. Take $b^2 = a^2(1 - e^2)$ and expand the terms within the parentheses:

$$b^2 = a^2(1-e)(1+e)$$

Since $A = a(1+e)$ and $P = a(1-e)$, there follows this interesting result:

$$b = \sqrt{AP}$$

6. As should be evident, an ellipse is determined in *shape* by its eccentricity and in *scale* by the length of its semi-major axis. Getting an accurate sense of an orbit's eccentricity is thus a key step to understanding, and ultimately predicting, an

object's path in the sky. If only the perihelion and semi-major axis of an object are known, the eccentricity can easily be calculated from this simple equation:

$$e = 1 - \frac{a}{P}$$

In that equation just two quantities, the perihelion distance and the semi-major axis, are all that are needed to construct the orbit. When just the semi-major and semi minor axes are known, eccentricity can be determined from this equation:

$$e = \sqrt{1 - \frac{b^2}{a^2}}$$

For example, in the case of Mars, $a = 1.5237$ AU and $b = 1.517$ AU. Hence with those as the known or given quantities, we could derive e from that equation:

$$e = \sqrt{1 - \frac{1.517^2}{1.5237^2}}$$

$$e = .093394$$

7. When the apsidal distances and semi-major axis are known, then we can use this equation:

$$e = \frac{\sqrt{1 - AP}}{a}$$

The reader is encouraged to derive the above equations on his or her own.

Deriving Kepler's Law of Areas

Problem Show how Kepler's Second Law, his so-called Law of Areas, flows naturally from the law of conservation of angular momentum.

Given

$L = mvr$	Angular momentum L at right angles to the radius, being the product of mass m, circular velocity v, and radius r.
$A = \frac{1}{2}bh$	The area of a triangle, being one-half the base times the height
θr	Distance along an arc, where θ is in radians and r is the radius vector. This distance approximates the base of a triangle when the angle is sufficiently small. This small angle we'll call $\Delta\theta$

Assumptions We will assume the law of conservation of angular momentum, a bedrock physical law, and the perfect "Keplerian" motion of a small planet (of

The Basic Geometry of the Ellipse

relatively insignificant mass compared to the Sun) orbiting the Sun in an elliptical path without being affected by the perturbing gravitational influences of other planets.

Method Consider two small, near-triangular slices of an ellipse. Imagine an object orbiting the Sun which is located at one focus, as in the accompanying figure.

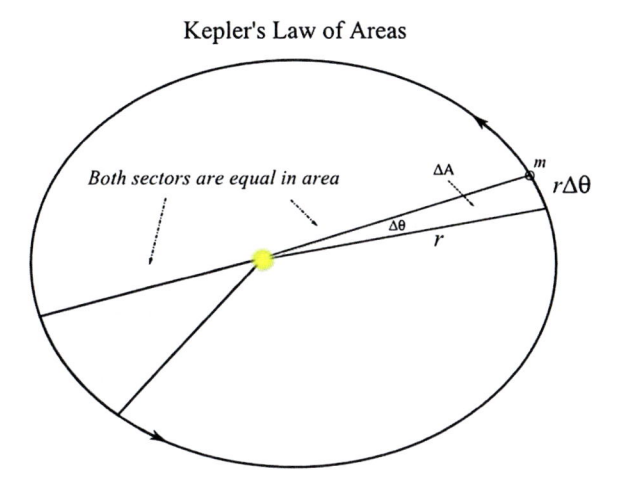

Kepler's Law of Areas

Both sectors are equal in area

Equal areas are swept out in equal times
$\Delta A / \Delta t = $ constant

At any small increment of time the angle (in radian measure) it sweeps out is $\Delta\theta$.[7] The small arc of a given small triangle, $r\Delta\theta$, represents the distance the planet travels along the ellipse in the same small amount of time, Δt. The component of velocity at right angles to the radius will be the distance traversed along the arc divided by the increment of time, $v = r\Delta\theta/\Delta t$.

This angular velocity notation will be useful in linking the constant of angular momentum with the areas of each sector. With appropriate subscripts, we can define two such small triangles. Their areas, ΔA_1 and ΔA_2, will be one-half their bases ($\Delta\theta_1 r_1$ and $\Delta\theta_2 r_2$) times their heights (r_1 and r_2) or $\Delta A_1 = \Delta\theta_1 r_1^2/2$ and $\Delta A_2 = \Delta\theta_2 r_2^2/2$.

Calculations The value of angular momentum at r_1 and r_2 is

$$L = mv_1 r_1 \quad \text{and} \quad L = mv_2 r_2$$

Substituting angular velocities into the angular momentum equations,

$$L = m\left(\frac{\Delta\theta_1}{\Delta t}\right) r_1^2 \quad \text{and} \quad L = m\left(\frac{\Delta\theta_2}{\Delta t}\right) r_2^2$$

---

[7] See Chap. 3 for a discussion of radian measure.

and acknowledging that angular momentum of the small mass is the same everywhere in the orbit, then

$$m\left(\frac{\Delta\theta_1}{\Delta t}\right)r_1^2 = m\left(\frac{\Delta\theta_2}{\Delta t}\right)r_2^2$$

Now multiply each side by $\Delta t/2m$:

$$\frac{1}{2}\Delta\theta_1 r_1^2 = \frac{1}{2}\Delta\theta_2 r_2^2$$

These equations define the areas of the respective (very small) sectors, and they are equal; thus,

$$\Delta A_1 = \Delta A_2$$

Both sectors therefore have equal areas swept out in the given equal times.

Observations

1. It is true that the arc along the curve does not form the rectilinear base of an exact triangle, and that the truer measure of area would be to use the chord length. But as we will see in the discussion of Newton's "evanescent arcs" in Chap. 5, the angle θ can be made as small as one chooses, and as it becomes very small, the chord length and the arc length ultimately approach equality, and the error disappears with infinitesimally small angles.
2. A more direct approach to the solution of this problem, though perhaps less intuitive, would be to see how, in any arbitrary triangular sector within the ellipse (again with its apex at the Sun), the change of its area with respect to time will always be constant. In other words, the area of any arbitrary sector being

$$\Delta A = \frac{1}{2}\Delta\theta r^2$$

The change of that area with respect to time is (dividing each side by Δt),

$$\frac{\Delta A}{\Delta t} = \frac{1}{2}\left(\frac{\Delta\theta}{\Delta t}\right)r^2$$

But since angular momentum is $L = m(\Delta\theta/\Delta t)r^2$ and is constant everywhere in the orbit, then

$$\frac{\Delta A}{\Delta t} = \frac{L}{2m} = \text{constant}$$

In other words, no matter what sector we pick, the change in its area with respect to time will be the same, which is Kepler's Second Law.

The Basic Geometry of the Ellipse

Determining the Semi-major Axis of the Asteroid Ceres

Problem Ceres, the first asteroid discovered (1803), has a period of 4.6 years. Using Kepler's Third Law, estimate the distance of Ceres from the Sun in AU.

Given

$P^2 = a^3$	Kepler's Third Law, where the units are years and AU

Assumptions We will disregard the gravitational effects of other planets or other perturbing forces.

Method Since we know the period, we simply solve for a.

Calculations

$$P^2 = a^3$$

$$a = P^{2/3}$$

$$a = (4.6)^{2/3}$$

$$a \simeq 2.77 \cdot \text{AU}$$

Observations

1. The result makes sense, since Ceres lies in the asteroid belt between Mars and Jupiter, which planets are respectively about one-and-a-half and five astronomical units out.
2. What if instead of astronomical units and years we used other units, such as kilometers and days? Since Kepler's Third Law is a ratio, the law holds for any consistent set of units. Expressing the law this way (where the symbol : indicates proportion),

$$P_1^2 : P_2^2 = a_1^3 : a_2^3$$

Then, for any single planet,

$$P^2 = ka^3$$

where k is a constant of proportionality, and $k = P^2/a^3$. Since the units in this law are years and astronomical units, for Earth, $k = 1$ (so that $P^2 = a^3$). That is the constant, using years and AU, applicable throughout the solar system. That is how we approached the Ceres problem above. But if the units were kilometers and days, then the applicable constant can be found, for example for Earth, at a mean distance of about 149.6 million kilometers:

$$k = \frac{P^2}{a^3}$$

$$k = \frac{365.24^2}{149,600,000^3}$$

$$k = 3.984 \times 10^{-20}$$

This is a rather inconvenient number! Let's test it on Mars, whose mean distance from the Sun is about 227.9 million kilometers. Its orbital period squared, using the constant, should be

$$P^2 = ka^3$$

$$P^2 = 3.984 \times 10^{-20} (2.279 \times 10^8)^3$$

$$P^2 = 4.72 \times 10^5$$

$$P \approx 687 \text{days}$$

This is the correct approximate value for orbital period of Mars. This labor reveals the great convenience of using years and AU, where the constant is just 1!

How Conic Sections May Be Generated by One Equation

Problem Using the accompanying figure, derive the general equation for the conic section.

Given The accompanying figure.

Assumptions Since this is an idealized mathematical construct, there are no physical assumptions that need to be made

Method Referring to the accompanying figure, the conic section has the property that all points on it have a constant ratio e (eccentricity) between the distance r from the focus F and a line known as the directrix.[8] Thus, $r/d = e$. The value of r is of great use in orbital problems. We can find r by ordinary trigonometry. We know DF is the distance d plus the distance r projected onto the x axis, or $r \cos \theta$. We can then put d in terms of r and e from the immediately preceding ratio, $d = r/e$, so the reference to the directrix disappears. Note that at $\theta = 90°$, or $\pi/2$ (assuming

---

[8] The derivation of the conic equation is standard in many textbooks, but the approach here was inspired by the particularly concise treatment given by Peter van de Kamp [3].

The Basic Geometry of the Ellipse

counterclockwise motion, as with solar system planetary orbits) the radius vector is vertical, and is equal to the parameter p. At that point $r = p = ed$.

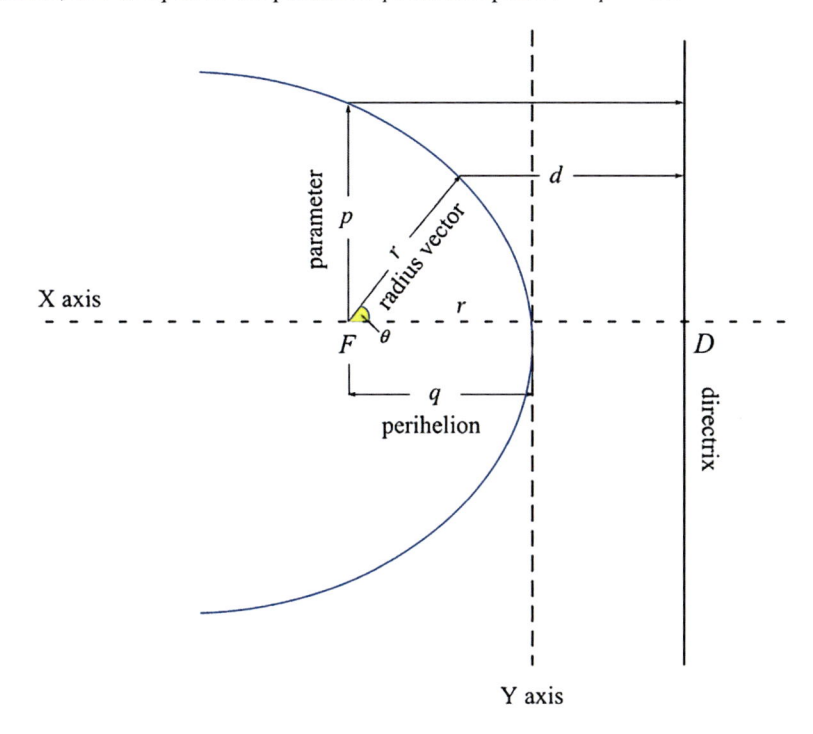

Defining the Conic Section

Calculations The distance from focus to directrix is:

$$DF = d + r\cos\theta$$

Since $d = r/e$, then

$$DF = \frac{r}{e} + r\cos\theta$$

$$DF = \frac{r}{e}(1 + e\cos\theta)$$

When $\theta = \pi/2$ (at 90°), then, at that point, $DF = d$. As noted, the parameter $p = ed$, so,

$$ed = r(1 + e\cos\theta)$$

$$p = r(1 + e\cos\theta)$$

And from this follows the very important relation which is the general equation for the conic section,

$$r = \frac{p}{1 + e \cos \theta}$$

Observations

1. Let us see what happens to the orbit when we vary eccentricity:

   If $e = 0$, then $r = p$ and the orbit is *circular*.

   If $e > 0$ but < 1, the parameter is smaller as e is larger (but never unity), and the orbit is an *ellipse*. That is, the width of the orbit becomes less than a circle. At $\theta = 0$, perihelion, then $r = p/(1 + e)$. The perihelion distance is always greater than half the parameter.

   If $e = 1$, the orbit is a *parabola*. At $\theta = 0$, $r = p/2$. The perihelion distance is always half the parameter.

   If $e > 1$, the orbit is a *hyperbola*. At $\theta = 0$, $r = p/(1 + e)$, the perihelion distance is always less than half the parameter. A comet, for example, with an eccentricity even slightly greater than unity will experience a perihelion passage very close to the Sun.

   We will encounter all of these orbital shapes in this book.

2. In the accompanying figure are the arcs of three conics.

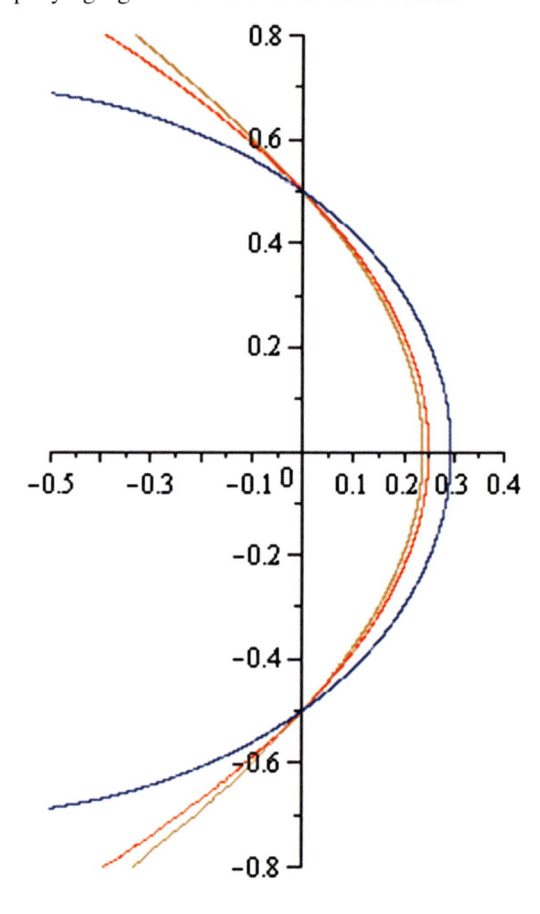

The Basic Geometry of the Ellipse

The plot is of the inner arcs of the equation for conics $r = p/(1 + e \cos \theta)$, for $e = .7$ (an ellipse, in blue), $e = 1.0$ (parabola, in red) and $e = 1.1$ (hyperbola, in gold). Notice the value of the radius vector at $\theta = \pi/2$, along the y axis is .5 for each curve, which is the parameter. Note that, at $\theta = \pi/2$, the cosine of the angle is zero and so the radius vector is equal to the parameter, $r = p$ for all conics at that angle.

3. It is important in the ellipse to be able to freely derive the semi-major axis from the parameter, and vice versa, when the eccentricity is known, or to estimate one of the unknowns when the other two are known or estimated. Since,

$$r_P = \frac{p}{1+e} \text{ for } \theta = 0 \text{ (perihelion)} \quad r_A = \frac{p}{1-e} \text{ for } \theta = \pi \text{ (aphelion)}$$

The semi-major axis is the average of these radii vector distances in the ellipse:

$$\frac{r_P + r_A}{2} = \frac{p}{1-e^2}$$

$$a = \frac{p}{1-e^2}$$

And,

$$p = a(1 - e^2)$$

This is a most useful relation. Let's try it again on Mars. With the information given above, find the parameter of the Martian orbit using AU as the units:

$$p = 1.5237(1 - .093394^2)$$

$$p = 1.5104 \text{ AU}$$

We have thus found the breadth of the Martian orbit by this simple means.

4. Do not confuse the semi-minor axis b equation with that for the parameter p:

$$p = a(1 - e^2) \qquad b = a\sqrt{1 - e^2}$$

5. If one takes the equation for the conic section and substitutes $a(1 - e^2)$ for p, then the resulting polar equation, independent of the parameter, is:

$$r = \frac{a(1 - e^2)}{1 + e \cos \theta}$$

6. One can use the polar equation to plot the orbit of Halley's comet, designated 1P for the first periodic comet. Its key orbital elements are $e = .967$ and $a = 17.834$ AU (values are rounded here).[9] With just these two numbers, and *Maple* software, the plot can be generated using the equation for the ellipse (numbers are AU):

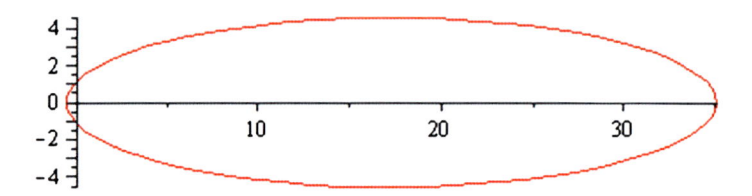

Orbit of Halley's Comet (aphelion at 35.7 AU)

Graphing the Orbit of Comet Schwassmann-Wachmann 3 Using Rectangular Coordinates

Above we derived the equation for the elliptical orbit in polar coordinates of r and θ. But there are many occasions when it is useful to work in the Cartesian, rectangular coordinates of x and y. There are some beautiful symmetries to appreciate too. The equation for the ellipse is,

$$\frac{x^2}{a^2} + \frac{y^2}{b^2} = 1$$

Where a and b are the respective semi-major and semi-minor axes of the ellipse, and the center of the ellipse is at the origin. We can see that if those axes are equal, such that $b = a$, then the equation becomes (by multiplying through by a^2) that of a circle of radius a:

$$x^2 + y^2 = a^2$$

Periodic comet Schwassmann-Wachmann 3, discovered in 1930 by astronomers Arnold Schwassmann and Arno Arthur Wachmann at Hamburg Observatory in Germany, is a regular visitor in our solar neighborhood, having an orbital period of a little less than five and a half years. As sometimes happens, so-called tidal forces exerted by the gravitational pull of other planets in the solar system, and by the Sun, can eventually cause a comet to break apart. Comet Schwassmann-Wachmann 3

---

[9] See the JPL Orbital Dynamics website, http://ssd.jpl.nasa.gov/sbdb.cgi?sstr=1P&orb=1#top

The Basic Geometry of the Ellipse

began to do so in 1995, but in May of 2006 it broke apart spectacularly: into first five then eight segments, and some of those pieces further disintegrated. As the comet passed by Earth in 2006 one could easily see with an amateur telescope (as the author did) the rare and beautiful sight of these multiple comets with their tails hovering against the backdrop of stars in the constellation Corona Borealis. The pieces were separated from each other after doubtless billions of years together and hundreds of millions of orbits. The problem below will presume we are working with the largest remnant, and likely parent of the breakaway comets.

Problem Given the information below,[10] graph the orbit of periodic comet 73P Schwassmann-Wachmann 3 in rectangular coordinates, with the long axis of the orbit on the x axis. Locate the foci and perihelion of the ellipse. Include the orbit of the Earth for reference.

Given

a	Semi-major axis of the comet: 3.063 AU
e	Eccentricity of the comet's orbit: .692
q	The comet's perihelion distance in AU: .9428
P	The comet's period in years: 5.36

Assumptions We will consider the comet's inclination of 11.39° and assume an orbit planar to the plane of the *ecliptic* (the Earth's orbital plane around the Sun being the reference plane). These orbital elements are the *osculating* elements taking into account the perturbations that have affected this comet's orbit as of the *epoch* (the time) when given. We assume the Earth's orbit is circular for this graph.

Method Use the equation for an ellipse in rectangular coordinates,

$$\frac{x^2}{a^2} + \frac{y^2}{b^2} = 1$$

Since the eccentricity and semi-major axis are given, the equation,

$$b = a\sqrt{1 - e^2}$$

will yield b. From the origin, or center of the graph, lay out the semi-major axis with ± a on each side of the origin. The maximum and minimum values of y will be ± b. To find the two foci, simply calculate ± F from either of these equations:

$$F = ae \qquad F = \sqrt{a^2 - b^2}$$

---

[10] The information may be found at the Minor Planet Center, website, http://www.minorplanetcenter.net/iau/Ephemerides/Comets/and from Elements and Ephemeris for 73P-C/Schwassmann-Wachmann at http://ubasti.cfa.harvard.edu/~cgi/ReturnPrepEph?d = c&o = 0073 P for epoch October 6, 2011.

80   4 The Geometry of the Solar System: Kepler's Laws of Planetary Motion

The Earth's orbit is a circle of radius 1 AU with the Sun (the focus) at the center. The graph may be done manually, by a graphing calculator, or by the use of math software, such as *Maple*, which produced the graph below.

Calculations First we calculate the semi-major axis:

$$b = a\sqrt{1 - e^2}$$

$$b = 3.063\sqrt{1 - .692^2}$$

$$b = 2.21 \text{ AU}$$

The foci are located at distance $\pm F$ from the origin:

$$F = \pm ae$$

$$F = \pm 2.1$$

The graph should look something like this:

**Orbit of Comet 73P Schwasssmann-Wachmann
(shown with Earth orbit as smaller circle)**

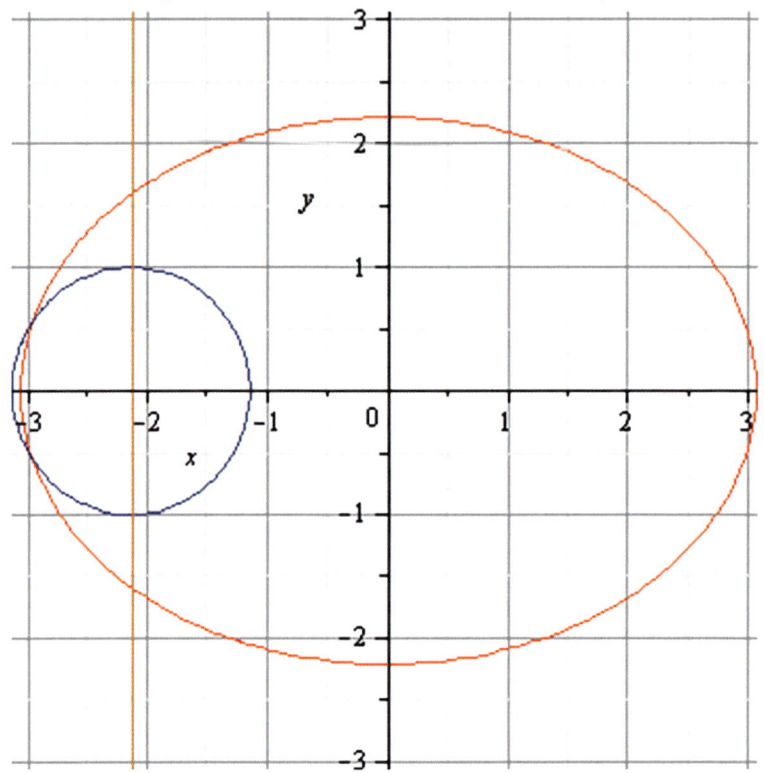

The Basic Geometry of the Ellipse

The Sun lies at one focus of the ellipse, at distance -2.1 AU from the origin. A vertical line has been drawn up through this focus. The empty focus is at $+2.1$ AU from the origin. The aphelion lies about 5 AU from the Sun, which is in the vicinity of Jupiter's orbit. (Recall that the aphelion is $2a - q \approx 5.2$ AU.) The tidal influences of Jupiter may have caused the breakup of this comet.

Observation As will be seen later, graphing an orbit in rectangular coordinates will prove to be highly useful in visualizing the position of the orbit in space.

Exercises: Applying Kepler's Laws to Comets, Asteroids, Spacecraft, and Moons The power and reach of Kepler's Laws is often surprising. These basic building blocks were empirically discovered by Kepler, and one of the first triumphs of Newton's work was to derive Kepler's laws theoretically from the principles of gravitation. Later we will discuss how Newton was able to do this. For now it is good just to practice using the laws, particularly Kepler's Third Law, on a variety of celestial objects.

Problems

1. Periodic comet Bennett C/1969 Y1 has a perihelion distance of $q = .5376$ and aphelion $Q = 281.89$ AU. Calculate the eccentricity of its orbit, its semi-major and semi-minor axes.
2. A spacecraft launched from Earth has a perihelion distance at Earth's orbit and an aphelion at Jupiter's orbit, at 5.2 AU. What is its period?
3. Distant, dwarf planet Sedna has an orbital period of 12,598.57 years. What is its semi major axis?
4. Periodic 2P/Encke has a semi-major axis of 2.21449 AU. What is its period in years?
5. The semi-major axis of Pluto's moon Charon is about 15.235 plutonian radii, and has an orbital eccentricity of .0022. What are its apsidal distances Q and q from Pluto?
6. Identify which of the following statements is incorrect. Assuming for a planet in the solar system the units are years and astronomical units: (a) the period is equal to the semi-major axis times the square root of its semi major axis; (b) the period is equal to the square root of the cube of the semi major axis; (c) the ratio of the period squared to the semi major axis cubed is a constant; (d) the period is equal to the semi major axis raised to the two thirds power.
7. Asteroid Eros has a semi-major axis of 1.45793 and a parameter of 1.38573. Find its period, its semi-minor axis, and its eccentricity.
8. Comet 1/P Halley and comet 13/P Olbers have periods of 75.32 and 69.52 years respectively and respective eccentricities of .967 and .93. What is the ratio of the semi-major and semi-minor axes of comet Halley to comet Olbers? Interpret your results.
9. Comet C/2007 Lulin has a nearly parabolic orbit with a parameter of 2.42458 AU and a perihelion distance of 1.212 AU. Sketch a graph of the orbit of this comet in relation to the orbits of the three inner planets.

10. Using the standard equation for the conic, draw a graph of the inner orbits (within 3 AU of the Sun) of three comets whose parameters are each .5 and whose eccentricities are .5, .8, and 1.0, respectively.

References

1. Koestler A (1985) The watershed: a biography of Johannes Kepler. University Press of America, Lanham, Md.
2. Rufus WC (1931) Johann Kepler, History of science society, Special publication no. 2. Waverly Press, Baltimore, pp 17–22
3. van de Kamp P (1964) Elements of astromechanics. W. H. Freeman, San Francisco, pp 8–9

Chapter 5
How the Moon Falls Toward the Earth (but Keeps Missing It)

Galileo had shown by experimentation and theory how things fall, and documented his findings in his 1638 *Dialog Concerning Two New Sciences*. He asked simple questions about everyday phenomena and tested them with experiment until he understood the general laws behind the phenomena of motion. Isaac Newton applied Galileo's and Kepler's laws of motion to the moon and planets and their satellites, and, with brilliant and subtle geometrical reasoning, developed the universal theory of gravitation that united earthly and heavenly phenomena. This he laid out with astonishing rigor in the first part of Newton's *Mathematical Principles of Natural Philosophy* (first published in Latin in 1687, and commonly known today by the first word of its Latin title, the *Principia*), perhaps the greatest masterwork of human thought in the history of science. Newton showed that the same force (gravitation) that caused the fall of an apple from a tree caused Jupiter's moons to orbit Jupiter, Earth's Moon to orbit Earth, and all the planets and moons together to orbit the Sun.

In his *Principia*, Newton dealt comprehensively with the implications of orbital motion under the influence of a centripetal force. We will see later how he demonstrated that the motions of the planets and moons under a hypothesized force that diminished inversely with the square of distance from the primary mass accurately matched the phenomena of the solar system. He proved in elegant prose and arcane mathematical language a vast body of propositions relating the motion of these revolving bodies, the tides, the orbits of comets, precession, and very much more. The *Principia* laid the foundation for a new kind of mathematics, the calculus, to deal with limits and areas. The mathematical concepts and proofs developed mainly in Book I of the *Principia* were applied to the observed phenomena of the solar system in Book III, which stunningly demonstrated their explanatory power.[1] Newton was able to deduce the inverse square law in a variety of ways, showing that it was sufficient to explain elliptical orbits (as well as motion along the other conic sections, hyperbolas and parabolas) where the center of force was at one focus, and he freshly derived Kepler's laws of planetary motion. Edmund Halley (1656–1742) had read some early drafts of Newton's scattered papers, saw their immense value, and had encouraged the reluctant Newton to publish the *Principia*. Halley in fact personally financed its publication. This alone has insured Halley a prominent place in the history of science. Halley is best known for using Newton's mathematics to predict the return of the great comet of 1682. Its appearance in 1758, exactly as predicted by Halley, triumphantly demonstrated the success of Newton's science, and forever gave Halley's name to the famous comet.

The Facts of Inertia: Newton's First Law

Let us return to some of the basic questions Newton asked as he was investigating the findings of Kepler and Galileo and the phenomena of motion. While still a young man, Newton was curious whether the same earthly force that caused heaviness, that made rocks to fall to the ground, also caused this planet's pull on the Moon. But why doesn't the Moon fall and hit the Earth in the same way that a rock falls to ground? Newton speculated that while the Moon does in fact *fall*, it is in effect falling *around* the Earth. Suppose there is a *centripetal* ("toward the center") force (not yet named gravity) pulling on the Moon. That force continually pulls it toward Earth. There is another inertial force, however, that makes it want to adhere to a straight line path. If you twirl an object fast at the end of a string, you feel a tug on the string, more the faster you twirl it. The pull on the string is the inertial tug on an object wanting (so to speak) always to go in a straight line. If the string is cut, the object will fly off in a straight-line tangent to the circle. Newton pondered such facts of experience. Wouldn't the Moon behave in the same way if not continually drawn back to Earth by some centripetal force originating from Earth? He realized the same thing that Galileo had intuited: that the tendency for things to move in a straight line, or if at rest, to remain at rest, was a fundamental quality of objects with mass. The principle became the basis for his first axiom or law of motion.

---

[1] Sir Isaac Newton [1]. (Short title) *Principia.* Book I (On the Motion of Bodies) and Book III (The System of the World) treat the motions of bodies and the consequences of gravitation. Book II of the *Principia*, however, deals with the motion of bodies in resisting mediums, of little concern in contemporary celestial mechanics. It may have been written chiefly to rebut Descartes theory of vortices. See, for example, the last Scholium in Book II, after Proposition LIII: "Hence it is manifest that the planets are not carried round in corporeal vortices; for, according to the *Copernican* hypothesis, the planets going round the sun revolve in ellipses, having the sun in their common focus...But the parts of a vortex can never revolve with such a motion." From Sir Isaac Newton [1]. All excerpts from the *Principia* discussed here are drawn from this translation.

The Facts of Inertia: Newton's First Law

Newton's First Law states that, in space, undisturbed by friction or other forces, a body at rest or in motion will remain in that state forever. If it is moving, it will continue its motion in a straight line for all eternity. It may be at rest[2] or in constant, uniform motion, but it will not accelerate or decelerate. The only thing that will change its course from an eternal straight-line path would be an external force applied to it. As Newton stated it in the *Axioms or Laws of Motion* to the *Principia*:

LAW I

EVERY BODY CONTINUES IN ITS STATE OF REST, OR OF UNIFORM MOTION IN A RIGHT LINE, UNLESS IT IS COMPELLED TO CHANGE THAT STATE BY FORCES IMPRESSED UPON IT.[3]

Such a force could be, for example, the influence of another planet, moon or star; friction; something hitting it; or ejections from it (such as the exhaust thrust in the case of a rocket, or an eruption from a comet's nucleus) to move it left or right or up or down, or to slow it down or speed it up.

If the Moon were moving elsewhere out in space, nowhere near another body, it would move in a straight line, forever. It would have *rectilinear* motion. But being near the Earth, the Earth draws it to it, away from its straight line path, into a *curvilinear* course. It experiences a centripetal force pulling it inward toward the center of the arc of its motion, but this force is not so great as to completely overcome the Moon's tendency to move away in a straight line tangent to that arc. There is a composition of forces: the inertial tendency of the Moon always to move in a straight line is just balanced by the centripetal force (not yet articulated as gravity by Newton in the early portions of the *Principia*) tending to make it fall to the Earth.

As mentioned in Chap. 1, forces can act in combination, and their component vectors combine to create a resultant motion. The rectilinear motion of the Moon is its inertial, straight-line vector component, at a velocity that just matches the centripetal vector component, continually "drawing off" the Moon into the resulting orbit, which is a path between the Earth's center and straight ahead. The schematic figure below illustrates this, but one must imagine the arc as being squeezed to an almost infinitesimally small size for each instant, so the path of the Moon at each conceived impulse of pull from the Earth deflects the Moon away from straight line motion at each corresponding instant.

---

[2] Being at "rest" in space was shown by Einstein in 1905 to be an entirely relative concept. For now we will use the terms "rest" and "motion" in the sense of absolute rest and motion relative to an absolute frame of reference, as Newton intended.

[3] *Principia*, 13.

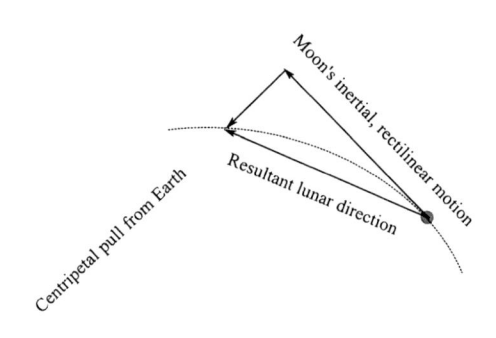

Components of Lunar Motion Every Instant
(Not to scale and arc size greatly exaggerated)

Of course the "impulses," like the arc lengths, must become correspondingly small so as to create, in our geometrical construction, an image of a *continuous* pull and a continuous deflection, as is the case in the physical world. This basic concept of vanishing arcs and small increments of time is fundamental to the calculus developed by Newton to analyze such motions.

If the Moon were somehow given a boost in its speed, the Earth's pull would, at the lunar distance, no longer be able to just balance its rectilinear motion: it would tend to fly off tangentially into space. Newton's example of a stone projected from the top of a mountain illustrates this. This famous diagram is from Book III of the *Principia* and accompanies his text.

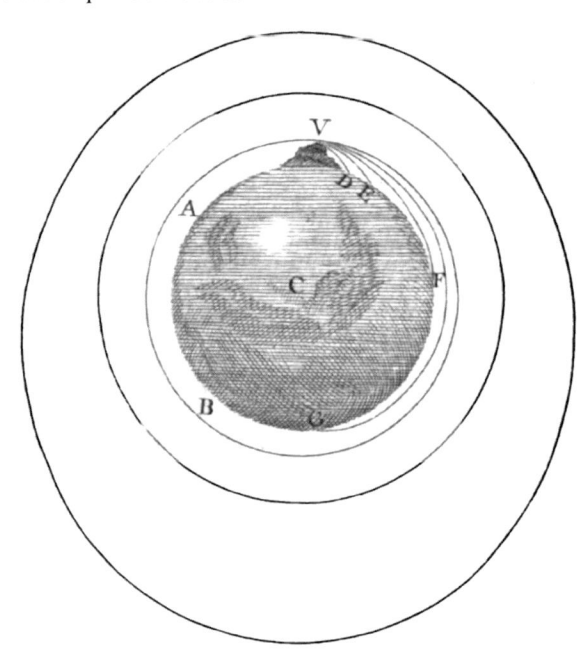

In the third part (Book III) of the *Principia,* called the *System of the World,* Newton lets us imagine a mountain, from which projectiles or stones are propelled horizontally with greater and greater force:

> [F]or a stone that is projected is by the pressure of its own weight forced out of the rectilinear path, which by the initial projection alone it should have pursued, and made to describe a curved line in the air; and through that crooked way is at last brought down to the ground; and the greater the velocity is with which it is projected, the farther it goes before it falls to earth. We may therefore suppose the velocity to be so increased that it would describe an arc of 1, 2, 5, 10, 100, 1000 miles before it arrived at the earth, till at last, exceeding the limits of earth, it should pass into space without touching it. ...
>
> [L]et us suppose there is no air about the earth...[I]f the velocity was still more and more augmented, it would reach at last quite beyond the circumference of the earth, and return to the mountain from which it was projected...and retaining the same velocity, it will describe the same curve, over and over.[4]

Newton has described, in so many words, how Earth satellites (not stones) would eventually be projected by rockets, and orbit the Earth, three centuries later.

What Is Centripetal Acceleration?

Any acceleration (change of motion) of an object is caused by a force acting on that object, in the direction of the induced motion. This should be true whether the object is at rest or already in motion (remember, motion is relative to the observer!). The force needed to accelerate a mass depends upon its mass, as we discussed in Chap. 1. The first clear exposition of these concepts came from Isaac Newton, in his Second Law of Motion. Here is how Newton put it, again from his *Axioms, or Laws of Motion*:

> LAW II
>
> THE CHANGE OF MOTION IS PROPORTIONAL TO THE MOTIVE FORCE IMPRESSED; AND IS MADE IN THE DIRECTION OF THE RIGHT LINE IN WHICH THAT FORCE IS IMPRESSED[5]

Our focus will be to apply this law first to find out about circular motion. (Understanding circular motion was Newton's first aim, and it led directly and naturally to understanding elliptical motion.) The Moon's orbit is nearly circular.[6] If the Moon is deflected from its course by a continual force pulling it towards the

---

[4] *Principia,* Book III, 551–52.

[5] *Principia,* Axioms, or Laws of Motion, 13.

[6] If the Moon's orbit were perfectly circular, its velocity at all points would be constant. It is assumed so for our instructional purposes, but any object in an elliptical orbit will move with velocities that vary depending on the nature of the ellipse, as it draws nearer or farther from the body orbited; its *mean* motion, however, will be constant. In the case of the Moon, too, the irregularities in the distribution of the mass of the Earth, the effects of the Sun's gravity, *etc.* all influence the shape of its orbit.

Earth, then there must be a continual acceleration acting in that direction.[7] Newton's Second Law states the equivalence of force (per unit of mass) and acceleration (commonly expressed by the equation $F = ma$).[8] This law is of great use in understanding the motions of the moons and planets, which orbit not in straight lines, but as if they are being continually drawn toward some center of force. Newton called the force acting *toward the center* the "centripetal force." He defined it this way:

DEFINITION V

A CENTRIPETAL FORCE IS THAT BY WHICH BODIES ARE DRAWN OR IMPELLED, OR ANY WAY TEND, TOWARDS A POINT AS TO A CENTRE[9]

This definition, at the beginning of the *Principia*, does not presume any particular hypothesis about what such a centripetal force might be. Newton does not assume that such a force is gravity, which was to be developed only later in the *Principia* as a unifying theory after he wove a meticulous and ornate tapestry of basic propositions. First, in Book I, he depicted geometrically how bodies tend to move. A body moving in a plane around the center of a circular orbit acts as if that body were drawn by a force pulling it toward the center. Only much later does he formally give this force a name and show how it applies to all the phenomena he presents in Book III.

Since there must be a force drawing an object into a curve, *all such motion along a curve must be accelerated motion*. And, according to Newton's earliest propositions, the accelerative force must be acting toward the center of the curve.

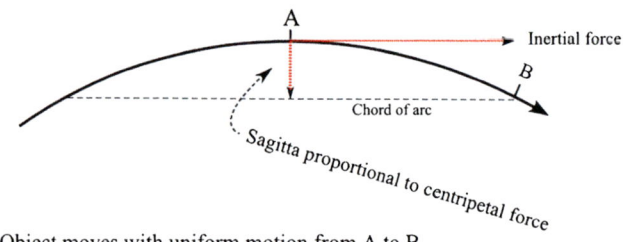

Object moves with uniform motion from A to B

In this figure, the object in orbit is moving (rather unusually for our solar system) clockwise, and the (not to scale) straight line vector toward the center of the arc

---

[7] A "change of motion" in Law II is a change in [mass × velocity]. In dealing with fixed masses, as is the case with most classical problems of celestial mechanics, Law II means that force = mass × change in velocity. Since a change in velocity is acceleration, then force = mass × acceleration. It can be said that the measure of mass is the force required to accelerate it. The equation also suits those situations where mass does change, such as in rockets whose expenditures of fuel continually reduce the mass of the rocket as it ascends.

[8] Sometimes acceleration and force will appear to be used interchangeably; this is because acceleration is force per unit mass.

[9] *Principia*, Definition V, 2.

What Is Centripetal Acceleration?

represents the direction of the centripetal force. The resultant motion along the arc of the orbit is again the composition of the inertial, straight-line (rectilinear) force of inertia wanting to keep the object going in a tangent to the curve, and the centripetal force pulling it inward. The vertical line, resembling a backward pointing arrow on a bow, is the bisector of the chord. It is called the *sagitta* (from the Latin word for arrow) *or versed sine* of the arc. We can think of it as the centripetal force vector whose length in each case will depend on the length and degree of curvature of the arc. Newton's object was to link the geometry of the curve to its centripetal force vector. This he did in a corollary to his first proposition of the *Principia*. Using his usual geometrical method of working in proportions, he compared in a corollary to the first Proposition of Book I, two hypothetical equal time arcs, relating the length of the versed sine to the centripetal force:

> The forces by which bodies, in spaces void of resistance, are drawn back from rectilinear motions, and turned into curvilinear orbits, are to each other as the versed sines [sagittae] of arcs described in equal times; which versed sines tend to the center of force, and bisect those chords when those arcs are diminished to infinity ...[10]

According to this, the centripetal forces of any two curved orbits will ultimately vary (that is, when the arc length is small at the limit where $\Delta t \to 0$) in the same proportion as the versed sines, or sagittae:[11] Intuitively, this makes sense. It seems natural that the centripetal forces on bodies moving in sharper curves with smaller radii would be greater than for more leisurely curves with larger radii. Let us suppose there are two moons revolving around a planet at different distances. Since they are in the vacuum of space, there will be no air resistance.

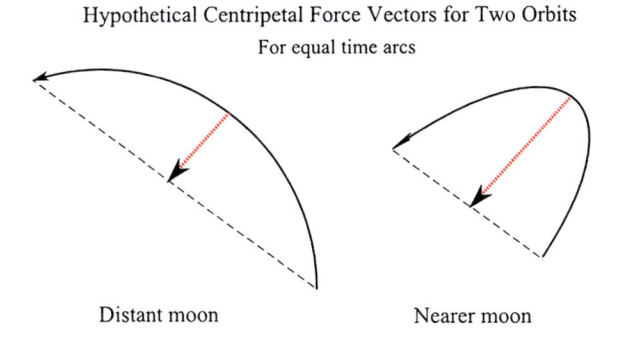

Hypothetical Centripetal Force Vectors for Two Orbits
For equal time arcs

Distant moon   Nearer moon

---

[10] *Principia*, Book I, Proposition I, Corollary 4, 42.

[11] While we cannot now take to time to prove this corollary to the proposition, it is an important one. It holds even for non-circular orbits. Newton creates parallelograms on the arcs of each orbit, with the chords forming the long axis of each parallelogram. For large arcs that are not circular, the line on each from A to the center of force may not be congruent (not match up) with the bisector of the chord. But as we take smaller and smaller units of time, the arcs too are diminished, the bisector lines converge with the force lines, and both point to the center of force. Thus, as we imagine the increments of time approaching zero (as $\Delta t \to 0$) the direction of the force vectors in each orbit ultimately coincide with the lines bisecting the chords.

Under the corollary just given, in equal times of measurement, the forces that pull the moons from straight lines into their own curved arcs will be proportional to the length of centripetal acceleration vector in each case. For the closer-in moon, whose orbit has a shorter radius, the arc of its orbit must logically bend more, so its sagitta will be longer, corresponding to the increased force on the nearer moon. For the more distant moon, feeling the centripetal force more weakly, its path will lie in a less curved line, and its versed sine will be smaller.

Comparing Centripetal Accelerations in Different Orbits

Velocity is always uniform along a circular orbit ("equable motion" in Newton's language), and sweeps out equal areas in equal times.[12] But if we change the arc and compare different circles – say, of different hypothetical moons in different orbits, how will this effect the centripetal forces acting on them?

In *Principia's* important Proposition IV of Book I, Newton dealt with the problem of relating the centripetal forces to velocities in circular orbits. He concluded that the centripetal forces depend on the squares of the arc distances covered in a given time, divided by the radii:

PROPOSITION IV, THEOREM IV (BOOK I)

> The centripetal forces of bodies, which by equable motions describe different circles, tend to the centers of the same circles; and are to each other as the squares of the arcs described in equal times divided respectively by the radii of the circles.[13]

---

[12] The Moon's orbit is not too far from circular, and so revolves around the Earth at a fairly constant velocity, running through its phases predictably in a cycle of a little more than 27 days. If an object is moving uniformly in a circular orbit, it will journey equal distances along any arc in equal times. It will also, according to Newton's very first proposition (and consistent with Kepler's Second Law discussed in Chap. 4), sweep out equal areas in equal times:

> The areas which revolving bodies describe by radii drawn to an immovable centre of force, do lie in the same immovable planes, and are proportional to the times in which they are described. (*Principia*, Book I, Proposition I, Theorem I.)

This too makes intuitive sense. Visualize each orbit composed of little triangles drawn from the center, where the sides are the radii and the bases are the distances traveled by the object in equal times. The length of each base thus depends on the velocity of the body. Since the area of each triangle is proportional to its base times height, where height – the radius – is constant, the areas of all triangles in a given orbit will be the same. The areas, too, will be proportional to the velocity, since, as we said, the object's motion is uniform along the arc. Now increase the number of triangles by making them smaller and smaller, until the triangles effectively merge into a circle, and the bases merge into the arc, and the conclusion is the same. In Newton's words: "Now let the number of those triangles be augmented, and their breadth be diminished *ad infinitum*; and... their ultimate perimeter...will be a curved line: and therefore the centripetal force, by which the body is continually drawn back from the tangent to this curve, will act continually..." (*Principia*, Book I, Proposition I, Theorem I, 41.) This line of argument is how Newton demonstrated his first proposition.

[13] *Principia*, Book I, Proposition IV, 45.

Comparing Centripetal Accelerations in Different Orbits

The centripetal forces on bodies in each circular orbit will be proportional to the arcs squared divided by the respective radii:

$$f_1 : f_2 :: \frac{\widehat{arc}_1^{\,2}}{r_1} : \frac{\widehat{arc}_2^{\,2}}{r_2}$$

This proposition has two parts. If during equal times we compare two bodies revolving in different circular orbits, the forces acting on them pull them toward the centers of their circles. This first idea we already supposed from the Definition V above. The key new concept here is in the second part of the proposition: that those forces are proportional to the squares of the *distances* along the arcs divided by the radius of each circle. With this result, Newton can take the next step and link force to an object's *velocity* in orbit.

The reason is this: since we know that the arc lengths represent velocities (as the motion is uniform), the squares of the arcs must also be proportional to the squares of the velocities. Thus, by positing that the centripetal forces are proportional to squares of the arcs divided by the radii, Newton must then logically conclude that the forces are also proportional to the velocities squared divided by the radii – a powerful, important conclusion. In mathematical symbolism, the centripetal force on each is proportional to v^2/r. Newton made this explicit in the first corollary to the proposition:

PROPOSITION IV, COROLLARY I (BOOK I)

Therefore, since those arcs are as the velocities of the bodies, the centripetal forces are as the squares of the velocities divided by the radii.[14]

The centripetal forces on bodies in each circular orbit will be proportional to the velocities squared divided by the respective radii:[15]

$$f_1 : f_2 :: \frac{v_1^2}{r_1} : \frac{v_2^2}{r_2}$$

And, in general, since accelerations are the result of forces, the centripetal accelerations are also proportional to:

$$v^2/r$$

These propositions set the stage for his famous moon experiment in his *System of the World,* which is Book III of the *Principia,* discussed in the next chapter, where he showed that the Moon's actual motion appeared to fit his geometrical conclusions.

---

[14] Ibid., Corollary 1.

[15] We use the symbol f to represent centripetal acceleration. We will maintain the reference here to forces, as Newton did, in describing the proportions, since acceleration is force per unit mass.

Proving *Principia's* Proposition IV: The Proportionality of Centripetal Forces in Circular Orbits

Problem Using the diagram below and principles of geometry, demonstrate the first corollary to Newton's Proposition IV that, for bodies in uniform circular motion, "the centripetal forces are as the squares of the velocities divided by the radii." That is, proportional in each case to v^2/r.

Given Uniform motion along circular orbits of any size, depicted below.
These basic theorems of Euclidean geometry:

1. An angle inscribed in a semicircle is a right angle. Thus *ABP* in the figure below is a right triangle.
2. If a line is tangent to a circle, it is perpendicular to the radius drawn to the point of contact. Hence *AD* is $\perp$ to *AP*.
3. In any right triangle, the perpendicular from the vertex of the right angle opposite the hypotenuse divides the triangle into two triangles, similar to each other and to the given triangle. Thus since $BC \perp AP$, then triangle *APB* is similar to triangle *ACB*. Also, $\angle CBA = \angle APB$.
4. A central angle is measured by its intercepted arc. Hence $\angle AEB = \overarc{AB}$.

Assumptions We will examine the special case of uniform circular motion. Because we are applying an idealized mathematical method, no physical assumptions need to be made now.

Method Draw a circle representing the orbit of some body, and inscribe a right triangle ABP along the axis, as shown below.

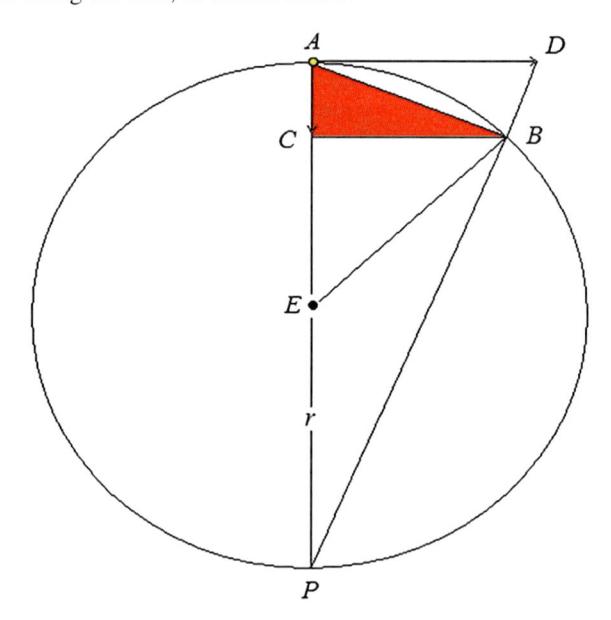

Now imagine the body is moving steadily along the arc $\overset{\frown}{AB}$. For the moment we will approximate the length of this path by the chord AB. Newton's Corollary IV to Proposition I in Book I stated that, the versed sine, which is the line AC, is proportional to the magnitude of the centripetal force pulling the object away from its rectilinear, tangential path along AD, where it would otherwise tend to go by virtue of inertia, according to Newton's First Law. Since equal areas are swept out in equal times, by Proposition I, the length of the arc along $\overset{\frown}{AB}$ is proportional to the velocity of the object.

For this problem, we need to find AC in relation to the radius of the circle by comparing triangle $\triangle CPB$ and the triangle $\triangle ACB$. They are similar, and the proportions of their sides will yield the relation between the line AC (the versed sine) and the diameter AP.

These triangles have straight lines, so how can we compare arc length AB with chord AB? Is it appropriate to approximate the length of the arc $\overset{\frown}{AB}$ by the chord AB? We will never have a convincing proof if we cannot obtain the true length of the arc $\overset{\frown}{AB}$. A core creative problem-solving insight in Newton's *Principia* derived from drawing geometrical constructions with straight lines, rectangles and triangles, then imagining the effect of creating ever smaller units of those until they approximated the curves in question. These rich and fruitful ideas, which are the foundation of the calculus, were introduced in Newton's Lemmas at the beginning of the *Principia*. We saw above in his proof of Proposition I how he divided a circle into triangles: then he increased their number as he diminished their size *ad infinitum*, so their ever smaller bases merged at last into a curve, and the conclusion he had demonstrated for each triangle was true for the circle. We can use that technique on the chord AB in the figure above. Imagine the point B in the figure being moved closer and closer to A, and as it approaches A, the arc $\overset{\frown}{AB}$ will diminish, becoming smaller and smaller. Similarly the tangents at the two points will converge. Eventually the length of the chord AB will approach the length of the "evanescent arc" $\overset{\frown}{AB}$, and become equal to it at the limit. To quote Newton, using a similar diagram, from the proof of his Lemma VII in Book I:

> Wherefore, the right lines AB, AD and the intermediate arc ...$[\overset{\frown}{AB}]$ (which are always proportional to the former), will vanish, and ultimately acquire the ratio of equality. (*Principia*, Book I, Lemma VII.)

Calculations Comparing triangles $\triangle CPB$ and $\triangle ACB$, for a circle of any size, and since triangle ABP is similar to triangle ACB:

$$\frac{AP}{AB} : \frac{AB}{AC}$$

$$AC : \frac{AB^2}{AP}$$

The above relationship holds true for any other circular orbit around this center of force. Now construct any other smaller or larger circle, labeled in the same way but with lower case letters.

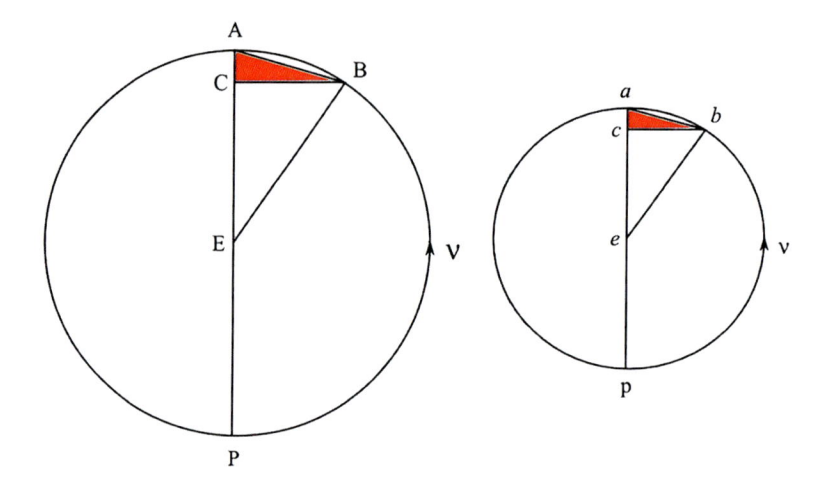

In this case too, for the right-hand circle,

$$ac : \frac{ab^2}{ap}$$

Then, for the two circles,

$$AC : \frac{AB^2}{AP} \text{ and } ac : \frac{ab^2}{ap}$$

AC and ac are as before the versed sines, which as we saw above (*Principia* Book I, Proposition I, Corollary IV) are ultimately proportional to the centripetal forces. We will call these forces F_c and f_c, and visualize them as force vectors pulling inward on each orbiting object. The problem here seeks the ratio of the *velocities* to those forces. This is straightforward, since as before, AB and ab are the arc lengths, and hence are the velocities of each object in a given period of time. Given that AP and ap are the respective diameters, or twice the radii, R and r, of the circles, the above ultimately reduces to these proportions:

$$F_c : \frac{V^2}{2R} \quad \text{and} \quad f_c : \frac{v^2}{2r}$$

$$F_c : f_c :: \frac{V^2}{R} : \frac{v^2}{r}$$

Hence the ratio of centripetal forces is as the squares of the velocities divided by the radii, which is the Q.E.D. of Newton's Proposition IV, Corollary I of his first book.

Observation In Chap. 7 we will explore the intriguing corollaries to Proposition IV, and how Newton richly mined the implications of this proposition to demonstrate that the inverse square law was consistent with any rotational scheme governed by Kepler's Third Law. For now we will address his final corollary to Proposition IV; its neat mathematical summary will be useful in Newton's famous Moon test in the next chapter:

COROLLARY IX TO PROPOSITION IV

From the same demonstration it likewise follows, that the arc which a body, uniformly revolving in a circle with a given centripetal force, describes in any time, is a mean proportional between the diameter of the circle, and the space which the same body falling by the same given force would describe in the same given time.[16]

The "mean proportional" from the above diagram is just arc $\widehat{AB}$ in the relation

$$\frac{AC}{\widehat{AB}} = \frac{\widehat{AB}}{AP}$$

Hence if we know distance traveled in a given time, AC, and the diameter of the orbit, AP, we can compute the fall distance in that time, AC.

Determining if Newton's Centripetal Acceleration for Circular Orbits Is Consistent with the Galileo's Distance-Time Squared Rule for Falling Bodies

The Galilean equations were derived from Galileo's observations of gravity's effects on falling bodies on the surface of the Earth. The logical question now is whether these equations, when applied to the motion of an object in space, imagined to be falling toward the Earth while at the same time moving in a straight line, will yield the same v^2/r relation for centripetal acceleration as Newton found.

Problem Show that the Galilean equations for a falling object under constant acceleration and for constant, rectilinear motion, are consistent with Newton's equation v^2/r for centripetal acceleration acting upon an object in a circular orbit in space.

Given The matters given for the above problem, and these Galilean equations:

$s = \frac{1}{2}at^2$	Equation for determining distance traversed under uniformly accelerated motion, where here s represents distance an object falls, a the acceleration due to gravity, and t is time elapsed during the fall
$s = vt$	Equation for distance s as a function of velocity v and time t

---

[16] *Principia*, Book I, Proposition IV, Corollary 9, 46.

5 How the Moon Falls Toward the Earth (but Keeps Missing It)

Assumptions Again, we are examining the special case of uniform circular motion, so the same assumptions apply here as in the preceding problem.

Method Here the goal is to see if the above-derived acceleration force, v^2/r, is consistent with Galileo's theorems, which he derived from experiment on the surface of the Earth. We must focus on the distance, AC, the "versed sine", since that is the alleged "fall" distance of an object as it moves from A to B along the arc of its circular orbit. We will assume that this AC distance is the distance s in the Galilean equation $s = \frac{1}{2} at^2$. As the Moon is falling, it is also moving in a projectile-like path in a straight line. The distance along the path AD is given by $s = vt$. The orbit is the resultant of these two motions. Since the fall along AC and the movement along AD are to be measured as of the same increment of time, we one may isolate the t in each equation and set the equations equal to each other. Then it is possible to solve for acceleration.

Calculations From the above diagram and the conclusion already demonstrated from the geometry of the circle,

$$AC = \frac{AB^2}{AP}$$

Calling r the radius of any such circle, and since $AP = 2r$,

$$2AC = \frac{AB^2}{r}$$

Put this on the shelf for now; it will be used later.

Now, since $s = \frac{1}{2}at^2$, it is assumed to be the distance the object, such as the Moon, "falls" along the line AC during the same time that it would move the distance $s = vt$ in the horizontal direction AD. Under this hypothesis, we can cast the equations into the geometrical terms of the diagram, then set each expression equal to t^2:

Distance along AC in time t	Distance along AD in time t
$s = \dfrac{1}{2}at^2$	$s = vt$
$AC = \dfrac{1}{2}at^2$	$AD = vt$
$t^2 = \dfrac{2AC}{a}$	$\dfrac{AD^2}{v^2} = t^2$

Since things equal to the same thing are equal to each other,

$$\frac{2AC}{a} = \frac{AD^2}{v^2}$$

Solving for acceleration, we have,

$$a = v^2 \frac{2AC}{AD^2}$$

But we know from above that $2AC = \frac{AB^2}{r}$. Making the substitution for $2AC$, we have,

$$a = \frac{v^2}{r} \frac{AB^2}{AD^2}$$

This still looks rather messy. But recall from Newton's Lemma VII that as we shrink the size of the arc ever smaller, by moving point B closer and closer to point A, "the right lines AB, AD and the intermediate arc... $\overset{\frown}{[AB]}$ (which are always proportional to the former), will vanish, and ultimately acquire the ratio of equality." Thus, as this shrinking occurs, we approach an ultimate limit,

$$\frac{AB^2}{AD^2} \to 1$$

and,

$$a = \frac{v^2}{r}$$

This is the equation for centripetal acceleration we sought, and we found it by using the theorems Galileo discovered from motion experiments on the surface of the Earth, and a small trick of calculus seen cleverly applied here by one of its inventors, Isaac Newton.

Deriving Galileo's Equation Geometrically from Newton's Equation for Centripetal Acceleration

Now we will work the converse of the above problem to see if we can derive Galileo's distance-time-squared relation from Newton's centripetal acceleration equation, using Newtonian geometrical techniques. These derivations help give physical reality to the somewhat abstract mathematical conclusion of Proposition IV and its first corollary.

Problem Derive the equation $s = \frac{1}{2}at^2$ geometrically using Newton's equation for centripetal acceleration.

Given The matters given for the above problem, and,

$s = \frac{1}{2}at^2$ The Galilean equation, where s represents distance an object falls, a the acceleration due to gravity, and t the time elapsed during the fall
---

Assumptions Same as in the above problem.

Method Further insight to the geometry of circular orbital motion is gained by going in the other direction, deriving from the geometry already given the Galilean equation for the fall distance of a uniformly accelerated object. Above we derived the centripetal acceleration equation $a = v^2/r$ from geometry. Here we derive the distance equation using the same basic geometrical principles.

Calculations Begin with the geometrical relationship we used above,

$$AC = \frac{AB^2}{AP}$$

Since $AP = 2r$ and the distance AC is the fall distance s in the given time, the relationship becomes,

$$s = \frac{1}{2}\frac{AB^2}{r}$$

The arc $\overset{\frown}{AB}$ is the distance traveled at uniform velocity, which for smaller and smaller times approaches the chord length AB. Thus, for any constant velocity along the arc, the length AB ultimately will be proportional to the time t. Since this length is measured by that uniform velocity times time, then,

$$s = \frac{1}{2}\frac{(vt)^2}{r}$$

$$s = \frac{1}{2}\left(\frac{v^2}{r}\right)t^2$$

Since v^2/r is centripetal acceleration, then, by substitution of acceleration a for the parenthetical term, we arrive at the Galilean equation,

$$s = \frac{1}{2}at^2$$

Observation We have shown that the distance a thing might fall along *AC* in the diagram would be consistent with the notion that it is accelerated continuously toward the center of the circle. Thus far Newton hasn't needed to say, "that's because of the force of gravity acting from the center which furnishes the acceleration." The relationships have so far still been geometrical and abstract. As noted above, Newton develops the mathematical principles in Book I of the *Principia* which he later, in Book III, ties together into a compelling narrative argument for gravity. In that book, he makes the physical connection between the Moon's fall and the fall of an object on Earth, which we will explore in the next chapter.

Reflections Upon Centripetal Acceleration

Suppose you are resting on the grass after a long bicycle ride. Your bike lies flat on the ground next to you. You lazily spin one of the horizontal tires. You notice a bug on the rim of the tire, and decide to give him a ride. By Newton's laws, even the bug on the rim of your rotating bicycle tire will experience acceleration outward. This is because its inertia tends to make it go in a straight line, whereas the tire it is on is rotating, pulling the bug constantly away from straightness. We want to know what acceleration the bug will experience if we spin the wheel at a constant velocity. We want to find the equation that will enable us to calculate the acceleration and thus the forces on that bug – and maybe even figure out how tightly its little feet adhere it to the rim. We have a hunch that if we find out, we could use that same equation to find the acceleration of any other orbiting or spinning thing, like particles in a cyclotron, molecules in a centrifuge, electrons in an atom or the planets and moons in our solar system.

How do we determine the acceleration on the bug? Consider what we know. The displacement away from straightness can be calculated if acceleration is known, by the $s = \frac{1}{2}at^2$ equation derived above. Inspecting that equation, we can see it is useful in another way: the acceleration can be calculated if the displacement is known. And that displacement can easily be determined by geometry.

Intuitively, we may suspect that the amount of this circular acceleration will be related to the velocity on the rim of the tire. If we turn the wheel fast enough, the bug will be flung off. It may also be related to the distance from the center. Let us derive the equation for circular acceleration using a simple diagram to illustrate the circular, clockwise motion of the tire. Assume the diagram represents a pizza-shaped slice of a spinning wheel. We know from Newton's laws that a particle (including a mass of negligible size, like a bug) on the rim of the circle would *tend* to travel along the straight line *d*. If it weren't dragged in by its attachment to the tire, it would go in a straight line at whatever velocity it has, which we'll call *v*. We turn the tire a little bit at a time, so the bug moves along the length of *d* in a little bit of time, which we'll represent by Δt. At the same time it is displaced away from straight by a distance *s*. (Each little triangle is repeated in each little bit of time Δt.) The bug is holding on with its feet to the rim, and the rim is circular, so it *cannot* go in a straight line, much as it wants to. The "away from straight" distance is s.

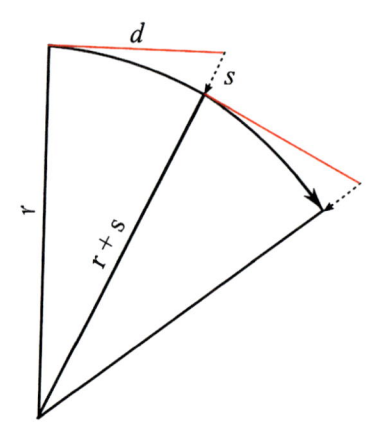

Object tends to move along tangent line d due to its inertia
Its actual circular path pulls it away at each instant, a distance s

According to Newton's Second Law, any deviation of a thing from a straight line path requires some force, which changes its velocity along an axis (from rest or otherwise), and thereby, by definition, accelerates. We know from the above discussion about velocity and acceleration that the displacement caused by acceleration is given by $s = \frac{1}{2}at^2$. For very small units of time, which we'll call Δt, we can write this equation as $s = \frac{1}{2}a\Delta t^2$. (Recall our purpose: if we can find the distance s, we can derive the acceleration from it.) We also know that, during the same small interval Δt the distance along the straight line d may be expressed by,

$$d = v\Delta t$$

Keeping these two equations in mind, we begin by looking at one of the small triangles on the wheel. Notice that because the line d is tangent to the rim, the angle between d and r is always a right angle. Using the Pythagorean Theorem, where in the diagram $r + s$ is the hypotenuse, we have,

$$r^2 + d^2 = (r + s)^2$$

$$r^2 + d^2 = r^2 + 2rs + s^2$$

$$d^2 = 2rs + s^2$$

Let us now assume that that arc along the rim is made smaller and smaller. In other words, our time intervals become shorter, and eventually are as short as we can make them. The distance s becomes smaller and smaller. But the distance s^2 becomes smaller much more rapidly than s does. In fact, when s is almost infinitesimally small, s^2 has become so much smaller that we can safely ignore the s^2 term.[17] The equation then becomes,

$$d^2 = 2rs$$

Since as we saw $d = v\Delta t$ and $s = \frac{1}{2}a\Delta t^2$, we can now substitute the right hand side of these equations for the distance terms in the above equation, which will yield an expression entirely in terms of velocity, acceleration and radius:

$$(v\Delta t)^2 = 2r\left(\frac{a\Delta t^2}{2}\right)$$

Solving for a results in the equation for *centripetal acceleration*:

$$a = \frac{v^2}{r}$$

This important equation again shows that the acceleration on the wheel will increase or decrease as the square of the velocity does, and inversely as the radius. If the wheel goes twice as fast, the acceleration increases four times. If at that new speed the bug goes halfway in toward the hub, he will experience double that acceleration – the turning curve is sharper. This relationship between acceleration and velocity and radius has *nothing to do with gravity*. It applies to any circular motion. It is a mechanical, geometrical fact of the inertial resistance to deflecting a thing from a straight line. That is the essence of circular acceleration. We saw above how Newton proved this using geometry and a few simple proportional arguments.

We could approach this derivation of centripetal acceleration in a still simpler way (there are indeed many ways to do this!). Recall the discussion on pendulums, the figure on pendulum geometry and the derivation of the fall distance in terms of radian measure. Beginning with the Galilean relation,

$$s = \frac{1}{2}at^2$$

then for small angles, and ultimately as we approach an infinitesimally small angle, we found that $s = r\theta^2/2$, where the angle is in radians. Making the substitution in the above equation, then isolating the acceleration term, yields

---

[17] For example, if $s = .01$, then $s^2 = .0001$; if $s = .0001$, then $s^2 = .0000001$.

$$\frac{r\theta^2}{2} = \frac{1}{2}at^2$$

$$a = \frac{r\theta^2}{t^2}$$

Since the velocity along the arc of motion may be given by $v = r\theta/t$, it is apparent that, when substitution is made of the square of this ($v^2 = r^2\theta^2/t^2$) divided by r, it takes us to this result,:

$$a = \frac{v^2}{r}$$

This is the same equation for centripetal acceleration we arrived at (and Newton arrived at) before, but through a slightly different path, using again the basic geometry of the circle, and some insights we had gained when working with the pendulum model.

Let us depart from derivations and try a practical example, to show the power of this simple equation. If this equation truly applies to any circular motion, why not try it on the Moon's (roughly) circular orbit around the Earth? We mentioned earlier that it travels in its orbit at the leisurely pace of about 1.02 km per second (1,020 m per second). Its mean distance is 384,400 km away from center of the Earth. This is 3.844×10^8 m away. From these facts alone, we can compute the circular acceleration acting on the Moon. The acceleration the Moon experiences in travelling along its arc, just like the bug on the wheel, is

$$a = \frac{(1020)^2}{3.844 \times 10^8} \simeq .0027 \ m/s^2$$

Now a fact of fundamental importance in the history of our understanding of the principles of celestial mechanics: this acceleration corresponds to the 2.7 mm per second per second gravitational acceleration you will recall we estimated from the inverse square law. Is this a coincidence? Or is it a hint of something fundamental about the nature of mass and orbital motion? It was a profound coincidence for Newton, as we shall see in Chap. 6, who explored its implications to the fullest. For now, however, it is sufficient to realize that we did not need to know anything about gravity to derive the equations that led us to the circular acceleration of the Moon. We used the model of a bug on a wheel. We then assumed the wheel was a third of a million kilometers in radius. The centripetal acceleration equation requires for its solution only the velocity and radius. We could have calculated the orbital speed from the Moon's distance from the Earth and the time it takes to make one orbit. The circular acceleration is a characteristic of inertia and the geometry of the orbit alone. Now we can take this a step farther and look at the actual forces on a circularly revolving or rotating object. To do this, we simply apply Newtons Second Law, which relates force to acceleration.

Inertial ("Centrifugal") Force

Suppose you are in your car going 10 m per second (about 22 mph) around a curve with a radius of 100 m (328 ft). By the above equation for centripetal acceleration, your acceleration will be one meter per second every second:

$$a = \frac{v^2}{r}$$

$$a = \frac{(10)^2}{100} = 1 \text{ m/s}^2$$

What *force* will you experience? By Newton's Second Law, the force will be equivalent to the mass times the acceleration, $F = ma$. Substituting centripetal acceleration, v^2/r, for the acceleration in Newton's Second Law, the equation for the force experienced by an object in circular motion becomes:

$$F_c = \frac{mv^2}{r}$$

In the above example, if you weigh about 60 kg (132 lb) you will feel an outward force of 60 N (the unit of measure for force). This, again, is the *inertial* force of you wanting to go in straight line, often (confusingly) called centrifugal force.[18]

A Useful Notation for Circular Motion

As we saw in the discussion of pendulums in Chap. 3, is convenient to utilize concepts of radial notation when dealing with problems in circular motion. Some interesting and useful relations follow, and it will be good to reinforce those here. Recall that an object moving in a circle covers 2π *radians* ("radiuses") every revolution, or an actual distance of $2\pi r$ meters. The velocity v is the time it takes the object to accomplish one such revolution, called the period of the object, which

---

[18] The phrase centrifugal force appears to have originated with Christiaan Huygens who stated his *Theorems on Centrifugal Force Arising from Circular Motion* as early as 1659, having developed them even earlier. He did not publish them, however, until 1673, where they appeared appended to his book on pendulums. See *Pendulum Clock*, 176–8.

we have here denoted by P. Therefore, as we have seen, $v = 2\pi r/P$. But if the velocity is expressed just in radians per second (think of radii per second) we use the symbol ω for *angular velocity*, or *angular frequency*, where $\omega = 2\pi/P$. For example, suppose you spin your bicycle tire, and it makes two revolutions (each of 2π radians) per second. Its angular frequency is $\omega = 2\pi/.5$ or 4π radii (radians) per second. If the tire is .75 m in diameter, the actual velocity of a bug on the rim is $\omega r = 2\pi r/P$ or about 4.7 m per second.

Since distance along a circular arc can be expressed as $r\theta$, the velocity of a point at distance r on a steadily rotating wheel (such as the rim of our tire) may be given by ωr where $\omega = \theta/t$, measured in radians per second. The equation for the angular acceleration a of a point on the rim is therefore:

$$a = \frac{v^2}{r}$$

$$a = \frac{(\omega r)^2}{r}$$

$$a = \omega^2 r$$

This is a simple and useful expression for centripetal acceleration. For constant circular motion, then, because acceleration a is the same symbol of the semi-major axis of an ellipse, we will typically use f to stand for acceleration. In sum, for velocity and acceleration:

$$v = \omega r$$

$$f = \omega^2 r$$

Similarly, centrifugal force may be expressed as:

$$F_c = m\omega^2 r$$

Exercises: The Fortunate Circle It has been argued that the very existence of uniform circular motion was a great bit of good fortune for Isaac Newton, for it allowed the successful development of his dynamics. Simple, rectilinear motion would not have permitted the needed growth of his theory, and elliptical and parabolic motion were too complex to permit the emergence of fundamental principles. The well-known Newton scholar, John Herival, stated: "...[T]he problem of uniform circular motion was at once not impossibly difficult and yet of sufficient complexity to call for a real advance in his concept of force and his method of applying it to motion in a curved path"[19] Newton not surprisingly

---

[19] John Herival [2].

A Useful Notation for Circular Motion

employed his creative imagination to go beyond the circle in order to study it. Herival explains:

> The first known discussion by Newton of the problem of circular motion is found at Axiom 20 of the *Waste Book*. The case considered is that of a ball moving on the interior of a hollow spherical surface. According to the principle of inertia there is a constant tendency for the ball to continue on in the direction of its motion at any point, i.e., along the tangent to the circle. And the fact that it does not so continue but moves instead in a circle argues the continuous action on it of a force. This force can only arise from the pressure between the ball and the surface. But if the surface presses the ball, the ball itself must press the surface. From which it follows that all bodies moved circularly have an endeavor from the centre about which they move.[20]

In considering cases where the ball was colliding against a many-sided polygon, Newton saw that the number of collisions would be reduced as the polygon (or circle if the number of sides goes to infinity) grew in size — as the circle's radius increased; but would increase (per unit time) as the velocity of the ball increased. He was thus on to quantifying centrifugal, and thus (by his Law III) centripetal force as some relation between velocity and the inverse of the radius.[21]

Problems

1. Identify any inapplicable answer(s): Centripetal force exists only: (a) where there is gravity; (b) where there is rotation; (c) where there is centrifugal force; (d) where there is inertia. Explain your reasoning.
2. Identify any inapplicable answer(s): Inertia is a tendency of a mass to: (a) move in a straight line; (b) stay still; (c) resist the force of gravity; (d) feel heavy. Explain your reasoning.
3. Identify any inapplicable answer(s): Motion along a curve: (a) will continue along a curve unless acted upon by an external force; (b) requires a force to make it happen; (c) requires gravitational force; (c) is always the composition of two forces. Explain your reasoning.
4. Explain Newton's comment that "versed sines tend to the center of force, and bisect those chords when those arcs are diminished to infinity..."
5. Explain the following connections, to the extent you understand such connections to exist: how, in circular motion (a) gravitational acceleration is related to centripetal acceleration; (b) centripetal acceleration is related to centrifugal force; (c) centrifugal force is related to inertia; (d) inertia is related to inertial force.
6. If the solar system were a large spinning disk and the planets were placed on that disk at the same relative intervals from each other (that is, for example, at distances from the center of about 1, 1.5, 5.2, and 9.6 units for the planets Earth Mars Jupiter and Saturn, respectively) what would be the ratio of the accelerations acting on those planets?

---

[20] Ibid.
[21] Newton developed this concept in the Scholium to his Proposition IV. *Principia*, Book I, Proposition IV, 47.

7. If two planetary orbits were to be compared, and the first planet was twice as far from the center of force but half as fast as the second planet, what would be the ratio of the centripetal accelerations acting upon them? Under these circumstances, if the first planet were half as massive as the second, what would be the ratio of their centrifugal forces be?
8. If you put a small steel ball weighing 1/10 of a kilogram in a centrifuge .5 m in radius, then turn this centrifuge to spin at a rate of 12,000 rpm, what will the centrifugal force be acting upon the steel ball? If the ball were suddenly to be ejected from the centrifuge, and ignoring friction and gravity, at what velocity will the ball travel? In what path? For how long?
9. Neptune's moon Naiad is about 48,227 km from the center of Neptune. It journeys in an almost circular orbit once in .294396 days. What is the centripetal acceleration acting upon Naiad? If you were informed that Naiad's mass is $.002 \times 10^{20}$ kg, what centrifugal force would you say it experiences?
10. Compare the centripetal accelerations and centrifugal forces acting upon Neptune's moon Naiad with Neptune's small moon Galatea. Galatea travels around Neptune at a distance of 61,953 km in a circular orbit with a period of .428745 days. Galatea's mass is $.04 \times 10^{20}$ kg. What are the ratios of the centripetal accelerations and centrifugal forces acting upon those two small moons of Neptune, Naiad to Galatea?

References

1. Newton I (1949) Mathematical principles of natural philosophy (trans: Motte A, 1729, revised by Cajori F). University of California Press, Berkeley, p 395
2. Herival J (1965) The background to Newton's principia, vol 7. Oxford University Press, London

Chapter 6
Newton's Moon Test

One of the most famous thought experiments in the history of science was made possible by a unique natural coincidence. The coincidence resides in the length of a second of time in relation to the minute, and the size of the Earth in relation to its distance from the Moon. In our system of keeping time, a second is one-sixtieth of a minute; the Earth's radius is also one sixtieth the distance to the Moon. Isaac Newton discovered that this coincidence of proportion enabled an easy test of his gravitational hypothesis that could be grasped by anyone.

About two decades before the publication of the *Principia*, Newton sought to prove that the force inducing objects, such as apples on a tree, to fall downward to Earth was the same force that kept the Moon enchained in its orbit, preventing it from going off into space. If the characteristic effects of this force were found to be the same for the Moon and the apple, then it would be evidence that the causes of the forces are the same. In particular, if the power of the force were found to vary with the inverse square of the distance all the way to the Moon, it would be a dazzling demonstration of the validity of that theory. It would suggest that the centripetal force emanating from the Earth varies so as to be strong at the Earth's surface, at one Earth radius, but attenuated by the inverse square of the number of radii distant from the Earth's center. The attenuation under such a law would be easily computable. It would be one-fourth as strong at two radii, one-ninth as strong at three radii and so forth. Because the mean lunar distance is about 60 Earth radii distant, it should by this reasoning make the force of the gravitational acceleration acting on each unit of mass there, only one thirty-six hundredth as strong as we experience on Earth. This fraction also happens to be the square of the number of seconds in a minute. Exactly how from these coincident facts Newton constructed an easily understandable demonstration of an inverse square law of gravity acting so far away as the Moon is best told by Newton himself in Book III of the *Principia*.

Newton did not discover this ready proof so easily, in fact was delayed about twenty years, mainly because in about 1666 the dimensions of the Earth were not accurately known.[1] William Whiston, who knew and succeeded Newton in his Lucasian professorship at Cambridge, described his conversation with Newton on this subject, and Newton's early frustration with his results:

> An Inclination came into Sir Isaac's Mind to try, whether the same Power did not keep the Moon in her Orbit, notwithstanding her projectile Velocity, which he knew always tended to go along a straight Line the Tangent of that Orbit, which makes Stone and all heavy Bodies with us fall downward, and which we call *Gravity*? Taking this Postulatum, which had been thought of before, that such power might decrease in a duplicate Proportion of the Distances from the Earth's Center. Upon Isaac's first Trial, When he took a Degree of a great Circle on the Earth's Surface, whence a Degree at the Distance of the moon was to be determined also, to be 60 measured miles only, according to the gross Measures then in use. He was, in some degree disappointed, and the Power that retained the Moon in her Orbit...appeared not to be quite the same that was to be expected, had it been the Power of gravity alone, by which the Moon was there influenc'd. Upon this Disappointment, which made Sir Isaac suspect that this Power was partly that of Gravity, and partly that of Cartesius's Vortices, he threw aside the Paper of his Calculation, and went on to other studies.[2]

It was years later, around 1675, that Newton came upon the more accurate value of a degree at the Earth's meridian. French astronomer Jean Picard measured its value to be 69.1 English statute miles. Newton read of this in the Royal Society's *Philosophical Transactions*, and is reported to have rushed to redo his calculations that he had set aside so many years before. This time he found, to his excitement, that his results agreed perfectly with his theory.[3] Here, in summary, is how Newton described his proposition and its proof in *Principia's* Book III, Proposition IV:

PROPOSITION IV. THEOREM IV (BOOK III)

That the moon gravitates towards the earth, and by the force of gravity is continually drawn off from a rectilinear motion, and retained in its orbit.

... Let us assume the mean distance [to the Moon] of 60 [semi]diameters in the syzygies; and suppose one revolution of the moon, in respect of the fixed stars, to be completed in $27^d.7^h.43^m$., as astronomers have determined; and the circumference of the earth to amount to 123249600 *Paris* feet, as the *French* have found by mensuration. And now imagine the moon, deprived of all motion, to be let go, so as to descend towards the earth with the impulse of all that force by which...it is retained in its orb, it will in the space of one minute of time, describe in its fall $15^{1/12}$ *Paris* feet...For the versed sine of that arc, which the moon, in the space of one minute of time, would by its mean motion describe at the distance of 60 semidiameters of the earth, is nearly $15^{1/12}$ *Paris* feet, or more

---

[1] A contrary explanation put forward for the delay was that there "were theoretical questions of great difficulty relating to the attraction of a sphere upon an external point – a problem which he [Newton] did not solve until 1684 or 1685..." It was only then that Newton proved that a planet or moon could be treated mathematically as if all its mass were concentrated in a point at its center. See Florian Cajori [1]. (Short title) *Principia*.

[2] This narrative was excerpted by Gale Christianson from *Memoirs of the Life and Writings of William Whiston*, London, 1749, I:36–37, and appears on page 79 of Gale E. Christianson [2]. The description of events here is drawn from pages 77–88 of that work.

[3] There is a great deal of literature on the early origins of Newton's Moon test. Some concise descriptions of his early musings during the plague years by people who knew Newton are presented in S. Chandrasekhar [3].

accurately 15 feet, 1 inch, and 1 line 4/9. Wherefore, since that force, in approaching to the earth, increases in the proportion of the inverse square of the distance, and, upon that account, at the surface of the earth is 60·60 times greater than at the moon, a body in our regions, falling with that force, ought in the space of one minute of time, to describe $60·60·15^{1/12}$ *Paris* feet; and, in the space of one second of time, to describe $15^{1/12}$ of those feet; or more accurately 15 feet, 1 inch, and 1 line 4/9. And with this very force we actually find that bodies here upon earth do really descend; for a pendulum oscillating seconds in the latitude of *Paris* will be 3 *Paris* feet, and 8 lines ½ in length, as Mr. *Huygens* has observed. And the space which a heavy body describes by falling in one second of time is to half the length of this pendulum as the square of the ratio of the circumference of a circle is to its diameter (as Mr. *Huygens* has also shown), and is therefore 15 *Paris* feet, 1 inch, 1 line 7/9. And therefore the force by which the moon is retained in its orbit becomes, at the very surface of the earth, equal to the force of gravity which we observe in heavy bodies there. And therefore...the force by which the moon is retained in its orbit is that very same force which we commonly call gravity...

Summary of Newton's Moon Test

It may be easier to understand this proof by dividing it into parts. We will explain each part below:

1. The Moon's fall in a *minute* at lunar distance is $15^{1/12}$ *Paris* feet;
2. The Moon's fall in a *second* on the surface of the Earth would be $15^{1/12}$ *Paris* feet;
3. The fall of a heavy body (such as a rock) in a second on Earth, as measured by a pendulum, is the same;
4. Therefore, the *force* pulling the Moon and the rock are one and the same: gravity

The First Part of the Test

Newton provides the initial data necessary for his proof, the Moon's distance and orbital period, and the Earth's circumference, then begins the first part of his test. He pictures the Moon stopped in its orbit and let go, to fall toward the Earth: "*And now imagine the moon, deprived of all motion, to be let go, so as to descend towards the earth ...*" His purpose is to find out how far it would fall in a minute. By using the geometrical methods of the last chapter, he determines that it would fall $15^{1/12}$ *Paris* feet in 1 min: "*it will in the space of one minute of time, describe in its fall $15^{1/12}$ Paris feet ...*" The problem below shows how he determined this. It is not important for us to know what a Paris foot is, because he uses the same units throughout.[4] In the end he shows that the results are all the same.

---

[4] One Paris foot is .32484 m (1.066 English feet).

The Second Part of the Test

In the second part of the test, Newton mentally removes the Moon (or any object at that distance) toward the surface of the Earth, and two things happen. The force increases by the square of the decreased distance (from 60 Earth radii to 1), and the fall distance decreases by the square of the reduced time (from 60 s to 1).[5] These two cancel out, so "*a body in our regions, falling with that force, ought ... in the space of one second of time, to describe $15^{1/12}$ of those feet ...*"

The Third Part of the Test

So far Newton has given us theoretical results, informing us how the Moon should be expected to behave if brought to Earth, assuming that an inverse square law governs: it *should* fall $15^{1/12}$ Paris feet in a second. Now he needs to compare theory with reality, and tell us how things *actually* behave on the surface of the Earth. Here Newton refers to experiments with the pendulum performed by the Dutchman Christiaan Huygens whom we encountered earlier. Recall that a pendulum "oscillating in seconds" is one where each swing is one second in duration. From such a pendulum the fall distance in one second can be calculated by the formula we derived earlier, $s = \frac{1}{2}\pi^2 l$, where s is the distance and l is the length of the pendulum. Huygens had determined the distance an object falls in 1 s on Earth; Newton discovered that it is the same as the fall distance computed above: "*And with this very force we actually find that bodies here upon earth do really descend...*"

The Fourth Part of the Test

Newton concludes with the famous pronouncement that the two phenomena must be caused by the same thing: "*And therefore the force by which the moon is retained in its orbit becomes, at the very surface of the earth, equal to the force of gravity which we observe in heavy bodies there.*"

Newton's conclusion that the distance the Moon falls toward the Earth in 1 min corresponds to the distance it would fall in a second if brought down to Earth was a brilliant exposition of the extension of the force of gravity all the way to the Moon. But was it an elegant coincidence? The result on its face did not necessarily mean

---

[5] This is because, by Galileo's theorem, for a given time, distance is proportional to acceleration, and by Newton's Second law, acceleration is proportional to force; thus the distance a thing falls is proportional to the force pulling it downward. By the same theorem, the distance something falls is also proportional to the square of time.

that gravity exists everywhere and universally. The Moon test was striking evidence that the inverse square law would explain why the Moon is held in its orbit, but it remained to be shown that there were not other factors at work that might apply elsewhere. In nature there are often complex conditions affecting results. Nor did this yet necessarily imply that all bodies attract all bodies. Nevertheless, this thought experiment was a major stepping stone in making all the pieces of the gravitational puzzle come together.

One may work out the mathematics of Newton's approach here, using the geometrical methods of Chap. 5. The problem below does just that.

Newton's Demonstration that "The Moon Gravitates Towards the Earth"

Problem Using the data provided by Newton in Proposition IV of Book III, and the additional information derived from it given below, check the accuracy of Newton's result that the distance the Moon, "deprived of all motion," would fall in 1 min from its orbit, is the same distance it would fall in 1 s at the surface of the Earth.

Given

123,249,600 Paris feet	Circumference of the Earth mentioned by Newton, from Picard's measurements
60 Earth radii	Mean distance from the center of the Earth to the Moon, estimated by Newton from the work of various astronomers
39,343 min	Period of Moon's orbital revolution, derived from $27^d.7^h.43^m$, as astronomers of the day had determined. This is about the modern value
$s = 1/2 \pi^2 l$	Equation found by Huygens for determining the distance s an object falls in one second from the length l of a pendulum which swings through its arc once a second

Assumptions We will assume, as Newton did, a circular lunar orbit with uniform lunar motion, and a distance to the Moon of 60 Earth diameters.[6] We will also assume a spherical Earth and Moon of uniform density, and no particular perturbing influences of other bodies, such as the Sun. Finally, it is an assumption that the value of g (Earth's gravitational acceleration) implicit in Huygens' pendulum results would be more or less universally applicable. In fact, the Earth is not a

---

[6] Newton begins Proposition IV in Book III of his *Principia* by summarizing the ancient and then current estimates for the lunar distance, in Earth radii. The values ranged from 59 (Ptolemy), 60 (Huygens), 60⅓ (Copernicus), $60^2/^5$ (Street), and 60½ (Tycho). He settles upon a value of 60. The true value, derived from the mean diameter of the Earth (calculated either as the average of the equatorial and polar radii, or as the mean volumetric radius), is about what Copernicus suggested.

uniform sphere, and spins, and the value of g does vary from place to place, so the pendulum lengths would not everywhere be precisely the same. But for the rather rough calculations and assumptions made here, it is safe to say the differences are not significant.

Method Consider the diagram below. Knowing how far the Moon falls in a minute (distance AC or DB in the diagram) in going from A to B requires knowing how far the moon travels in 1 min along the arc $\widehat{AB}$.[7]

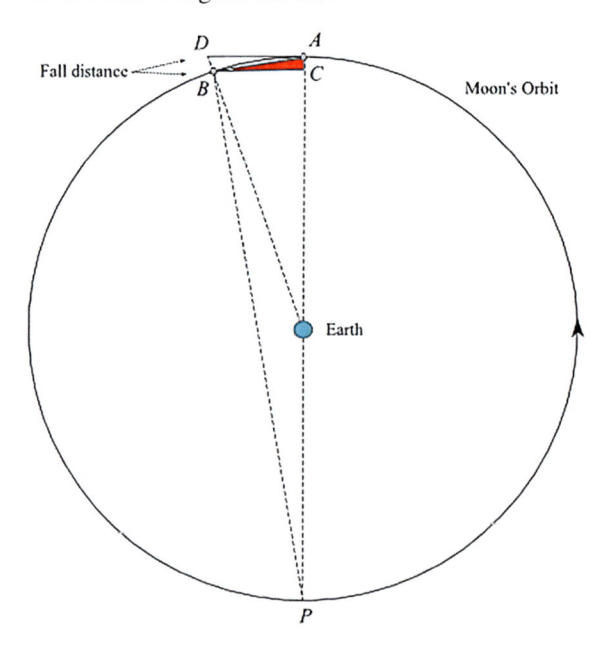

To compute the fall distance AC, we recall from Chap. 5 that, from Newton's Proposition IV, Corollary IX, the arc $\widehat{AB}$ is the "mean proportional between the diameter of the circle, and the space which the same body falling by the same given force would describe in the same given time":

$$\frac{AC}{\widehat{AB}} = \frac{\widehat{AB}}{AP}$$

This is a useful proportion since the fall distance AC is what we seek. The other unknowns are the arc length and the diameter of the lunar orbit. The distance to the Moon is given in Earth radii. From the given circumference we can find the radius

---

[7] If one views the lunar orbit from northern hemisphere of the Earth, it moves counterclockwise. If viewed from the southern hemisphere, the path is clockwise. In either case, the mathematical principles are the same.

The Fourth Part of the Test

of the Earth. Given that the Moon is given as 60 Earth radii distant, one can then compute the two unknowns in the above proportion, and arrive at the AC distance the Moon falls in 1 min. The steps are as follows: First, using the diameter of the Earth, calculate the diameter of the lunar orbit AP. From this, find the circumference of the lunar orbit. This latter, when divided by the period in minutes, yields the arc length between B and A travelled by the Moon in 1 min. Since that arc length is the mean proportional between AC and AP, distance AC can readily be computed from the equation just given.

The next part of the analysis entails the application of the inverse square law. By Newton's imaginary relocation of the Moon to the Earth's surface, the strength of the centripetal force is increased by the square of the reduced distance. This yields a new result for a minute of fall time. To find the fall in just 1 s on Earth, we apply the Galilean relation that, for a constant acceleration, distance is proportional to the square of the time. For the empirical part of the test, it is necessary to translate the length of Huygens' seconds pendulum into fall distance on Earth by the equation $s = \frac{1}{2}\pi^2 l$. Then empirical and theoretical results may be compared.

Calculations *First part: finding the fall distance of the Moon in its orbit.* First we calculate the diameter of the Earth so we can find the diameter of the lunar orbit. For conceptual consistency, we will use the units and number of digits Newton did. The diameter of the Earth will be its circumference C_E given by Newton divided by π:

Diameter of Earth: $D = C_E/\pi = 123{,}249{,}600/3.14. = 39{,}231{,}566$ Paris feet
Diameter of Moon's orbit: $AP = 60 \cdot D = 2{,}353{,}893{,}968$ Paris feet

The distance $\widehat{AB}$ is the portion of the Moon's orbital circumference C_M traversed by the Moon in 1 min.

Circumference of Moon's orbit $C_M = \pi \cdot AP = 7{,}394{,}976{,}000$ Paris feet
Period of one lunar revolution in orbit $P = 39{,}343$ min
Distance along arc $\widehat{AB}$ traversed in one minute $= C_M/P = 187{,}962$ Paris feet

This is the velocity of the Moon in its orbit, in Paris feet per minute. From that number we can determine the fall distance of the Moon. Since

$$\frac{AC}{\widehat{AB}} = \frac{\widehat{AB}}{AP}$$

Then the distance AC will be

$$AC = \frac{\widehat{AB}^2}{AP}$$

$$AC = \frac{187{,}962^2}{2{,}353{,}893{,}968}$$

$$AC = 15\frac{1}{120} \text{ Paris feet}$$

But Newton states, "it will in the space of one minute of time, describe in its fall $15^{1/12}$ *Paris* feet ..." The value we derived of 15 and 1/120 Paris feet (in decimals, about 15.01) is slightly less than the 15 and 1/12 Paris feet (about 15.1) stated in the proof of the Proposition and found, as we shall see in the third part of the test, by the pendulum experiment. The late Indian physicist and Nobel laureate Subrahmanyan Chandrasekhar speculated that Newton wrote the value $15^{1/12}$ Paris feet from memory from an earlier calculation.[8] On the other hand, Dana Densmore believes the difference can be accounted for by subtracting the force of the Sun's influence, a correction she assumes Newton computed but did not state in the proof of this famous proposition.[9]

Second part: finding the fall distance of the Moon as if it were on the Earth's surface. The next part of the proposition states that "since that force, in approaching to the Earth, increases in the proportion of the inverse square of the distance, and, upon that account, at the surface of the earth is 60 · 60 times greater than at the moon, a body in our regions, falling with that force, ought in the space of 1 min of time, to describe 60 · 60 · $15^{1/12}$ *Paris* feet; and, in the space of 1 s of time, to describe $15^{1/12}$ of those feet..." That is, if we take as our supposition that the inverse square law applies, then the acceleration on Earth should be 3,600 times stronger at 60 Earth radii closer. Doing the calculation as Newton suggests, we find the Moon's fall closer to home in 1 min of time would be:

$$60 \times 60 \times 15.01 = 54{,}032 \text{ Paris feet in one minute on Earth}$$

Since the fall distance at this Earthly acceleration is proportional to time squared, by Galileo's theorem ($s \propto at^2$), then 54,032 Paris feet is the distance the Moon would fall on Earth in t^2, or in 60 s squared, or in 3,600 s. To find the fall distance in 1 s, we must divide by 3,600:

$$\frac{54{,}032}{3{,}600} = 15\frac{1}{120} \text{ Paris feet in one second on Earth}$$

---

[8] See S. Chandrasekhar, *Newton's Principia for the Common Reader*, 358, where Chandrasekhar states in a note to his result of $15^{1/120}$ Paris feet: "Newton gives instead $15^{1/12}$ *Paris* feet. Perhaps he wrote this value from memory (?) from an earlier computation with different parameters from those listed in this proposition."

[9] D. Densmore [4]: "...[T]he reason his [Newton's] number $15^{1/12}$ Paris feet, differs from ours, 15.01 Paris feet, is that he has taken into account the effect of the sun's force on the moon." Densmore cites and calculates this difference from Newton's Proposition 45, Corollary 2, though Newton did not himself refer to that Proposition or corollary for support in Proposition IV of Book III discussed here. Moreover, the first part of the proof is itself entirely geometrical, and does not rely upon a presupposition of actual gravitational forces.

As noted, Newton has accepted the value of 15 and 1/12 Paris feet, "*or more accurately 15 ft, 1 in., and 1 line 4/9.*" Since a line is a twelfth of an inch, then the fall of the Moon if brought to Earth is,

$$s_{Moon} = 15 + \frac{1}{12} + \frac{1}{144}\left(1 + \frac{4}{9}\right) \simeq 15.1 \text{ Paris feet (4.9 meters)}$$

Third part: comparing this result with the fall of an object, such as a rock, on Earth. Newton says, "And with this very force we actually find that bodies here upon earth do really descend..." Newton's refers to the pendulum studies of Christiaan Huygens who in 1673 published *Horologium Oscillatorium* (the *Penduluum Clock*).[10] This work was available to Newton. Huygens' reported value for the distance a pendulum would fall in a second matches what Newton calculated.[11] Huygens found, consistent with Newton's report, as follows:

> ...[I]f the Parisian foot is given, then we would say that the simple pendulum, whose oscillations mark off seconds in an hour, has a length equal to three of these feet, plus eight and one-half lines.[12]

A line is a twelfth of a Parisian inch. Huygens' pendulum length is therefore about 3.06 Paris feet. Putting that as the value of l in the equation $s = \frac{1}{2}\pi^2 l$ we have,

$$s = \frac{\pi^2 l}{2}$$

$$s = \frac{3.14^2(3.06)}{2}$$

$$s = 15\frac{1.15}{12} \text{ Paris feet (4.9 meters)}$$

This result is compatible with what Huygens found, and Newton concludes that the fall in 1 s according to the pendulum "*is therefore 15 Paris feet, 1 in., 1 line 7/9.*"

$$s_{Pendulum} = 15 + \frac{1}{12} + \frac{1}{144}\left(1 + \frac{7}{9}\right) \simeq 15.1 \text{ Paris feet (4.9 meters)}$$

---

[10] For a modern translation by Richard J. Blackwell of Christiaan Huygens' 1673 work, See C. Huygens [5] (Short title) *The Pendulum Clock*.

[11] For an analysis of the pendulum result in contemporary mathematics, see S. Chandrasekhar, *Newton's Principia for the Common Reader*, 359–360.

[12] *The Pendulum Clock*, 168. This is found in Huygens' Proposition XXV, "*A method of establishing a universal and perpetual measure.*"

Observation The realization that the force of gravity on Earth, when attenuated by the inverse square law, matches actual fall of the Moon in its orbit was a stunning insight. Newton revealed that gravity's reach extends at least to the Moon. We will show later how he deduced that it extends to the heavens.

A Simple Confirmation of the Inverse Square Law Between Earth and Moon Using Modern Data

It is instructive to work through a short-hand version of Newton's Moon test using more modern quantitative methods. We will use the computed acceleration of the Moon in its orbit, its distance from Earth, and the value of g on Earth to confirm Newton's Moon test and his Proposition IV of Book III of the *Principia*.

Problem Given the Moon's mean distance and orbital velocity below, and using the accelerations of the Moon toward the Earth, show how far the Moon falls toward the Earth in 1 min and compare that result with how far an object on Earth falls in 1 s. Show that the result is consistent with the inverse square law.

Given

384,400 km	Moon's mean distance from Earth, which is approximately 60 Earth radii away
1.023 km/s	Moon's mean velocity in its orbit
$s = \frac{1}{2} at^2$	Galilean distance equation for constant acceleration, with the same variables as before
9.8 m/s^2	Value of gravitational acceleration at the surface of the Earth
$a = v^2/r$	Equation for centripetal acceleration of a circularly revolving object

Assumptions The same assumptions as in the previous problem. In particular, note the uniform circular motion of the Moon, which again is an acceptable approximation for our instructional purposes, as it was here for Newton.[13] This assumption thus neglects the fact that the Moon's distance from Earth varies some in the course of its orbit, as does its velocity. We also neglect the influence of the Sun on the lunar orbit.

Method The length AC is the distance the Moon falls in a given increment of time (we have not yet placed any scale on the above diagram). Set this distance equal to s in the Galilean equation. That equation contains an unknown – the acceleration term. Substitute Newton's expression for centripetal acceleration v^2/r, whose derivation we discussed in the last chapter, for the term a, then solve for distance s. That will yield the distance the Moon falls in time t. Since the time given in the problem is 60 s, the result will be our answer. For Earth, use the value 9.8 m/s^2 for acceleration.

---

[13] Its actual eccentricity is roughly .05. The lunar orbit not as circular as Venus's orbit, whose eccentricity is less than .01, or Earth, at roughly .02, but more circular than Mars's .09 eccentricity.

Calculations Begin with the distance equation, $s = at^2/2$. Since centripetal acceleration is given by $a = v^2/r$, the right-hand side of this expression may be substituted for a in the distance equation:

$$s = \frac{1}{2}\left(\frac{v^2}{r}\right)t^2$$

$$s = \frac{1}{2}\left(\frac{1.023^2}{384,400}\right)60^2$$

Since the above units are in kilometers, the result will be in kilometers. In meters, the distance (rounded) the Moon descends in its orbit in 1 min is,

$$\underbrace{s = 4.9 \text{ meters}}_{\text{Moon's fall in its orbit in one minute}}$$

Incidentally, this is a very leisurely descent for this great object! It is about 16 ft. Now again compare this with the descent of objects closer to home, on the surface of the Earth. Again begin with Galileo's equation:

$$s = \frac{1}{2}at^2$$

On Earth, in 1 s, an apple falls,

$$s = \frac{1}{2}(9.8)(1)^2$$

$$\underbrace{s = 4.9 \text{ meters}}_{\text{An apple's fall on Earth in one second}}$$

Observation Newton elsewhere in the *Principia* showed the applicability of the inverse square law to elliptical orbits, consistent with the first two laws of planetary motion discovered by Kepler, published in 1609 in *Astonomia Nova*. Newton's Section III of Book I of the *Principia* dealt with "The motion of bodies in eccentric conic sections." The first entry in that section, Proposition XI, Problem VI, stated the case for elliptical motion: "If a body revolves in an ellipse; it is required to find the law of centripetal force tending to the focus of the ellipse." He proved that the centripetal force on a body is inversely as the distance from the object to the focus of the ellipse, a key conclusion making his theory of gravity consistent with ellipses and all other conic sections.

A Geometric Approximation of the Moon's Period (and Other Diversions)

It can be rewarding to explore some mathematical variations on Newton's conclusions. Below are some exercises that may deepen understanding of the Moon's orbital geometry.

Problem Suppose the distance to the Moon were unknown. With just the Moon's velocity along its orbit and the calculated downward fall toward the Earth in 1 min, and using *radian* measure, geometrically approximate the period of the Moon.

Given

4.9 m	Distance the Moon's falls toward Earth in one minute, along line AC in the diagram
1.023 km/s	Moon's mean velocity in its orbit, along arc $\stackrel{\frown}{AB}$ in the diagram, which for short arcs approximates the length of the chord AB

These further basic theorems of Euclidean geometry:

1. A central angle is measured by its intercepted arc. Hence $\angle AEB = \stackrel{\frown}{AB}$
2. An inscribed angle is measured by one half its intercepted arc. Hence $\angle APB = \frac{1}{2} \stackrel{\frown}{AB}$, and $\angle APB = \frac{1}{2} \angle AEB$. Likewise $\angle APB = \angle CBA$.

Assumptions We again approximate the lunar orbit by a circle.

Method Consider the following diagram of the Moon's orbit around the Earth.

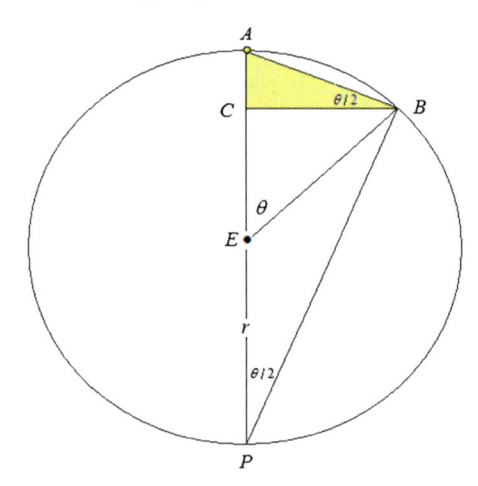

For the sake of interest and variety, we take a southern hemisphere view of the lunar orbit, and the Moon is revolving clockwise. (Mathematically, of course, there is no difference from the reverse depiction; the sagitta, or fall distance AC, has the same geometrical and physical meaning as before.) Our goal is to find the

A Geometric Approximation of the Moon's Period (and Other Diversions)

(unrealistically exaggerated) central angle $\angle AEB$, labeled θ in the diagram. If we know what part of a circle the Moon travels the angle θ in 1 min, we can readily calculate how long it takes to go all the way around its circumference, which is its period.

The measure of distance along arc $\overset{\frown}{AB}$ is the central angle of that arc (here, θ) in *radians* times the radius r of the circle. But if as stated in the problem we are to assume the radius is unknown, then we have to rely upon the information provided about the smaller triangle $\triangle ACB$. We are helped in this by the given geometrical principle that the angle $\angle CBA$ (marked $\theta/2$) is half of the central angle. To see how this smaller triangle may be useful, assume that point B is the center of a smaller circle whose radius is approximately AB. Then we can find $\theta/2$ in radians by the ratio of AC/AB (since in radians, $\theta/2$ times the little radius AB yields AC; hence $AC/AB = \theta/2$). Since AC and AB are both given, we can determine $\theta/2$ and thus θ. Then we simply divide this central radian angle into 2π, the number of radians in a circle, to yield an approximate period for the lunar orbit.

Calculations The moon falls 4.9 m in 1 min along the line AC (Newton's versed sine, or sagitta), but moves in its orbit 1.023 km in 1 s along the arc $\overset{\frown}{AB}$. It therefore moves $1.023 \times 1{,}000 \times 60 = 61{,}380$ m in 1 min along the arc $\overset{\frown}{AB}$. Consistent with Newton's Lemma VII, we can approximate for very small angles the chord distance AB by the arc $\overset{\frown}{AB}$. At any such small angle, the ratio of the fall to the arc distance in one full minute is then,

$$\angle CBA = \frac{\theta}{2} \approx \frac{AC}{AB} = \frac{4.9}{6.138 \times 10^4} = .00007983 \text{ radians}$$

The approximated "slope" of this arc (the ratio AC/AB) is thus about 1 m of fall for every 12,500 m (about seven and three quarter miles) gained along its orbital arc (which the Moon traverses in a little more than 12 s). This ratio is equivalent to the angle $\theta/2$ in radians. Therefore, the measure of angle θ ($\angle AEB$) in radians is twice that or .00015966. We divide this into 2π, which is the circumference of the circle in radians, to find the number of minutes in the mean orbital period:

$$\theta = .00015966 \text{ radians}$$

$$P_{min} = \frac{2\pi}{.00015966}$$

$$P_{min} = 39{,}353$$

To find the mean orbital period in days, divide P_{min} by 1,440, the number of minutes in a day:

$$P_{days} \approx 27.3 \text{ days}$$

Observations

1. The above is reasonably close to modern values. The published mean lunar period is 27.3217 days, or about 39,343 min.[14]
2. We can also geometrically approximate the orbital period P of the Moon using its velocity in orbit and its mean distance from Earth (about 384,400 km). Again, because the lunar orbit is not a perfect circle, the results again will only be approximate. Recall that the length of the arc $\widehat{AB}$ will be θ in radians times the radius: $r\theta_{rad} = \widehat{AB}$. Hence:

$$\theta_{rad} = \frac{\widehat{AB}}{r}$$

$$\theta_{rad} = \frac{6.138 \times 10^4}{3.844 \times 10^8}$$

$$\theta_{rad} = .000159677$$

This is the angular measure of revolution in 1 min. To find the orbital period, it is required to find how many of these measures fill an entire circle. To do this we divide it into 2π:

$$P_{min} = \frac{2\pi}{.000159677}$$

$$P_{min} = 39,349$$

$$P_{days} \approx 27.3 \text{ days}$$

3. Above we geometrically derived the orbital period from the velocity and fall distance, then from velocity and radius distance. Now we derive the fall distance AC from the angular velocity and period alone. In other words, can we just look at the velocity along the *orbital arc alone* and from it compute the distance the Moon falls in 1 min? If we know only its orbital period and velocity, we can indeed geometrically calculate the slope and the fall distance. The Moon's actual orbital period is about 27.3217 days, which is about 39,343 min. We'll use radian measure to determine its angular velocity in radians per minute.

Let's review the origin of the relation between the half-angle $\theta/2$ in the triangle $\triangle CBA$, and its sides, AC and AB. Since we know from before that,

$$\frac{AC}{AB} = \frac{AB}{AP}$$

---

[14] See NASA's site, http://nssdc.gsfc.nasa.gov/planetary/factsheet/moonfact.html.

If we again imagine the arc $\widehat{AB}$ and therefore θ to become smaller, as described in Newton's Lemma VII, then chord AB is approximated by $r \times \theta_{rad}$,

$$\frac{AC}{AB} \approx \frac{r\theta_{rad}}{2r}$$

and ultimately,

$$\frac{AC}{AB} = \frac{\theta_{rad}}{2}$$

consistent with our earlier conclusions. Here we are taking the fractional part of the orbit to be one part in 39,343, so it is safe enough for our purposes to approximate the chord length with the arc length. We called the ratio AC/AB the "slope" of the orbital arc. The angular velocity of the Moon in radians per minute along its arc is,

$$\theta_{rad} = 2\pi/P$$

$$\theta_{rad} = 2\pi/39,343$$

$$\theta_{rad} = .0001597 \text{ radians per minute}$$

Substituting this value into the above equation, we have,

$$\frac{AC}{AB} = \frac{.0001597}{2} = .00007985 \text{ radians}$$

Since the length AB is the lunar velocity (1.023 km/s) times the time increment we have chosen (1 min), its length is, again, $1.023 \times 1,000 \times 60 = 6.138 \times 10^4$ m. Solving for AC, we have,

$$AC = (7.985 \times 10^{-5})(6.138 \times 10^4)$$

$$AC \simeq 4.9 \text{ meters}$$

The above result again is the distance in meters the Moon in its orbit falls toward Earth in 1 min under the assumptions given above.

4. The *acceleration* will be twice this. This is because, recapitulating from above, in the ultimate case,

$$\frac{AC}{AB} = \frac{r\theta_{rad}}{2r}$$

and, since $\theta_{rad} = 2\pi/P$,

$$\frac{AC}{AB} = \frac{2\pi r}{2rP}$$

But $2\pi r/P$ is also the velocity in orbit, so

$$\frac{AC}{AB} = \frac{v}{2r}$$

Solving for AC we have,

$$AC = \frac{ABv}{2r}$$

Since AB in unit time is also the velocity,

$$2AC = \frac{v^2}{r}$$

Thus the Moon, which falls about 4.9 m in 1 min, experiences a centripetal acceleration toward the Earth of twice this in the first increment of time. Since $s = \frac{1}{2}at^2$, where in our case $s = AC$, then $a = 2AC/t^2$. Thus, using seconds as the units of time, the Moon's acceleration toward the Earth is:

$$a = \frac{2AC}{t^2}$$

$$a = \frac{2(4.9)}{60^2}$$

$$\underbrace{a = .0027 \text{ m/s}^2}$$

The Moon's centripetal acceleration toward the Earth

This is 2.7 mm per second, every second. Whereas the pull of Earth's gravity on the surface of the Earth at one Earth radius, is 9.8 m per second, every second, at 60 Earth radii removed from the center of the Earth, the "g" force is slightly less than one over 60^2 or one thirty-six hundredth of that value. Yet that diminished force is still enough to hold the Moon in its gentle orbit around us.

A Pendulum in Space

Newton showed that the distance the Moon falls in 1 min equals the distance an object on Earth falls in 1 s. A simple and intuitive way of comparing the effects of the inverse square law of gravity on Earth and at lunar distance is to imagine two

A Pendulum in Space

pendulums, one on Earth and the other somehow suspended in space at the distance of the Moon. The following problem explores the result.

Problem Take two imaginary pendulums of equal length, one on Earth and the other somehow suspended in space at the distance of the Moon, oriented so that it hangs in the direction of the Earth. Given the relation below for determining the period of oscillation of a pendulum, show by proportions that the period of pendulum at lunar distance, about 60 Earth radii from the center of the Earth, is one-sixtieth the period of a pendulum on Earth.

Given

$P \propto \sqrt{\frac{l}{g}}$	Period relation for pendulums with small amplitudes
l	The length of the pendulums, here assumed to be one unit length
g	The gravitational acceleration acting on the pendulum mass

Assumptions Because this is a thought experiment, we do not need to establish the physical possibility of supporting a pendulum in space hovering at lunar distance. That is the beauty of thought experiments! We will assume that the pendulum is stationary in space with respect to the Earth, at constant distance, and ignore the effects of the Sun, Moon, and other planets. Finally, we will assume the pendulums are each swung through small angles, so the above period relation holds.

Method We will take the value of g at the Earth's surface as one unit. It is "one g," and can be denoted g_e. The value of g at the distance of the Moon can be denoted by g_m. By applying the inverse square law through a distance of 60 Earth radii, we can see the effect of the change of g on the period of the pendulum.

Calculations Take the period of the Earth-bound pendulum to be P_e and the one at lunar distance to be P_m. The ratio of the periods will be,

$$\frac{P_e}{P_m} = \frac{\sqrt{\frac{l}{g_e}}}{\sqrt{\frac{l}{g_m}}}$$

Since the pendulum length in each case is one unit, and g_e is also one unit, the ratios reduce to this,

$$\frac{P_e}{P_m} = \frac{1}{\sqrt{\frac{1}{g_m}}}$$

If we apply an inverse square law to the Earth's gravity, the gravitational acceleration g_m at lunar distance is $1/60^2$ times g_e, or 1/3600th that of Earth's gravity at its surface. Given this, the ratio then becomes,

$$\frac{P_e}{P_m} = \frac{1}{\sqrt{\left(\frac{1}{3600}\right)}}$$

or,

$$\frac{P_e}{P_m} = \frac{1}{60}$$

Thus, the space pendulum at 60 Earth radii has a period that is 60 times longer than the period of the Earth pendulum: $P_m = 60 \times P_e$. If the Earth pendulum swings once a second, the space pendulum at the Moon's distance will swing once in a minute, exactly the result of Newton's Moon test but obtained by a different method.

Observation This result should not be surprising, since it is consistent with Galileo's discovery that the forces making an object fall are independent of forces perpendicular to the direction of fall. For the first part of its swing, a pendulum is, essentially, a falling object, and continues through the remainder of its swing through inertia. That it is subjected to a transverse force pulling it inward in its swing is independent of the force pulling it downward, just as Galileo's rolling ball on the inclined plane of his experiments still "fell" though a vertical distance, independently of the force from the plane making it go transversely.

Exercises: Working with Proportions One of the pleasures of reading Newton is to absorb his proportional way of thinking. Before instruments had quantified many things with great accuracy, he, like those before him, would derive fundamental results about the universe in stepping-stone fashion, by considering the ratio of this to that, and then from that to one other. By comparing the inverse squares of forces (called taking the "subduplicate ratio" by Newton) at different distances, he demonstrated basic principles without having to know the value of constants, such as the gravitational constant G (which we will see later). His Moon test was a classic illustration of the method. The beauty of it is that it is usually simple and intuitive. Courses in the physical sciences do not seem to practice it, preferring to jump right into the direct, numerical solution of problems. But if you can become comfortable with proportional thinking, it will greatly enhance your insights into the heart of the subject. Most or all of the problems below can be solved by the use of simple proportional thinking.

Problems

1. If the moon's tangential motion were suddenly stopped, the moon would: (a) remain at rest in space in accordance with Newton's first law of motion; (b) begin falling to earth with constant acceleration; (c) continue to accompany the Earth in its orbit around the sun; (d) impact the Earth with an acceleration equal to 9.8 m/s squared; (e) none of the above.
2. Given that the moon is about 60 earth radii distant, what would its period be at half that distance?

3. The orbit of Venus is .723 astronomical units from the sun. What gravitational pull from the Sun does it experience in relation to the Earth?
 4. The planets Mars, Jupiter, and Saturn are 1.6, 5.2 and 9.8 astronomical units from the Sun, respectively. In proportion to the planet Jupiter, what is the Sun's gravitational force per unit mass acting on Mars and Saturn?
 5. The Moon travels in its orbit at a velocity of about 1 km/s. The Earth orbits the Sun, which is 390 times more distant than the Moon, at about 30 km/s. What is the gravitational acceleration of the Earth acting on the Moon in proportion to the Sun acting on the Earth?
 6. Mars is 1.6 astronomical units from the sun. Using just this information, determine the ratio of the velocity of earth to the velocity of Mars.
 7. Saturn, at 9 AU from the Sun, orbits it at about 9.69 km/s. What is Saturn's fall toward the Sun in 1 min of time? One AU is about 150 million kilometers.
 8. Calculate the Earth's fall toward the Sun in 1 min of time as it orbits it at about 30 km/s. Using this information, what is the Sun's gravitational force per-unit mass at the distance of Saturn's orbit? How does this compare with your answer to the previous problem?
 9. Assuming the Sun is 333,000 times more massive than the Earth, and is 390 times more distant than the Moon, what is the Sun's gravitational acceleration acting on the Moon in relation to the Earth's gravitational acceleration acting on the Moon?
10. The Earth is about 81 times more massive than the Moon, which is about 60 Earth radii distant. What is the Moon's gravitational pull on a 1 kg mass on the Earth's surface, in relation to the Earth's gravitational pull on that mass?

References

1. Cajori F (1949) Principia in modern English: Sir Isaac Newton's mathematical principles of natural philosophy (trans: Motte's Revised), 663, App. note 40. University of California Press, Berkeley
2. Christianson GE (1984) In the presence of the creator: Isaac Newton and his times. The Free Press, New York
3. Chandrasekhar S (1995) Newton's principia for the common reader. Clarendon Press, Oxford, pp 1–6
4. Densmore D (1996) Newton's principia: the central argument. Green Lion Press, Santa Fe, 303
5. Huygens C (1986) The pendulum clock or geometrical demonstrations concerning the motion of pendula as applied to clocks. University of Iowa Press, Ames

Chapter 7
Newton Demonstrates How an Inverse Square Law Could Explain Planetary Motions

By relating the fall of the Moon in space to the descent of objects at the surface of the Earth, Newton showed the inverse square law applies out so far as the Moon. But could he prove that the inverse square law applied everywhere else in the solar system, as appeared to be the case with Kepler's laws? Because the Earth does not have two moons, Newton could not directly determine whether the Keplerian proportions held true for the Moon, and thereby empirically link an inverse square law with that law. But he could inquire mathematically whether an inverse square law could account for Keplerian motion. Certainly, a theoretical connection between the two laws would more convincingly establish the universality of gravitation. This he did in the remarkably brief and intriguing corollaries to *Principia's* Proposition IV of his first book.

Clearing the Mathematical Path to the Inverse Square Law

With his characteristically concise, elegant, and rather cryptic approach, Newton explored a range of mathematical possibilities in the corollaries to Proposition IV of Book I of the *Principia*. They can be looked at in two ways. They can first be viewed as the development of useful mathematical relationships to apply in systems where the given conditions appear to hold true, for instance in the comparison of the periods of two satellites in the same system, or in other contexts when comparing centripetal forces among systems. They can be seen more profoundly, consistent with his overall plan of developing a case for his gravitational theory, as a series of systematic tests of hypotheses. By what we may infer as a thought-experiment, he allowed orbital periods to vary with distance in both Keplerian and non-Keplerian ways, to see if the centripetal forces implied by those hypotheses would vary inversely with the square of the distance from the force. If a hypothesis implied forces contrary to experience, it could be ruled out. Newton did not formally connect the mathematics to the physical world until Book III, but the result in Book I was apparent: the only hypothesis that fit was where periods and distances are related by

Kepler's Third Law; then and only then must the centripetal forces be governed by an inverse square law. And conversely, only with an assumed inverse square law are periods and distances related in the way described by Kepler's Third law.

Corollary I *Relating Arcs to Velocities*

As we saw from Proposition IV, when we compare the motion of bodies in two circles, the centripetal forces on each tend to the centers of the circles, and are in the ratio of the arcs traversed in equal times divided by the respective radii. Taking the respective arcs to be $\widehat{AB}_1$ and $\widehat{AB}_2$ and letting f with appropriate subscripts represent acceleration (force per unit mass), then

$$f_1 : f_2 :: \frac{\widehat{AB}_1^2}{r_1} : \frac{\widehat{AB}_2^2}{r_2}$$

In Corollary I to Proposition IV, Newton made the easy conversion of the arcs-squared to velocities squared. This holds since the times elapsed along the arcs are the same, so the velocities are proportional to the distances. Thus,

$$f_1 : f_2 :: \frac{v_1^2}{r_1} : \frac{v_2^2}{r_2}$$

And generally,

$$f \propto \frac{v^2}{r}$$

which is the all-important centripetal acceleration relation. This was Newton's revealing conclusion in Book I, Proposition IV and its first corollary. As we saw, it was a vital connecting link between orbital motion and Galileo's equation for falling objects on Earth.

Corollary II *The Relation of Centripetal Forces to Periods*

In Corollary II to the same proposition, Newton shows that "since the periodic times are as the radii divided by the velocities," the centripetal forces are proportional to the radius divided by the square of the period:

$$f_1 : f_2 = \frac{r_1}{P_1^2} : \frac{r_2}{P_2^2}$$

$$f \propto \frac{r}{P^2}$$

This result follows simply from the equation for centripetal acceleration developed in Corollary I and the fact that in each circle the velocity is equal to the circumference divided by period of revolution, $v = 2\pi r/P$ (mathematically equivalent to Newton's statement that the periods are proportional to the radii

Clearing the Mathematical Path to the Inverse Square Law

divided by the velocities); that is, velocities are proportional to the radius divided by periods:

$$\underbrace{f \propto \frac{v^2}{r}}_{\text{Centripetal acceleration}} \quad \text{and} \quad \underbrace{v \propto \frac{r}{P}}_{\text{Circular velocity}} \Rightarrow \underbrace{f \propto \frac{r}{P^2}}_{\text{Centripetal acceleration}}$$

This is an extremely useful relation which we will visit several times more in this book.

Corollaries III to VI *The "What-if" Corollaries*

After these straightforward conclusions, the remaining corollaries of Proposition IV of *Principia's* Book I use these results to explore "what ifs", considering what the centripetal forces would be if the relationship between the periods and the distances from center of force varied in certain ways. The variations would be mathematically true, but would the world they described conform to experience? We will algebraically demonstrate the "what if" Corollaries III through VI of Proposition IV, something Newton left to the reader. We will do it consistently with the Newtonian way of proportional thinking.

Corollary III

Whence if the periodic times are equal, and the velocities therefore as the radii, the centripetal forces will be also as the radii; and conversely. (*Principia*, Corollary III to Book I, Proposition IV.)

This corollary tests the centripetal forces if the periods are *equal*, independent of the radii. This would be the case if, for example, the solar system revolved around the sun like a record on a turntable or a merry-go-round, where the outer edges rotate once in the same amount of time as the inner surface. If this were the case with the planets, all planets at their distances would complete one revolution around the Sun at the same time. The implication as to centripetal force (shown below) is that the forces would vary as the radii. In such a system, orbits at twice the distance from the center of force would imply twice the inward-pulling centripetal force. This outcome would resemble the way centripetal force works with springs and pendulums, where the greater the distance from the center (or the longer the stretch), the stronger the force is tending to pull it back. If in this case the planets had equal periods, they would be compelled back toward the center with forces increasing as one went outward. The planets of solar system (if indeed the solar system could even have formed under such a regime) would very soon have collapsed into the Sun. The result (of course!) does not conform to experience or to Kepler's Third Law.

Proof of Corollary III

Corollary II (shown above) can be expressed proportionally in this way, using Newton's comparison of circles of radii r_1 and r_2 respectively:

$$f_1 : f_2 = \frac{r_1}{P_1^2} : \frac{r_2}{P_2^2}$$

If as stated the periodic times are equal, then $f_1 : f_2 = r_1 : r_2$ and generally,

$$f \propto r$$

Corollary IV

If the periodic times and the velocities are both as the square roots of the radii, the centripetal forces will be equal among themselves; and conversely. (Principia, Corollary IV to Book I, Proposition IV)

This corollary tests the centripetal forces if the periods (and velocities) are proportional to the *square roots* of the radii. The result: constant centripetal force. This would mean that the Sun would attract all planets equally anywhere in the solar system – no inverse square law would apply in this case.

Proof of Corollary IV

We have from Corollary II above:

$$f_1 : f_2 = \frac{r_1}{P_1^2} : \frac{r_2}{P_2^2} \quad \text{and} \quad f_1 : f_2 = \frac{v_1^2}{r_1} : \frac{v_2^2}{r_2}$$

Rearranging the proportions in terms of the squares of the times and velocities yields,

$$P_1^2 : P_2^2 = \frac{r_1}{f_1} : \frac{r_2}{f_2} \quad \text{and} \quad v_1^2 : v_2^2 = f_1 r_1 : f_2 r_2$$

Thus,

$$\frac{P_1}{P_2} = \frac{\sqrt{r_1}\sqrt{f_2}}{\sqrt{r_2}\sqrt{f_1}} \quad \text{and} \quad \frac{v_1}{v_2} = \frac{\sqrt{r_1}\sqrt{f_1}}{\sqrt{r_2}\sqrt{f_2}}$$

Now if $P_1 : P_2 = v_1 : v_2$, then,

$$\frac{\sqrt{r_1}\sqrt{f_2}}{\sqrt{r_2}\sqrt{f_1}} = \frac{\sqrt{r_1}\sqrt{f_1}}{\sqrt{r_2}\sqrt{f_2}}$$

$$f_2 = f_1$$

Corollary V

If the periodic times are as the radii, and therefore the velocities equal, the centripetal forces will be inversely as the radii; and conversely. (*Principia*, Corollary V to Book I, Proposition IV)

Clearing the Mathematical Path to the Inverse Square Law

In this corollary, we are to assume that the periods will be proportional to the radii. In such a hypothetical case, a planet, such as Saturn, nine astronomical units out from the Sun would be presumed to take 9 years to revolve around the Sun (not the 29 years Saturn actually takes!). The result: the greater the distance, the less the centripetal force. But this is a simple inverse relationship, not an inverse square relationship. So far, none of the hypotheses that have deviated from Kepler's Third Law have come even close to an inverse square relationship of force to distance.

Proof of Corollary V

If the periodic times are as the radii, the velocities will be equal, since from Corollary II above, $P_1 : P_2 = \frac{r_1}{v_1} : \frac{r_2}{v_2}$, therefore $P_1 : P_2 = r_1 : r_2$. To find the relationship between centripetal force, the radii and velocities, we can use the expression given in Corollary I:

$$f_1 : f_2 = \frac{v_1^2}{r_1} : \frac{v_2^2}{r_2} \quad \text{so} \quad \frac{f_1}{f_2} = \frac{v_1^2 r_2}{v_2^2 r_1}$$

Where velocities are equal,

$$\frac{f_1}{f_2} = \frac{r_2}{r_1}$$

And generally,

$$f \propto \frac{1}{r}$$

Corollary VI

If the periodic times are as the 3/2th powers of the radii, and therefore the velocities inversely as the square roots of the radii, the centripetal forces will be inversely as the squares of the radii; and conversely. (*Principia,* Corollary VI to Book I, Proposition IV)

At last Newton tests the period-distance relationship of Kepler's Third law. This is the empirically verified relationship of the planets in the solar system, although Newton is not yet necessarily making a connection to the physical world: he is exploring a mathematical relationship that, if proven, will be useful in Book III where he does connect these results with observations of real things in the heavens. Now, what will the hypotheses in this corollary imply as to centripetal force? Result: An inverse square relationship, which conforms to the observed fall of the Moon discussed above, and as we will see later, the motion of bodies in elliptical orbits.

Proof of Corollary VI

It is given that $P_1 : P_2 :: r_1^{3/2} : r_2^{3/2}$, which was the important empirical relationship among the orbiting planets found by Kepler. Since we know from Corollary II above that $P_1 : P_2 : r_1/v_1 : r_2/v_2$, we can make the appropriate substitutions to obtain relations between the radii and velocities only. Thus,

$$r_1^{3/2} : r_2^{3/2} = \frac{r_1}{v_1} : \frac{r_2}{v_2} \quad \text{so} \quad \frac{r_1 v_2}{r_2 v_1} = \frac{r_1^{3/2}}{r_2^{3/2}}$$

Remembering that $1/r = r^{-1}$, we have,

$$\frac{v_2}{v_1} = \frac{r_1^{3/2} r_1^{-1}}{r_2^{3/2} r_2^{-1}} \quad \text{which becomes,} \quad \frac{v_2}{v_1} = \frac{r_1^{1/2}}{r_2^{1/2}}$$

So here, the velocities are inversely as the square roots of the radii. Now, from above we know,

$$\frac{f_1}{f_2} = \frac{v_1^2 r_2}{v_2^2 r_1} \quad \text{and} \quad \frac{v_2^2}{v_1^2} = \frac{r_1}{r_2}$$

The first equation comes from Corollary V above, and the second we obtained from squaring the velocity expression. The velocity squared is common to both equations, so we can make the substitution, eliminating the velocity terms and solving for the force terms:

$$\frac{f_1}{f_2} = \frac{r_2^2}{r_1^2}$$

And generally,

$$f \propto \frac{1}{r^2}$$

This is Newton's inverse square law of gravitation: The force acting on a body in a circular orbit is inversely proportional to its distance from the center. It was a momentous result that came directly from Kepler's Third Law.

And Conversely...

The corollaries end with the phrase, "and conversely" meaning that the principle works in reverse. If A leads to B, then conversely, B leads to A. Newton does not prove the converse either (again leaving that to us); however the implication of the converse of Corollary VI is powerful:

- Kepler's Third Law relating orbital periods to distances necessarily implies an inverse square law of centripetal force; and conversely,
- An inverse square law of centripetal force necessarily leads to orbital periods and distances consistent with Kepler's Third Law.

Summary of Above Corollaries to Proposition IV

The above corollaries to Proposition IV of Book I may be summarized in this way:

Corollary to fourth proposition	If periodic time P varied with the distance r from the center of force in this way...	Which means P varies as the following power of r	How would the centripetal force vary?	Does the result conform to experience?
III	P is constant (does not vary with the radius)	r^0	$f \propto r$	No
IV	P varies as $\sqrt{r}$	$r^{1/2}$	f is constant	No
V	P varies as r	r^1	$f \propto 1/r$	No
VI	P varies as $r^{3/2}$	$r^{3/2}$	$f \propto 1/r^2$	Yes!

The most important relation is thus found in the sixth corollary, where the inverse square law arises out Kepler's Third Law.

A Small Lesson

Newton appended a "Scholium," or small lesson, to his Proposition IV, as he did with various propositions throughout the *Principia*, showing his awareness of the implications of what he had discovered:

> The case of the sixth Corollary [the inverse square law] obtains in the celestial bodies (as Sir *Christopher Wren,* Dr. *Hooke,* and Dr. *Halley* have severally observed); and therefore in what follows, I intend to treat more at large of those things which relate to centripetal force decreasing as the squares of the distances from the centres.
> Moreover, by means of the preceding Proposition and its Corollaries, we may discover the proportion of a centripetal force to any other known force, such as that of gravity. For if a body by means of its gravity revolves in a circle concentric to Earth, this gravity is the centripetal force of that body. But from the descent of heavy bodies, the time of one entire revolution, as well as the arc described in any given time, is given ...[1]

Of Circles and Ellipses

Newton later in *Principia* proved that circularity was not essential for the truth of these propositions; Kepler stated that his Third Law applied to the planets which by his Second law were shown to moving in ellipses. Newton generalized what Kepler had discovered from Tycho's observation of the planets as applicable both to circular and elliptical motion:

> The same things being supposed [from the previous Proposition XIV, i.e., revolution about a center, and centripetal force varying inversely as the distance from the center], I say, that the periodic times in ellipses are as the 3/2th power of their greater axes.[2]

---

[1] Florian Cajori [1]. (Short title) *Principia*.
[2] *Principia*, Book I, Proposition XV, Theorem VII, 62.

Therefore the periodic times in ellipses are the same as in circles whose diameters are equal to the greater axes of the ellipses.[3]

Evidence from Distant Worlds: The Moons of Jupiter and Saturn Obey Kepler's Third Law, and Therefore the Inverse Square Law

The inverse square law of centripetal force certainly appeared to explain the Moon's orbital motion and the planetary orbits in the solar system. The mathematics of the *Principia* united Kepler's Third Law with the v^2/r centripetal acceleration to produce, remarkably, the inverse square relationship. But skeptics may have deemed those circumstances unique in their own effects. It was one thing to demonstrate that the Moon's fall around the Earth was consistent with a theory that considered the Earth as the source of some mysterious force keeping the Moon from flying off into space. It was another to credit to such an unknown force the motions of all bodies *everywhere*. While the mathematical conclusions were based on the assumption that Kepler's Third Law applied everywhere, this was not necessarily self-evident. Nothing about motion in the celestial sphere in the 1,600s could be taken for granted, and there were competing theories everywhere: Kepler's mystical cosmology, Descartes' swirling vortices, the Tychonic and Copernican solar systems, to name a few. Earlier in the same century, Galileo had been put under house arrest for supporting a heliocentric view of the universe.

Newton had prepared a powerful rebuttal to these alternatives in his one simple theory of gravitation. The identical effects of the Sun and Earth on their orbiting bodies was strong evidence, but showing *empirically* that the inverse square law also applied to distant Jupiter and Saturn and their satellites, even as they went around the Sun, would greatly tighten his case for gravity. This would mean showing clearly that Kepler's Third Law applied even to remote Jupiter and Saturn.

But how would that be proof? Newton always had in mind his "Rules of Reasoning in Philosophy," which begin Book III of the *Principia*, the first two of which state:

RULE I

We are to admit no more causes of natural things than such as are both true and sufficient to explain their appearances.[4]

RULE II

Therefore to the same natural effects we must, as far as possible, assign the same causes.[5]

---

[3] Ibid., Corollary.
[4] *Principia*, Book III, 398.
[5] Ibid.

Applying these rules to the present case, one might frame the matter this way: If many types of orbital motion (around the Earth, the Sun, and outer planets) appear to be governed by a centripetal force, and that force in each investigated circumstance is governed by the inverse square law, then it is scientifically reasonable to conclude that the centripetal forces tending to the center of each of these bodies are caused by the same thing: a property of matter which in every case in nature will diminish in an inverse square manner with distance from the center.

The Motion of Jupiter's Satellites as a Test for the Universality of the Inverse Square Law

Around the time of Kepler's discoveries, Galileo had fashioned improved versions of a telescope, newly invented by Dutch lens makers. His first telescopic observations of Jupiter revealed a large orange orb surrounded by four little whirling moons. Indeed, it looked like a miniature solar system. John Flamsteed (1646–1719), the first Astronomer Royal of Greenwich Observatory in England, carefully measured the Jovian satellite motions and studied their distances. In letters to Isaac Newton, Flamsteed noted that the little moons followed Kepler's Third Law as exactly as the accuracy of his observations would permit. This was important information because, as Newton showed, the inverse square law flows naturally from Kepler's Third Law. One is the mathematical cousin of the other. If Newton could show that the inverse square law applied even to the remotest bodies then known in the solar system, it would be strong evidence for the universality of the gravitation.

In Book III of the *Principia*, Newton tested Kepler's Third Law against actual observations of Jupiter's and Saturn's moons and the known planets of the solar system.[6] The book begins with the presentation of "Phenomena," which are just systematically organized statements of known facts. From these factual foundations

---

[6] There is a 1728 translation of *The System of the World* (or *Treatise on the System of the World*). It is believed to be an earlier and more accessible draft of what eventually became Book III of the *Principia*. The translator of this is unknown, but may have been Andrew Motte, who also translated the *Principia*. An online version of the 1728 *System of the World* is at http://books.google.com/books?id=rEYUAAAAQAAJ&pg=PR1#v=onepage&q&f=false.

Newton had famously contentious disputes with Flamsteed. Interestingly, in this earlier account, Newton gives abundant credit to Flamsteed for his Jupiter observations and his insightful application of Kepler's Third Law to them. But in Book III of the *Principia*, there is no mention whatsoever of Flamsteed's contribution on this matter. Compare page 401 (*Phenomenon I*) of Book III of the *Principia* to pages 555–556 (section [6.]) of that publication. Here is what Newton said in the earlier version:

> This proportion [Kepler's law] has long ago been observed in those satellites [of Jupiter]; and Mr. *Flamsteed*, who had often measured their distances from Jupiter by the micrometer, and by the eclipses of the satellites, wrote to me, that it holds to all the accuracy that possibly can be discerned by our senses. And he sent me the dimensions of their orbits taken by the micrometer, and deduced the mean distance of Jupiter from the earth, or from the sun, together with the times of their revolutions. ... Ibid., 555.

he drew conclusions in the propositions that followed. Those propositions as we saw refer for their mathematical support to the propositions, corollaries, and lemmas of Book I. The first two phenomena pertain respectively to the moons (here called "planets") of Jupiter and Saturn. Using data obtained by Flamsteed and other astronomers to confirm the expected result that the moons of Jupiter and Saturn obeyed Kepler's Third Law, Newton arrived at the result that those moons acted under the influence of a centripetal force with their host planet as its center, which force fell off with the inverse square of the distance from that center:

PHENOMENON I

That the circumjovial planets, by radii drawn to Jupiter's centre, describe areas proportional to the times of description; and that their periodic times, the fixed stars being at rest, are as the 3/2 th power of their distances from its centre.[7]

PHENOMENON II

That the circumsaturnal planets, by radii drawn to Saturn's centre, describe areas proportional to the times of description; and that their periodic times, the fixed stars being at rest, are as the 3/2 th power of their distances from its centre.[8]

In the text accompanying Phenomenon I, Newton informs us that the orbits of Jupiter's moons "differ but insensibly from circles concentric to its centre; and their motions in those circles are found to be uniform." If the moons of Jupiter appeared at unequal distances at their greatest elongations from Jupiter, or traversed different parts of their orbits at different speeds, this would be evidence of non-circular orbits. Since their orbits appear circular, however, his work is made simple: it is easy to conclude that they describe areas proportional to their times, consistent with Kepler's Second Law. The same result follows for Saturn in Phenomenon II.

The last part of the Phenomena states that the satellite motions are consistent with Kepler's Third Law, which is a key stepping stone to the Propositions that follow. The data on the accompanying table is taken as it is presented in the *Principia* from Book III, Phenomenon I. The table shows how Newton found the period-squared, distance-cubed relationship in Jupiter's moons, the same ratio of period to orbital distance that is Kepler's Third Law and that governs the motions of planets around the Sun.[9] As can be seen from the table, the bottom row showing Newton's *calculated* distances (in Jovian radii) of the satellites' orbits, derived from the periodic times and the application of Kepler's Third Law, match closely the distances derived from observations, in the rows above it.[10] His table for Saturn

---

[7] *Principia*, Book III, Phenomenon I, 401.

[8] *Principia*, Book III, Phenomenon II, 402.

[9] *Principia*, Book III, Phenomenon I, 401.

[10] The names of the satellites in order, from Jupiter outward are: Io, Europa, Ganymede, and Callisto. Having been discovered by Galileo, they are referred to as the "Galilean" moons of Jupiter. Jupiter has over 60 moons; but these are the brightest and the most beautiful to see, being visible even in binoculars. The text here is replicated from the *Principia*, using Newton's notational form. The reader should be cautioned about the notation: a number such as $8^{2/3}$ does not mean 8 raised to the 2/3 power, but is 8 and the fraction 2/3.

in Phenomenon II reaches the same conclusion with respect to the four observed moons of that planet.

The periodic times of the satellites of Jupiter

$1^d.18^h.27^m$	$3^d.13^h.42^s$	$7^d.3^h.42^m.36^s$	$16^d.16^h.32^m.9^s$

The distances of the satellites from Jupiter's center

	1	2	3	4	
From the observations of:					
Borelli............	$5^{2/3}$	$8^{2/3}$	14	$24^{2/3}$	
Townly by the micrometer.........	5.52	8.78	13.47	24.72	*Semi-*
Cassini by the telescope...........	5	8	13	23	*diameter*
Cassini by the eclipse of satellites............	$5^{2/3}$	9	$14^{23/60}$	$25^{3/10}$	*of Jupiter*
From the periodic times.................	5.667	9.017	14.384	25.299	

So far, Newton has shown, albeit with admirable thoroughness, what was already known about these moons: that they appear to be governed by Kepler's Third Law. The crucial step in making his case for gravity comes next, after the Phenomena, in the first proposition of Book III:

BOOK III PROPOSITION I

That the forces by which the circumjovial planets are continually drawn off from rectilinear motions, and retained in their proper orbits, tend to Jupiter's centre; and are inversely as the squares of the distances of the places of those planets from that centre ... The same thing we are to understand of the planets which encompass Saturn ...[11]

This proposition has two parts. The first states that the "forces ... tend to Jupiter's centre." This is a straightforward conclusion following from Newton's earliest propositions in Book I: that any body moving along a curve in a plane, and which by a radius drawn to a point describes areas proportional to times, is "urged by a centripetal force directed to that point."[12] Thus Jupiter's moons are "urged" to the center of Jupiter. The second part, the conclusion we have been waiting for, states that this urging force *obeys the inverse square law*. To prove it, Newton calls upon his workhorse Corollary VI in Book I's Proposition IV: "If the periodic times are as the 3/2th powers of the radii ... the centripetal forces will be inversely as the squares of the radii; and conversely." In other words, if the orbital periods of the moons of Jupiter and Saturn obey Kepler's Third Law, as observation shows, then those planets hold them with a force that varies inversely as the distance from the planet, a marvelously simple and far-reaching conclusion!

---

[11] *Principia*, Book III, Proposition I, Theorem I, 406.

[12] *Principia*, Book I, Proposition II, 42. The actual text of Book I, Proposition II, Theorem II, reads: "Every body that moves in any curved line described in a plane, and by a radius drawn to a point either immovable, or moving forwards with an uniform rectilinear motion, describes about that point areas proportional to the times, is urged by a centripetal force directed to that point." *Principia*, at 42.

Newton then extrapolates this conclusion logically to the primary planets:

BOOK III, PROPOSITION II, THEOREM I

That the forces by which the primary planets are continually drawn off from rectilinear motions, and retained in their proper orbits, tend to the sun; and are inversely as the squares of the distances of the places of those planets from the sun's centre.[13]

In other words, the inverse square law is observed to apply in the satellite systems of Jupiter and Saturn and indeed all the planets of the solar system.

Finding the Keplerian Proportionality Constant in the Jovian Satellite System

Newton's development of many of his key propositions employed a kind of proportional thinking that is powerful and intuitively clear. Proportional thinking is reasoning on the basis of proportions, such as by saying the ratio of period squared is proportional to distance cubed among all of the planetary orbits. The proposition above and all the corollaries and their proofs discussed in the last chapter, for example, are presented by Newton in the proportional manner. The problem below and those that follow in this chapter are more or less exercises in that tradition. Here again we begin with Jupiter's moons, then will apply proportional analysis to some contemporary situations.

Problem Using the observational data above gathered by Newton from Book III of the *Principia*: Find the period-squared/distance-cubed proportionality constant for the Jovian satellites. Then, assuming the constant was derived from just one satellite, derive the bottom row of Newton's table, where he determines and compares the distances of all the satellites from the periodic times. The units to be used in these proportions are days and Jovian radii.

Given Newton's Jovian satellite data as shown in the above table.

Assumptions We will assume such accuracy of the data as was available to Newton. As to the distances, assume that Cassini's data from the eclipses of the satellites is the most accurate. Newton seemed to suggest that this was the more accurate method.[14]

Method For the first part of the problem, begin by finding the periodic times in decimal parts of a day. This is the period, P, for each satellite. Then we will find the ratio of those times squared to the decimal distances (in Jovian radii) a, cubed. The ratio will be the proportionality constant. Since $P^2 = ka^3$, it is then possible to calculate for each satellite the ratio $k = P^2/a^3$. For the second part of the problem,

---

[13] *Principia*, Book III, Proposition I, Theorem I, 406.

[14] See, for example, *The System of the World*, id., 556.

Finding the Keplerian Proportionality Constant in the Jovian Satellite System

the constant being given, the task is to solve this equation for a, for each of the periodic times.

Calculations The respective periodic times in seconds for each of the four satellites are:

Io	Europa	Ganymede	Callisto
1.769 days	3.55 days	7.155 days	16.689 days

The *squares* of these times are, respectively:

3.129	12.603	51.194	278.523

The distances (using the last entry for Cassini) are:

5.667	9.0	14.383	25.3

The distances *cubed* are:

181.995	729	2,975.42	16,194.277

The ratios of each periodic time squared to each distance cubed are therefore, respectively:

.0172	.0173	.0172	.0172

The ratios are all nearly the same, consistent with Kepler's Third Law.

The average of these ratios is .0172. This is the ratio of periods squared over distances cubed, holding true for each satellite. Thus the constant of proportionality among Jupiter's satellites, using Newton's data and the units of days and Jovian radii, is:

$$k = .0172$$

For the second part of the problem, we assume that this proportionality constant could have been derived from the observations of any one of the Galilean satellites. The task now is to take the periodic times of the remaining satellites and derive their distances from Jupiter. We'll see how the data match up with the last row in Newton's table.

Obtaining the following relationships again from $P^2 = ka^3$, we isolate the distance term:

$$a^3 = \frac{P^2}{k} \to a = \sqrt[3]{\frac{P^2}{k}} \to a = \sqrt[3]{\frac{P^2}{.0172}}$$

The resulting derived distances, in Jovian radii, for each respective period are:

5.666	9.015	14.385	25.3

These are in good agreement with Newton's tabulated distances derived from the periodic times.

Observations

1. It is interesting to compare the radii and periods Newton used with the modern values. As evident from the chart below, the periods were more accurately determined in Newton's day than were the distances from the planet. This is no doubt because then it was far easier to time satellite occultations and eclipses of satellites than it was to measure the tiny moons' distances against the planet's bright image in wobbly air through imperfect optics. Even though micrometers were in use, the telescopes were still by modern measures crude.

Constant	Io	Europa	Ganymede	Callisto
Orbital Period (Newton et al.)	1.769 days	3.55 days	7.155 days	16.689 days
Orbital Period (modern)	1.769	3.551	7.155	16.689
Distance (radii) from Jupiter (Newton et al.)	5.667	9.0	14.383	25.3
Distance (radii) from Jupiter (modern)	5.91	9.4	14.97	26.33

Using the modern values, a proportionality constant of .0152 would apply using units of days and Jovian radii.

2. The equation $P^2 = a^3$ applies to the solar system, since as we saw,

$$P_1^2 : P_2^2 = a_1^3 : a_2^3$$

holds for all the planets, where the units are astronomical units and years. In other words, the proportionality constant implied in the equation $P^2 = ka^3$ is 1, which works if we use the Earth's orbital distance as our measuring rod for the solar system, and years are the unit of time. Recall from Chap. 4 that if we take period in days and distance in kilometers, the constant is far from one. In fact in that case $k = 3.984 \times 10^{-20}$, an awkward number.

Using a Modern Jovian Proportionality Constant to Find the Periodic Time of Jupiter's Inner Satellite Amalthea

Problem Amalthea is a Jovian moon located only about 181,400 km from the center of Jupiter, almost as close to Jupiter' center as our Moon is to Earth. But Jupiter's equatorial radius is a lofty 71,492 km, over 11 times the radius of the Earth. Amalthea thus flies across the sky only about 110,000 km above Jupiter's cloudtops. Given these facts, what is the periodic time of Amalthea likely to be?

Finding the Keplerian Proportionality Constant in the Jovian Satellite System

Given

$P^2 = ka^3$	Kepler's Third Law in its simplest form, where k is a proportionality constant whose value will depend on the gravitational system we are observing (in our case Jupiter's) and our choice of units
k	The proportionality constant .0152 derived above for the Jovian system using modern data, where the units are days and Jovian radii
a	The number of Jovian radii to the Amalthea: 2.537

Assumptions We assume a circular orbit of Amalthea about Jupiter. We can ignore their true revolution about their common center of mass, since Jupiter is so overwhelmingly more massive than its tiny companion. We ignore all "non-Keplerian" effects (such as perturbations by Jupiter's other satellites, and by other bodies in the solar system). We also assume the planet and its satellite are spheres of uniform density and the sufficient accuracy of the above numbers for our illustrative purposes.

Method Solve the Kepler's Third Law for P, and the result for the period will be in days.

Calculations Since $P^2 = ka^3$,

$$P = \sqrt{ka^3}$$

$$P = \sqrt{(.0152)(2.537)^3}$$

$$P = .498 \text{ days}$$

$$P \approx 12 \text{ hours}$$

Observations

1. The published period of Amalthea is .498179 days.[15] It does in fact have a nearly circular orbit, with an eccentricity of .003.
2. We can estimate Amalthea's mean velocity by using the circular velocity equation (using a for the radius) $v = 2\pi a/P$; however, if we want the result in kilometers per second, we will have to use kilometers for the semi-major axis (here being the radial distance from Jupiter's center since we assume a circular orbit) and convert the period to seconds. The period of .498 days is about 43,000 s, and its given distance is 181,400 km. Its orbital speed is thus,

$$v = \frac{2\pi(1.814 \times 10^5)}{4.3 \times 10^4}$$

---

[15] See, http://nssdc.gsfc.nasa.gov/planetary/factsheet/joviansatfact.html.

$$v = 26.5 \text{km/s}$$

This is extremely fast for a satellite. Our Moon moves gracefully through its realm at about a kilometer a second. Amalthea moves about as swiftly in its tiny orbit as does Mars at perihelion in its orbit around the Sun.

Applying the Proportionality Constant in Kepler's Third Law for the Earth–Moon System to Find the Distance to a Geosynchronous Satellite

Problem Find the proportionality constant in Kepler's Third Law for the Earth–Moon system, and use it to approximate the distance from Earth's center to a satellite with a period of exactly 1 day, in Earth radii, then in kilometers.

Given

$P^2 = ka^3$	Kepler's Third Law, where again k is a proportionality constant whose value will depend on the gravitational system we are observing (here, the Earth–Moon system) and our choice of units
P	The lunar period of about 27.32 days
a	The number of Earth radii to the Moon: 60.27
6,378.1	Earth's mean equatorial radius in kilometers

Assumptions We assume a circular orbit of the Moon about the Earth, ignoring their true revolution about their common center of mass. More importantly, we ignore the Moon's own mass here. We also ignore all "non-Keplerian" effects (such as perturbations by the Sun and other planets). We also again assume the Earth and Moon are spheres of uniform density and the sufficient accuracy of the above numbers for our illustrative purposes.

Method We are no longer here considering the Jovian system, so the Keplerian proportionally constant for that system cannot be used. The reader is invited to articulate why this is so. We must here ascertain from data what the proportionality constant would be for the Earth–Moon system. To do this, we may solve the Kepler equation for k, then using that value, set the period equal to one and solve for a. Units will be Earth radii. Multiplying by Earth's radius gives the value in kilometers.

Calculations Since $P^2 = ka^3$, then

$$k = \frac{P^2}{a^3}$$

$$k = \frac{(27.32)^2}{(60.27)^3}$$

$$k = .00341$$

This is the proportionality constant for things revolving around the Earth, *using days and Earth radii as units*. What is the distance for a satellite to have a 1 day period? Since $P^2 = ka^3$,

$$a = \sqrt[3]{\frac{P^2}{k}}$$

$$a = \frac{1}{\sqrt[3]{.00341}}$$

$$a \simeq 6.44 \text{ Earth radii}$$

To find the approximate distance in kilometers, we multiply the above result by the Earth's radius in kilometers:

$$6378.1 \times 6.44 = 42{,}378 \text{km}$$

Observation

1. If we could see it, a satellite in an equatorial orbit at this distance will appear to hover over one spot. It will be a *geosynchronous* satellite (in synch with Earth's rotation). Many communication satellites are in such orbits.
2. Here the units we used above were radii and days. We could easily find other constants using other units for other problems. We could, for example, call the lunar distance one "lunar unit" and keep the period at days. So to find the period of something orbiting Earth halfway to the Moon, we would again find the proportionality constant for those units (in that gravitational system) then work out the simple calculation. Taking that example, with a being one *lunar unit* (in the equation $k = P^2/a^3$), the constant becomes $k = 27.32^2$ or 746.38. The period at the halfway point becomes:

$$P = \sqrt{ka^3} \rightarrow P = \sqrt{(746.38)(.5)^3} \simeq 9.66 \text{ days}$$

At half the lunar distance the period thus drops to a little more than a third of the lunar period.
3. The method of using the proportionality constant for Kepler's Third Law for a given orbital system assists in the intuitive understanding of orbital relationships. But as will be seen later, there are many other ways to derive the

key orbital quantities. Typically, for well-observed orbiting systems, such as satellites, planets, the period will be known or determinable from observation, and Kepler's Third law is then used to deduce the mass either of the primary, where the secondary's mass is insignificant, or the combined masses of the system where it is not. For slow-moving double stars, or newly discovered comets or exoplanets, the period may need to be derived, again from Kepler's Third Law, a tool of amazing utility.

Exercises: Fear and Panic Around Mars In his 1882 book, *Popular Astronomy*, Simon Newcomb described the discovery of the satellites of Mars 5 years earlier:

> On the night of August 11th, 1877, Professor Asaph Hall, while scrutinizing the neighborhood of Mars with the great equatorial of the Washington Observatory, found a small object about 80 seconds east of the planet. Cloudy weather prevented further observation at that time; but on the night of the 16th it was again found, and two hours' observation showed that it followed the planet in its orbital motion. Still, fearing that it might be a small planet which chanced to be in the neighborhood, Professor Hall waited for another observation before announcing his discovery. A rough calculation from the observed elongation of the satellite and the known mass of Mars showed that the period of revolution would probably be not far from 29 hours, and that, if the object were a satellite, it would be hidden during most of the following night, but would reappear near its original position towards morning. This prediction was exactly fulfilled, the satellite emerging from the planet about four o'clock on the morning of August 18th.
>
> But this was not all. The reappearance of the satellite was followed by the appearance of another object, much closer to the planet, which proved to be a second and inner satellite. ... The most extraordinary feature of the two satellites is the proximity of the inner one to the planet, and the rapidity of its revolution. The shortest period hitherto known is that of the inner satellite of Saturn – 22 hours 37 minutes. But the inner satellite of Mars goes round in 7 hours 38 minutes. Its distance from the center of the planet is about 6000 miles, and from the surface less than 4000. If there are any astronomers on Mars with telescopes and eyes like ours, they can readily find out whether this satellite is inhabited, the distance being less than one sixtieth that of the moon from us.[16]

The outer and inner satellites were named, respectively, Deimos ("Fear") and Phobos ("Panic"), from the *Illiad*. Agnes Clerke, an historian of Nineteenth Century astronomy, described the rapidity of the inner moon this way:

> This is the only known instance of a satellite circulating faster than its primary rotates, and is a circumstance of some importance as regards theories of planetary development. To a Martian spectator the curious effect would ensue of a celestial object, seemingly exempt from the general motion of the sphere, rising in the west, setting in the east, and culminating twice, or even thrice a day; which, moreover, in latitudes above 69° north or south, would be permanently and altogether hidden by the intervening curvature of the globe.[17]

---

[16] Simon Newcomb [2].
[17] Agnes M. Clerke [3].

Mars is on the inner edge of the asteroid belt, and it is likely that the tiny moons of Mars are captured asteroids. Deimos, the outer satellite sighted first by Professor Hall, actually orbits Mars in slightly more time than given in this early account. Modern data[18] has Deimos orbiting once in 1.2624 days, or a little more than 30 h and 18 min. Its distance from the center of the planet is 23,458 km (14,576 miles), about a sixteenth the distance from the center of our Earth to our Moon — a little more than half the distance to Earth's geosynchronous satellites. Phobos is much closer to Mars, only 9,376 km (5,826 miles) from its center. In the following problems we will assume the orbits of these moons are circular, which they very nearly are, and explore these unusual objects.

Problems

1. Given Deimos' period and distance, find the Keplerian proportionality constant and approximate period of Phobos by this method.
2. Phobos' actual period is .3189 days. Find the ratio of the respective centripetal accelerations of Phobos to Deimos toward the center of Mars.
3. Use two methods from Newton's corollaries to his *Principia,* Proposition IV (Book I) to find the ratio of the velocities of Phobos to Deimos.
4. What are the satellites' mean circular velocities in kilometers per day, and in kilometers per second?
5. Use Newton's v^2/r equation for centripetal acceleration to calculate the actual centripetal accelerations of each of Mars' satellites in SI units (meters per second and meters), and their ratio.
6. If Mars had a third satellite four times as far away from it as Deimos is, how much longer, proportionately, would its period be?
7. Referring to the above problem, using one or more of Newton's corollaries in his *Principia,* Proposition IV (Book I), tell how much less, proportionately, would the acceleration acting on the hypothetical third satellite be? Would the same ratio hold true if the comparison were to Phobos?
8. How much slower would be hypothetical third satellite's velocity be? Develop and explain your answer again using one or more of Newton's corollaries to his Proposition IV.
9. The radius of Phobos' orbit is 2.76 Martian semi-diameters. If a hypothetical third satellite were orbiting at the very surface of Mars (but not touching it, and ignoring atmospheric friction) what would the centripetal acceleration acting upon it be? Use any of Newton's corollaries to his Proposition IV to find your answer.
10. Would you expect the results of your previous answer to be the same or different from the surface gravity of Mars (i.e., the value of the Martian g at its surface)? Explain your answer.

---

[18] From the NASA/JPL link, http://ssd.jpl.nasa.gov/?sat_elem.

References

1. Cajori F (1949) Principia in modern English: Sir Isaac Newton's mathematical principles of natural philosophy, Book I, Proposition IV, Theorem IV, Scholium, 46 (trans: Motte's Revised). University of California Press, Berkeley
2. Newcomb S (1882) Popular astronomy. Harper & Brothers, Washington, pp 329–330
3. Clerke AM (2003) A popular history of astronomy during the nineteenth century. Sattre Press, Decorah, p 283; first ed., 1902

Chapter 8
Newton's Master Stroke: The Universal Law of Gravitation

If Kepler's laws applied to planets revolving around the Sun, perhaps it was due to some as-yet (in 1619) undiscovered force that caused such motions. Kepler was convinced that such a force originated in the Sun, operating at one focus of the ellipses described by the planets. He speculated that the Sun's force might be akin to magnetism. Civilization would have to wait another 70 years before Isaac Newton would unveil his theory of gravitation.

Newton developed his theory in the *Principia* in slow and careful steps. Before Book III, Newton makes no assumptions whatsoever about causes of phenomena in the physical world, since his purpose was to derive mathematical principles. In the very first proposition of Book I, Newton describes curved motion in a plane, and concludes that if bodies move in a curve about a center of force, then they describe areas proportional to the times described, echoing Kepler's Second Law:

BOOK I, PROPOSITION I. THEOREM I

The areas which revolving bodies describe by radii drawn to an immovable centre of force do lie in the same immovable planes, and are proportional to the times in which they are described.[1]

There is nothing in this concise and powerful proposition that suggests what the force is to which such bodies are drawn, or what law governs the force, inverse square or otherwise. Nor does it state whether there is even a mass at the center causing the force, or if the force is variable or constant. In Proposition II, Newton tells us the converse of Proposition I, that if a body is found in that situation to describe areas proportional to the times, then it is urged by a centripetal force:

---

[1] Florian Cajori [1], (Short title) *Principia*.

BOOK I, PROPOSITION II. THEOREM II

Any body that moves in any curved line described in a plane, and by a radius drawn to a point either immovable, or moving forwards with an uniform rectilinear motion, describes about that point areas proportional to the times, is urged by a centripetal force directed to that point.[2]

Beginning with these two propositions and the application of some geometry, Newton derived the v^2/r proportion for centripetal acceleration. Once Newton had found it, he applied it to the 3/2th power law of Kepler's Third Law of planetary motion. This he did in the sixth corollary to Book I, Proposition IV, and it led straight to the door of the inverse square law. Since, as we saw in the second corollary to that proposition, centripetal acceleration v^2/r reduces to r/P^2, the connection appears easy:

$$\underbrace{f \propto \frac{r}{P^2}}_{\text{Centripetal acceleration}} \quad \text{and} \quad \underbrace{P^2 \propto r^3}_{\text{Kepler's Third Law}} \quad \Rightarrow \quad \underbrace{f \propto \frac{1}{r^2}}_{\text{Inverse square law}}$$

Newton was able to test this result empirically on the moons of Jupiter and Saturn. He found the inverse square law applied to those moons. Those moons "describe areas proportional to the times of description" around those planets, so the planets themselves must be the source of the centripetal force, consistent with Book I's Proposition II. With it, and the careful scaffolding of laws and propositions constructed in Book I of the *Principia*, Newton linked the mathematics with the evidence in Book III. There he stated that the centripetal force that has figured in so many of his propositions is *gravity*, and is dependent only upon the mass of the object. This Newton saw as *universal*, affecting the Moon and planets just as it affected things like apples falling on Earth – although in 1687 it was unknown to anybody other than Newton:

BOOK III PROPOSITION VIII, THEOREM VII

That there is a power of gravity pertaining to all bodies, proportional to the several quantities of matter which they contain.[3]

This understated conclusion had been approached with the utmost caution, but was the apex of his work; the idea of action at a distance was deeply profound and counter-intuitive, presenting a challenge to its acceptance.

We have been discussing mainly centripetal accelerations (forces acting on unit masses) and how they relate proportionally to other factors of orbital motion. But now we need to delve deeper into attractive forces of quantities of masses, not limiting ourselves to proportions of accelerations. It will be instructive in this regard to employ Newton's Second Law to derive the inverse square law of forces, then construct Newton's gravity equation.

---

[2] Ibid., Book I, Proposition II, Theorem II, 42.
[3] Ibid., Book III, Proposition VIII, Theorem VIII, 414.

# Working with Forces: Deriving the Inverse Square Law from Newton's Second Law

Problem Using the equation for circular centripetal acceleration, Newton's Second Law and Kepler's Third Law, show that the centripetal force acting on a circularly orbiting object is inversely proportional to the square of the distance from the center of force.

Given

$f_c = v^2/r$	Centripetal acceleration f_c in circular orbital motion, where v is the velocity of an orbiting object and r is the radius of the orbit
$F = mf$	Newton's Second Law, where m is mass and f is acceleration[a]
$v = 2\pi r/P$	Circular orbital velocity, where P is the period and r is the radius
$P^2 = kr^3$	Kepler's Third Law. For elliptical orbits, $P^2 = ka^3$ where a is the semi-major axis of the ellipse

[a] It is common to use the letter a to stand for acceleration when writing Newton's Second Law. But since that letter is also used to stand for semi-major axis of the ellipse, to avoid confusion we here use the notation f to represent acceleration, f_c centripetal acceleration, and F_c centripetal force

Assumptions We assume that each object, planet, moon or sun is a homogeneous spherical object with uniform density, which may be correctly mathematically represented as if all its mass were concentrated at its center. We also assume revolution in a circular orbit, unaffected by other forces.

Method Noticing that the terms for acceleration, radius, velocity and period all appear more than once in the equations, fruitful combinations and substitutions may be made. Since we know by Newton's Second Law that force equals mass times acceleration, $F = mf$, and that the centripetal acceleration of a body in circular orbit is its velocity squared divided by the radius, $f_c = v^2/r$, it follows that the force acting on an object in circular orbit will be the product of its mass and centripetal acceleration. It is then possible to substitute the velocity and period terms to obtain an expression solely in terms of the force and radius. Note that centripetal force per unit mass is equivalent to centripetal acceleration.

Calculations Since $F = mf$ and $f_c = v^2/r$, then, making the substitution for acceleration,

$$F_c = m\left(\frac{v^2}{r}\right)$$

This is the force that is needed to draw a mass m into a circular orbital path against its inertial tendency to go in a straight line. A body in circular motion moves with a velocity given by the circumference of the circle divided by the period, $v = 2\pi r/P$. Substituting the right-hand side of this velocity equation into the above force equation, we have,

$$F_c = m\left(\frac{4\pi^2}{P^2}\right)r$$

This is still the centripetal force acting on a circular moving body of mass m at distance r from the center of revolution, whether it be a car on a curve, a skater on ice, a bug on a wheel, or a satellite around a planet. There is no suggestion here of what keeps the mass m in a curve of radius r. Now we consider it to apply to bodies in space orbiting with a period P around a center of force, such as the Sun, using Kepler's Third Law, $P^2 = kr^3$, to eliminate the period term from the equation:

$$F_c = m\left(\frac{4\pi^2}{kr^3}\right)r$$

$$F_c = m\left(\frac{4\pi^2}{k}\right)\frac{1}{r^2}$$

Since all the terms except radius on the right side of the equation are constants, the inverse square proportional relationship emerges clearly:

$$F_c \propto \frac{1}{r^2}$$

Observation We are beginning to get at quantitative, rather than purely proportional, relationships. In the above equation,

$$F_c = m\left(\frac{4\pi^2}{k}\right)\frac{1}{r^2}$$

the terms in parenthesis are constants, yet k is unknown; these terms will be the same for every planetary orbit. Here we will introduce a proportionality constant and call it G, making it equal to the $4\pi^2/k$ term. The new equation is

$$F_c = \frac{Gm}{r^2}$$

This is still the centripetal force acting upon a mass m at distance r from the center of the orbit, varying as the inverse square of the distance from the center, but without hypothesis as to cause, or knowledge of G.

Constructing Newton's Gravity Equation

A moving mass m, such as the Moon, tends by its inertia ever to go in a straight line, and it resists deflection into a curved path. If another force tends to pull it into a circular arc, then that force must be proportional to this:

$$F_c \propto \frac{m}{r^2}$$

Constructing Newton's Gravity Equation

where r is the radius of the circular arc. This is the force resisting the pull of the mass into curvature. The fundamental question before Newton was, how can this force be related to the force *causing* the pull of the mass into curvature? Newton concluded that the Moon's fall to the Earth is governed by the same force that makes an apple fall to the ground. The rate of the Moon's fall matched just what would be expected if the Earth's gravity decreases with the inverse square of the distance from the center of the Earth, and acts in its more diminished strength even as far away as the Moon. Using M to represent the Earth's mass, the *gravitational* force originating from the Earth's mass must therefore be expressed this way:

$$F_g \propto \frac{M}{r^2}$$

There is a clear invitation in these two proportions, F_c and F_g, to relate them to each other: the force causing the deflection into an orbit must be equal to the force resisting it. And if the force of Earth's gravity acts on the Moon, then the Moon's gravity must similarly act on the Earth, forcing it into an orbit as well. How did Newton more generally link these two expressions of force reciprocally acting on the Moon and Earth to a postulated force, of gravity, originating from the Earth and Moon, respectively? The key is in Newton's Third Law:

LAW III

TO EVERY ACTION THERE IS ALWAYS OPPOSED AN EQUAL REACTION: OR, THE MUTUAL ACTIONS OF TWO BODIES UPON EACH OTHER ARE ALWAYS EQUAL, AND DIRECTED TO CONTRARY PARTS.[4]

Whatever draws or presses another is as much drawn or pressed by that other. If you press a stone with your finger, the finger is also pressed by the stone. If a horse draws a stone tied to a rope, the horse (if I may so say) will be equally drawn back towards the stone... This law takes place also in attractions...

This law helps us understand the mutual effect of forces. Thus the body in the center of the orbit, which we call M, experiences an equal but opposite force from that experienced by the mass m orbiting it. The acceleration of the Moon toward the Earth would be proportional to the Earth's mass attracting it divided by the square of the distance to the center of the Earth (the radius of the orbit); the acceleration of the Earth toward the Moon would likewise be proportional to the Moon's mass attracting it divided by the square of the radius of its orbit. Each pull on the respective mass, at a given distance from the other, is correspondingly resisted by a centripetal force proportional to its own mass.

These are intriguing relationships. Assuming again m is the mass of the Moon, M is the mass of the Earth, and r is their separation measured from the center of the Earth, what is the force acting upon the *Earth*, M, toward the Moon? It appears that it is Gm/r^2. What is the force acting upon the *Moon*, m, toward the Earth? It likewise appears to be GM/r^2.

---

[4] *Principia*, 13–14.

152   8 Newton's Master Stroke: The Universal Law of Gravitation

To state this mathematically, we may again use the symbol f to represent the respective accelerative forces acting on each body, with the subscripts M and m to signify the Earth and Moon masses respectively:

$f_M = \frac{Gm}{r^2}$	The Moon's pull on the Earth (f_M) causing its moonward acceleration
$f_m = \frac{GM}{r^2}$	The Earth's pull on the Moon (f_m) causing its earthward acceleration

In sum, we have assumed as Newton did that the attractive force of a given mass is proportional to its quantity, and that this force obeys the inverse square law. From this follows the conclusion that each of the forces acting on the Moon and Earth, respectively, must be proportional to the masses of the Earth and Moon, respectively.

Problem Use the Earth – Moon system to show how Newton's Universal Law of Gravitation flows naturally from the above relations, $f_M = Gm/r^2$ and $f_m = GM/r^2$, where m and M are again the masses of the Moon and Earth respectively, and r is the distance between their centers.

Given

$f_M = \frac{Gm}{r^2}$	Centripetal acceleration of Earth towards the Moon
$f_m = \frac{GM}{r^2}$	Centripetal acceleration of Moon towards the Earth

Method To visualize the relationships among the forces, we may imagine the relatively small acceleration f_M of the Earth toward the Moon (the latter shown by the smaller "mass" in the diagram below), as a line or vector drawn toward the Moon.

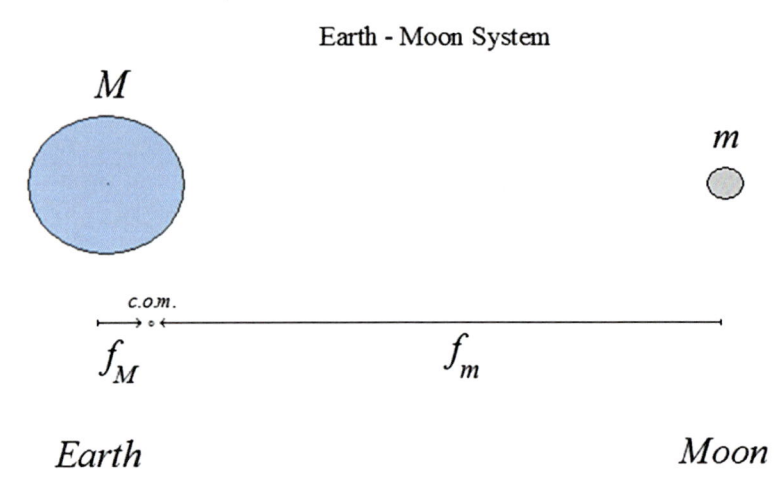

Earth - Moon System

M

m

c.o.m.

f_M f_m

Earth   Moon

f_M = Force of Moon's pull on Earth
f_m = Force of Earth's pull on Moon

Constructing Newton's Gravity Equation

Because the line is drawn as originating from the Earth does not mean that this acceleration f_M originates from the Earth – it originates from the Moon's mass and is shown as an accelerative pull upon the Earth toward Moon. The length of this not-to-scale short line is intended to be proportional to the amount of the acceleration induced by m, the mass of the Moon. Similarly, the greater acceleration f_m of the Moon toward the Earth, induced by the Earth's greater mass M is shown as a vector toward the Earth. Its longer line is intended to be proportional to the Earth-induced acceleration f_m in the direction of the Earth.

Assumptions We again assume that each object, or planet, moon or sun is a spherical object with uniform density, which may be represented as if all its mass were concentrated at its center (a key conclusion of Newton's work). We also assume circular orbits in space, and the absence of any other forces (including friction of air, or the perturbing forces of other planets or the Sun) affecting their motion.

Calculations The accelerative force pulling on each body will depend on the mass of the attracting body. Therefore the accelerative force f_M acting on Earth is to the accelerative force f_m pulling on the Moon, as the mass of the Moon m is to the mass of the Earth M:

$$\frac{f_M}{f_m} = \frac{m}{M}$$

$$M f_M = m f_m$$

The accelerations acting on each body, given in terms of mass and distance are,

$$f_M = \frac{Gm}{r^2} \text{ and } f_m = \frac{GM}{r^2}$$

Substituting the right-hand side of these expressions into the previous equation, we have,

$$\frac{GMm}{r^2} = \frac{GMm}{r^2}$$

Thus the absolute forces *acting on each body* are equal. The force experienced by any body is,

$$F_g = \frac{GMm}{r^2}$$

Observations

1. This is Newton's Universal Law of Gravitation: the gravitational force attracting two bodies varies as the *product* of each of their masses, and inversely as

the square of their distance apart. The constant G is called the *gravitational constant*, and has been determined to equal 6.674×10^{-11} N·m^2/kg^2 from experiment.[5]

2. We could have just used Newton's Second Law, $F = mf$, to say that the acceleration acting on the Moon, induced by the Earth's mass, $f = GM/r^2$, when multiplied by the Moon's mass, m, gives the force acting on the Moon:

$$F = mf$$

$$F = m\left(\frac{GM}{r^2}\right)$$

$$F = \frac{GMm}{r^2}$$

Similarly, the acceleration acting on the Earth, induced by the Moon's mass, $f = Gm/r^2$ when multiplied by the Earth's mass, M, is gives the force acting on the Earth:

$$F = M\left(\frac{Gm}{r^2}\right)$$

$$F = \frac{GMm}{r^2}$$

And they are equal to one another.

3. The conclusion about the proportionality of mass and mutual gravitational acceleration is intuitive, since a force applies to every part of a mass equally. In Newton's words, from Book III of *Principia*:

> [S]ince all the parts of any planet A gravitate towards any other planet B; and the gravity of every part is to the gravity of the whole as the matter of the part is to the matter of the whole; and (by Law III) to every action corresponds an equal reaction; therefore the planet B will, on the other hand, gravitate to all parts of the planet A; and its gravity towards any one part will be to the gravity towards the whole as the matter of the part to the matter of the whole.[6]

---

[5] The Committee on Data for Science and Technology (CODATA) maintains standards for international uses. See http://physics.nist.gov/cuu/Constants/international.html. The full value for G given by CODATA is 6.67384×10^{-11}. http://physics.nist.gov/cgi-bin/cuu/Value?bgl search_for=G. m^3 kg^{-1} s^{-2}.

[6] *Principia*, Book III, Proposition VII, Theorem VII, 414–15.

4. We know the force acting on a standing person by the Earth's gravity, since by Newton's Second Law $W = mg$, where W is her weight and g is the acceleration of gravity induced by the mass of the Earth. If the mass of a person is doubled, the force acting on her body – her weight – is doubled: $2W = 2mg$: the force acting on her is proportional to her mass: $F::m$. If the Earth's mass were doubled (doubling g), her weight would again be doubled. It is intuitive that the force acting on a person thus varies as the product of each of the masses.
5. If a 60 kg (132 lb) student is standing on the surface of the Earth, the force on him will be,

$$\frac{GMm}{r^2} = \frac{(6.674 \times 10^{-11})(5.98 \times 10^{24})(60)}{(6.38 \times 10^6)^2} \simeq 588 \text{ newtons}$$

Force is measured in newtons (abbreviated N), in SI units (from $F = mf$) of kg·m/s². Note that the distance r is taken from the center of the Earth to the surface, i.e., it is the radius of the Earth. Now imagine he has climbed Mt. Everest. What force will be exerted on him then? Remember the inverse square law, which has taken up residence in the denominator of Newton's gravitation equation. Since he is climbing in altitude, and so moving farther away from the center of the Earth, the r^2 term will be slightly larger, and we should expect the force on the climber to be slightly smaller. Mt. Everest is about 6.39×10^6 km from the center of the Earth, so he will experience a force at its summit of 586 N, somewhat less than back down on the surface. This too will be the force he exerts upon the Earth, by virtue of Law III.

6. It often convenient to use the symbol μ for the product GM. When doing geocentric calculations, $GM = 3.99 \times 10^{14}$ N·m²/kg² where the primary mass is the Earth. It is a good number to remember.
7. Recall the formula for the acceleration felt by a small object in the gravitational field of the larger body, $f = GM/r^2$. For the situation of the Earth, we can write it as $g = \mu/r^2$ and check to see if we get an expected result. At the Earth's surface the value should be $3.99 \times 10^{14}/4.07 \times 10^{13} = 9.8$ m/s² which is correct.

Reflections upon the Equilibrium of Gravitational and Inertial (Centrifugal) Forces of a Mass in Orbit

We will take a moment to step back and review one implication of Newton's gravitation law. We consider the case where the primary mass is much larger than the secondary. In such cases it is safe to assume the small mass orbits the center of the larger mass. (For Earth satellites and most of the smaller planets and satellites of the solar system this is very nearly so, though in fact in all binary systems the masses orbit their common center of mass, as we will see below.) Let's call this little mass m_2 to distinguish it from the primary mass m_1 around which it orbits. The force needed to deflect the secondary mass from a rectilinear path is given by $F = m_2 f$, Newton's Second law. Any acceleration f of this mass, along any axis, bending it away from a straight line, will be *resisted* by the inertia of the mass m_2. Newton also found that centripetal acceleration is a function of the velocity and the radius of the circular orbit: $f = v^2/r$. Thus we again link centripetal acceleration in circular orbits with his second law. Substituting the expression for circular (centripetal) acceleration into that law gives the force needed to bend the object off of its stubborn tendency to go in a straight line:

$$\underbrace{F = m_2 f}_{\text{Newton's Second Law}} \quad \text{and} \quad \underbrace{f = \frac{v^2}{r}}_{\text{Circular or centripetal acceleration}} \rightarrow F = m_2\left(\frac{v^2}{r}\right) \quad \begin{array}{l} \textit{Newton's Second Law and} \\ \textit{equation for centripetal} \\ \textit{acceleration are combined} \\ \textit{to show the force resisting} \\ \textit{deflection from a rectilinear path} \end{array}$$

The resulting force (often called centrifugal force) will increase dramatically (by the square) as we pick up the velocity. But if we lengthen the radius of spin, the force is diminished. A heavier mass, too, will result in a greater force than a lighter one in the equation (double the mass, double the force).

But the force that bends the path of our wants-to-go-in-a-straight-line mass is not identified in the above equation. It could be anything. When we twirl a weight on a string, the force is the pull of the string, making sure it doesn't fly off. But for objects in space, the force bending the path from straight is *gravity*. It is given by Newton's gravity equation $F = Gm_1m_2/r^2$, where the force is dependent on the product of the masses and is reduced by the square of the distance r between them. Now we can see how, in a stable circular orbit, the interplay of two physical attributes: first, the gravitational force, represented by this equation,

$$\underbrace{F_g = \frac{Gm_1m_2}{r^2}}_{\text{The gravitational force pulling } m_1 \text{ and } m_2 \text{ together}}$$

and second, the straight line, inertial tendency of a mass to resist the centripetal deflection into an orbital arc, represented by this equation,

$$\underbrace{F_c = \frac{m_2 v^2}{r}}_{\text{Force on } m_2 \text{ needed to deflect its path toward } m_1 \text{ (from } F=mf\text{)}}$$

The force needed to bend a mass into an orbit is supplied exactly by the gravitational pull of the other mass. The forces are therefore set equal to each other:

$$\frac{Gm_1m_2}{r^2} = \frac{m_2 v^2}{r}$$

The respective accelerations are:

$$\underbrace{\frac{Gm_1}{r^2} = \frac{v^2}{r}}_{\text{Gravitational and circular accelerations are equivalent}}$$

The two accelerations are equivalent. This shows, as a condition to maintaining the circular orbit, the gravitational acceleration (left-hand term) induced by the primary mass m_1 at the distance r, is equivalent to the centripetal acceleration acting on the secondary mass m_2 at that distance. The greater the mass m_2 of the orbiting object, the greater is the force needed to overcome its inertia to move it, but equivalently so is the gravitational force acting on it; so the secondary mass terms cancel out.

Recall that this is why dropped objects of any mass hit the ground at the same time. Friction aside, the velocity of dropped objects on Earth is dependent only on the height from which they are dropped. Farther from Earth, the distance from Earth's center starts to make a difference, and the inverse square law weakens the Gm_1 gravitational force. But at any given distance, no matter how far, the "fall" toward the Earth of an object of any mass, be it a small artificial satellite or the Moon, will by the same principle have the same velocity. This velocity can be calculated, as we have already seen. From the above equation, solve for v to find the velocity of any m_2 in a circular orbit, where (again $m_2 \ll m_1$):

$$v = \sqrt{\frac{Gm_1}{r}}$$

The physical insight upon which an understanding of orbital motion rests is that there must be an exact match-up of gravitational force and orbital velocity, for a given radius, for any object to exist in a stable orbit. If the velocity is too low in relation to the gravitational force, the object will succumb to gravity and perhaps crash into the primary mass. If the velocity is too great in relation to the attractive pull of the primary mass, the secondary mass may not hold it, and the object will move away.

The mating of the two equations above yields circular orbital velocity in terms of mass and orbital radius. It links radius with velocity for any orbited mass. It is the "just-right" ("Goldilocks") velocity for an orbiting thing at that distance to stay around a mass the size of m_1. If for example we know the mass of the Earth and the distance to the Moon, we can determine the Moon's "perfect" velocity through space (about 1 km/s) to keep it where it is. The equation tells us that the greater is the radius, the lesser must be the velocity to maintain orbit. In past times when the Moon was orbiting closer to the Earth, we know the lunar month must have been shorter.

Kepler's Third Law as Modified by Newton

A common assumption in finding the period of a circular orbiting thing is that of an insignificantly small object orbiting a much more massive one. In the solar system, especially with the inner planets, this is a safe assumption. The Earth for example is 1/333,000 the mass of the Sun, and Mars about one tenth even of that. But what is the picture like when the secondary mass is not so small? Double stars of often quite comparable masses are frequently found orbiting one another around their common center of mass. Even in our solar system, Jupiter's mass is roughly 1/1,000 that of the Sun, nothing to be ignored for accurate computation. And certainly the Moon's mass, at about 1/81 the mass of the Earth, cannot be discounted as insignificantly small for accurate determination of its period. How can we find the orbital period in such cases? We return to our original constructed model, but this time imagine the objects as spinning around each other, like two dancers holding each other's arms as they turn.

Suppose we consider any two orbiting bodies where the secondary mass is non-trivial in relation to the primary. Let r_1 and r_2 be the respective distances between the center of mass or "balance point" of the two masses m_1 and m_2. Think of the centripetal force of each spinning body as a function of its distance from the center of mass, around which it is spinning, just as the case of the wheel we discussed earlier, only each mass of the unequal pair is spinning in a circle of different radius.

Center of Mass of a Binary System

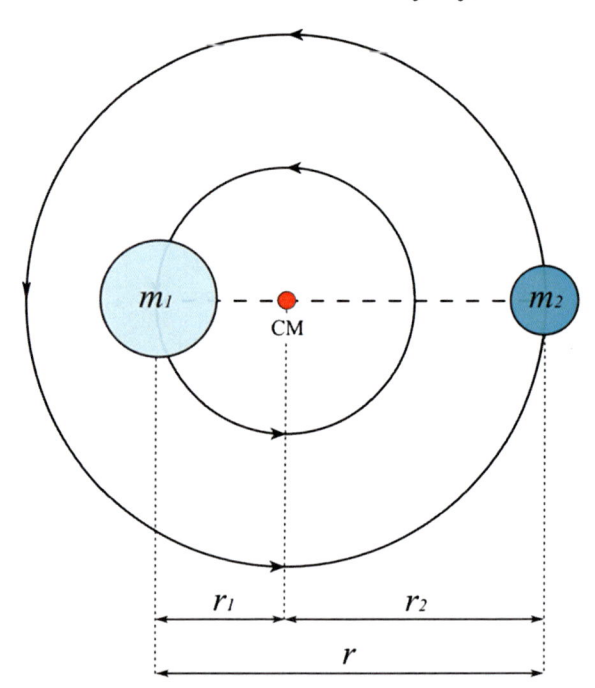

Kepler's Third Law as Modified by Newton

This again is the force resisting change of motion (acceleration) inward, away from straight line motion. This resisting force, from Newton's Second Law, can be expressed as $F = m_2\omega^2 r_2$ for the secondary mass, and $F = m_1\omega^2 r_1$ for the primary.[7] But their gravitational attraction is a function of their own distances apart r, consistent with Newton's gravitation equation. Their distance to each other is the sum of their respective distances to the center of mass, so $r = r_1 + r_2$ (see the figure). Writing the equations for each body, we have:

$$\underbrace{\frac{Gm_1 m_2}{r^2}}_{\text{Mutual gravitational force attracting } m_1 \text{ and } m_2} = \underbrace{m_2\omega^2 r_2}_{\text{Force experienced by } m_2 \text{ as it orbits center of mass at distance } r_2}$$

$$\underbrace{\frac{Gm_1 m_2}{r^2}}_{\text{Mutual gravitational force attracting } m_1 \text{ and } m_2} = \underbrace{m_1\omega^2 r_1}_{\text{Force experienced by } m_1 \text{ as it orbits center of mass at distance } r_1}$$

Canceling the common terms on each side of the equations reveals the balance of gravitational and centripetal accelerations on each body:

$$\underbrace{\frac{Gm_1}{r^2} = \omega^2 r_2}_{\text{Balance of accelerations acting on } m_2}$$

$$\underbrace{\frac{Gm_2}{r^2} = \omega^2 r_1}_{\text{Balance of accelerations acting on } m_1}$$

Our goal is to put everything in terms of r, which is the sum of r_1 and r_2. We take the pair of expressions just found, each representing the distance to the center of mass,

$$r_2 = \frac{Gm_1}{\omega^2 r^2} \quad r_1 = \frac{Gm_2}{\omega^2 r^2}$$

and add them,

$$r_2 + r_1 = \frac{Gm_1}{\omega^2 r^2} + \frac{Gm_2}{\omega^2 r^2}$$

Now, since $r = r_1 + r_2$,

$$r = \frac{Gm_1}{\omega^2 r^2} + \frac{Gm_2}{\omega^2 r^2}$$

---

[7] Recall that in the angular notation for acceleration, $a = \omega^2 r$ is equivalent to $a = v^2/r$.

We can simplify the whole expression by multiplying each side by $\omega^2 r$ and taking the square root. The result is the relative velocity in a circular orbit (since $v^2 = \omega^2 r^2$):

$$\omega r = v = \sqrt{\frac{G(m_1 + m_2)}{r}}$$

Or relative acceleration:

$$\omega^2 r = f_r = \frac{G(m_1 + m_2)}{r^2}$$

This shows rather simply that the individual centripetal forces caused by the respective revolutions of each mass at different radii from the center of mass is equivalent to combining virtually all of the mass in one place, as if there were one small particle only orbiting at distance r from their centers.

Rearranging the above equation for r gives,

$$\omega^2 r^3 = G(m_1 + m_2)$$

This is Kepler's Third Law for circular orbits, in angular notation. Recalling that $\omega = 2\pi/P$, we may substitute it for ω^2 and solve for period, yielding the customary expression of Kepler's Third Law in terms of the period:

$$\frac{4\pi^2}{P^2} r^3 = G(m_1 + m_2)$$

$$P = \frac{2\pi}{\sqrt{G(m_1 + m_2)}} r^{3/2}$$

This relationship is Newton's modification of Kepler's Third Law. It is sometimes referred to as the Kepler-Newton Law of orbital motion. It applies equally to ellipses, where the distance term is the semi-major axis of the ellipse a rather than the radius r.

Notice that the period will be the same regardless of how the total mass of the system is allocated between the two masses. If we do a thought experiment and mentally shift quantities of mass back and forth between the two masses m_1 and m_2, making one mass bigger at the expense of the other, then shifting it back, the period equation tells us that the period will remain the same, since the *sum* of the masses will be the same. That is, as long as the number in the parenthesis $(m_1 + m_2)$ is unchanged, the period will be unchanged. The only consequence will be to shift the location of the center of mass closer to the bigger mass, and (as we will see below) alter the velocities of each mass around the center of mass. In fact we could mentally reduce the mass of, say m_2 to almost infinitesimal size in relation to m_1

without affecting the period. In that case the center of mass would be (as it is with our Sun in the solar system) virtually identical with, or at least very close to the center of m_1.

Why, After Newton, It Became Evident that Kepler's Proportions Are Not Strictly Satisfied for Large Secondary Masses

Kepler derived his Third Law from observations of Mars, and then saw how neat the fit was with the other planets. The giant planets, Jupiter and Saturn, were the outermost planets known in Kepler's day. Their slow movement through the sky made Mars the logical choice for Kepler's research. Mars' relatively swift movement had enabled Tycho Brahe, upon whose data Kepler relied, to obtain a great deal of data over many orbits. Newton was interested in re-deriving Kepler's Third Law not from observation, but from his theory of gravitation. The question that presented itself to Isaac Newton was how the concept of gravitation meshed with Kepler's Third Law.

The Strict Proportionality of Kepler's Third Law

Kepler found that for the planets in the solar system the squares of the periods in years are proportional the cubes of the semi-major axes in astronomical units, that is, this relationship is a constant. Thus,

$$\frac{P_1^2}{a_1^3} = \frac{P_2^2}{a_2^3} \ldots = \frac{P_n^2}{a_n^3} = 1$$

Or, put another way, with the same meaning,

$$P^2 = ka^3$$

Where $k = 1$ if the units are in years and AU. As far as the measurements of Tycho Brahe using the non-telescopic instruments of the day could suggest to Kepler, there was nothing in the data to challenge this relationship. When analyzed by Newton, the movements of Jupiter's moons too were completely in accord with law; the proportion yielded the same constant for all the Galilean moons. Applying Newton's law of gravitation to the solar system naturally produced this version of

Kepler's Third Law, where the constant of proportionality arose from the mass of the Sun and the gravitational constant:

$$P^2 = \frac{4\pi^2}{GM} a^3$$

where M is the Sun's mass and G the gravitational constant. From this, it is apparent that while the proportionality constant may not be one, there is still a constant proportion, for any period and semi-major axis:

$$\frac{P_1^2}{a_1^3} = \frac{P_2^2}{a_2^3} = \frac{P_n^2}{a_n^3} = \frac{4\pi^2}{GM}$$

The Keplerian proportionality of period squared to semi-major axis cubed is still retained. And, as we will see later, with the appropriate choice of units and value of G, we can make this proportion again equal one.

The Deviation from Strict Keplerian Proportionality

But when Newton considered that orbiting masses actually revolved about their common center of mass (including even the Sun and planets in the solar system), the full expression for Kepler's Third Law became this,

$$P^2 = \frac{4\pi^2}{G(M+m)} a^3$$

where again M is the mass of the Sun and m is the mass of a planet (or satellite). The resulting Keplerian proportionality became this,

$$\frac{P^2}{a^3} = \underbrace{\frac{4\pi^2}{G(M+m)}}_{\text{This term is not a constant, but varies with each mass } m}$$

and is no longer technically a constant applicable to all sets of orbits around the Sun, since each planet has a different mass, albeit small in relation to the mass of the Sun (or other primary). Comparing the orbits of two or more planets (or satellites), we see the problem more clearly:

$$\frac{P_1^2}{a_1^3} = \frac{4\pi^2}{G(M+m_1)} \ldots \frac{P_2^2}{a_2^3} = \frac{4\pi^2}{G(M+m_2)} \ldots$$

The constant proportionality has disappeared, because the little mass terms m_1 and m_2 etc. will be different for each planet, and alter the beautiful proportionality that existed when we ignored it. One might even say that Kepler's Third "Law," as he framed it has been destroyed. We can see this more plainly if we cancel out the common terms, and set it this way:

$$\frac{P_1^2}{a_1^3} : \frac{P_2^2}{a_2^3} = \frac{1}{M+m_1} : \frac{1}{M+m_2}$$

The ratios are not constant. This is quite different from Kepler's Third Law as he presented it. But if we think of the little planetary masses m for the most part being overwhelmingly smaller than the vastly larger solar mass M, then $M + m \approx M$, and Kepler's simple proportions are retained.

Why did Kepler not notice what Newton discovered? Because Mars' mass is incomparably smaller than the Sun, about one three-millionth of its mass, Kepler was unaware of any small anomalies in Mars' period that would have perplexed the development of his third law. The empirical law he devised based on Mars' movements appeared to be completely accurate because the mass of Mars was insignificant in relation to the overwhelming mass of the Sun. And even the giant planets Jupiter and Saturn seemed to fit the scheme, because their masses are about one and three-and-a half thousandths the solar mass, respectively, and any errors attributable to ignoring those masses were not within the limits of detection of the non-telescopic instruments of the day.

Comparing Periods with and Without the Secondary Mass

Let us compare the periods in the two versions of Kepler's Third Law to see the practical consequences of neglecting the secondary mass (where we assume $m \approx 0$) We can say that P' is the period obtained by disregarding the secondary mass, and P is the true period, where both masses are taken into account. Compare the periods in these equations, without and with the secondary mass accounted for:

$$P'^2 = \frac{4\pi^2}{GM} a^3 \quad P^2 = \frac{4\pi^2}{G(M+m)} a^3$$

If we assume the mass units are *solar masses* such that $M = 1$ and m is expressed in fractions of a solar mass, and that distance units are astronomical units, such that $a = 1$, then the ratio of the periods is:

$$\frac{P'^2}{P^2} = 1 + m$$

$$P' = P\sqrt{1+m}$$

Neglecting the secondary mass in calculations therefore naturally results in slightly longer values for the periods by the factor $\sqrt{(1+m)}$.

Example: The Significant Consequences of Disregarding the Mass of Jupiter

The Sun's mass is huge; is the secondary mass worth worrying about? How can we get a sense of the practical difference? We can get an idea of the upper range of the difference by examining the case of Jupiter, the solar system's most massive planet. The ratio of the planet's mass to the Sun's mass (taken as unity) is about 0.000954. We would expect the difference in the period by neglecting this mass to be $\sqrt{(1+m)}$, or $\sqrt{(1+.000954)}$, or about 1.000477 times longer than the true published period of about 4334.82 days. This is about 2 full day's difference in period, a large error that would result from neglecting that planet's mass in the period calculation. On the other hand, the error factors in period by neglecting the mass of smaller Uranus (another gas giant planet) and Venus (a small rocky planet) are only about 1.00002 and 1.000001 respectively.

When one is considering satellites or spacecraft in orbit around the Earth or other body, the mass of the satellite may be safely ignored. For example, if a 10,000 kg satellite is orbiting the Earth, its mass in relation to the Earth's mass is about 1.7×10^{-21} to 1.

Exercises: The Intriguing Case of Eugenia and Petit Prince Eugenia was the 45th asteroid discovered. It was found in Paris on June 27, 1857 by H Goldschmidt. The asteroid is 215 km in diameter, and is about 2.7 AU distant from the Sun, orbiting it in the swarm of small bodies between Mars and Jupiter known as the asteroid belt. A century and a half later, in 1999, astronomers using new, adaptive optics technology on the Canada-France-Hawaii Telescope on Mauna Kea, Hawaii, discovered that Eugenia has a companion: a tiny moon orbiting in a nearly circular orbit at a distance from Eugenia not much greater than the distance between Paris and Rome. This was a stunning achievement, to resolve such a faint, close object at over 400 million km away. The little moon was named Petit-Prince ("The Little Prince") after the children's book of that name by Antoine de Saint-Exupery, which is about a prince who escapes to an asteroid. Petit Prince is perhaps 13 km in diameter. A second companion, S/2004 (45) 1, and even closer to Eugenia, was discovered in 2007. Eugenia is thus a triple asteroid system, one of only a handful known. The following problems explore some of the dynamics of Eugenia and Petit Prince.

Problems

1. Eugenia is 2.723537 AU from the Sun. Beginning with the mass of the Sun and the given distance to the asteroid, find the velocity of Eugenia in its orbit around the Sun, in kilometers per second. Take the mass of the Sun to be 1.9891 x 10^{30} kg. and the length of an astronomical unit to be 149,598,000 km.[8]
2. Find period as if Eugenia's orbit were a perfect circle (which it almost is) and compare it with the method of finding velocity you used in the previous problem.
3. Find Eugenia's period in years using Kepler's Third Law and the SI system
4. Find Eugenia's period in years and days using Kepler's Third Law and astronomical units.
5. What is the centripetal acceleration acting on Eugenia, resulting from its orbital motion? Express your result to six significant digits in m/s^2.
6. Using the result from problem two above, compute the gravitational acceleration induced by the Sun keeping Eugenia in orbit. Express your result to six significant digits in m/s^2 and compare it with your previous result.
7. Compute the gravitational force in newtons (kg-m/s^2) between the Sun and Eugenia. Assume the mass of Eugenia is 5.8 × 10^{18} kg.
8. Compute the centrifugal force in Newtons to four significant digits acting on Eugenia resulting from its orbital motion, and compare it with your previous result.
9. The orbit of Eugenia's moon Petit Prince has a semi-major axis of 1,184 km. Find its period in days by Kepler's Third Law. For the moment, ignore Petit Prince's mass.
10. The orbital period of Petit Prince has been estimated from photographs to be 4.766 days. From this, determine the combined mass of the asteroid and the satellite around the center of mass. Compare with the published masses of about 5.8 × 10^{18} for Eugenia and 1.2 × 10^{15} for Petit Prince. What percent contribution to the total mass of the Eugenia–Petit Prince system is Petit Prince's mass?

Reference

1. Cajori (1949) Principia in modern English: Sir Isaac Newton's mathematical principles of natural philosophy. Book I, Proposition I, Theorem I (Motte's Translation Revised). University of California Press, Berkeley, p 40

---

[8] The precise length of an astronomical unit is 149,597,870.691 km. See http://neo.jpl.nasa.gov/glossary/au.html.

Chapter 9
Gravity on This and Other Worlds

An intuitive way of understanding gravitational force is to think of it as the "g" force of a planet or moon. This can be especially helpful when dealing with small orbiting objects such as artificial satellites. Force per unit mass is acceleration,[1] so a one kilogram ball (a "unit mass") on the surface of a planet will experience a gravitational acceleration, or g force, that is proportional to the quantity of mass in the planet and the inverse square of the distance of the ball from the center of the planet. It is a useful way of looking at gravity and comparing the gravity of different bodies.

How Planetary Radius and Density Affect g

Will two planets of exactly the same amount of particles of matter (mass) have the same surface g? The answer is no. Planets may be of different sizes, so the surface of one may be much farther from the center than the surface of another, diminishing the relative value of g of the bigger planet according to the inverse square law. If you are asking how a large planet can have the same amount of matter as a small one, you will recall that different masses may have different *densities*. The amount of matter per unit volume of one planet or satellite may be, and often is, dramatically larger or smaller than another. A large low-density body may have the same g at its surface as a smaller, higher density body. Compare our little rocky Earth with the giant beach ball that is Saturn. The Earth's mean density is about 5.5 grams per cubic centimeter.[2] To put this in context, basalt is about 3 g/cm^3 and iron about 7.9 g/cm^3.

---

[1] Recall that by Newton's Second Law, force equals mass times acceleration. Acceleration is therefore the same as force per unit mass, and this is how gravitational force is usually expressed, as acceleration. Some authors use the term "accelerative force" to covey this concept clearly.

[2] We will use the c/m^3 convention here since that is most commonly used in the astronomical literature. Earth's density in the mks system is 5,513.4 kilograms per cubic meter.

The Earth's mean density thus lies roughly between basalt and iron. Saturn's mean density, however, is only .687 grams per cubic centimeter (687.1 kilograms per cubic meter), [3] or one eighth that of Earth. Being less than the 1 g/cm^3 reference density of water, it would float. Cork is about .25 g/m^3 and ice is about .92 g/cm^3. Saturn's density is somewhere between cork and ice.

Saturn is indeed large. Its mean radius is 58,232 km, about nine times the Earth's, but the value of g on Saturn's surface is only about 10.44 m/s^2, slightly more than Earth's surface g. We know this from the equation we saw in the last chapter: $f = Gm/r^2$, where f is the accelerative force (per unit mass) of gravity at distance r from the center of the mass in consideration. Since we are calling this force the g force, we may write that $g = Gm/r^2$.

Let's try an experiment. Suppose we could enlarge Saturn a little while keeping the quantity of mass within it the same. We will inflate it until the value of g on its surface is the same 9.8 m/s^2 as the Earth's g at its surface. To do that, we turn the equation on its head and solve for radius:

$$r = \sqrt{\frac{Gm}{g}}$$

Inserting values of Saturn's mass (5.683×10^{26} kg) and Earth's g, we get an enlarged radius of 62,212 km, a 107% increase over its former radius. What must then happen to its density? Since the mass is the same, it must decrease with the expanded volume. The mass of a spherical object of uniform density is $m = 4/3\pi r^3 \rho$ where ρ is the density, so the density must be given by,

$$\rho = \frac{3m}{4\pi r^3}$$

We insert the new radius into the equation and the new density of our enlarged Saturn is about .564 g/cm^3, or 82% of its former value.

Now what would the result if we *shrank* Saturn so that its mean radius was the same as the Earth's (6,371 km)? What would its surface g and new density be then? Using $g = Gm/r^2$ with the Earth's radius (in meters – remember your units!) we get almost 935 m/s^2! Its new, enormous density has become 763.6 g/cm^3! This is about 100 times denser than iron. Keeping these effects of density and distance in mind, we will turn to planet Earth and consider what its g really means.

---

[3] See http://ssd.jpl.nasa.gov/?planet_phys_par. Densities in general vary with temperature and pressure.

All the Mass of a Body May Be Regarded as Concentrated in a Single Point

Newton proved that, for a sphere of uniform density, the gravitational attraction of a mass may be regarded as emanating from a single point at its center. He labored for years to figure this out, but he did finally prove it. With these assumptions, it is mathematically safe to regard the gravitational force of the Earth as concentrated in a point at its center, and that gravitational acceleration induced by the Earth's (point) mass is dependent on the distance to the surface of the Earth from that point, in other words, the radius. One must look at the radial distance from the center of the Earth (or other body) to its surface to find its g acceleration.

The Earth Spins

One of the ancient arguments made by Ptolemy and others against the rotation of the Earth, and for a geocentric model of the universe, was that a spinning Earth would make things fly off its surface. The Earth does rotate on its axis, and that rotation generates inertial (so-called centrifugal) forces[4] that counteract the force of gravity. Suppose you could control the spin of the Earth with a large dial. Turn it up to make the Earth spin much more rapidly and we all feel lighter; spin it fast enough and it flings us into space, like children flying off a too-speedy turntable on a playground. Spin it with an immense velocity and the Earth's oceans depart into space. Faster still and the Earth itself begins to come apart. The centrifugal acceleration that could make these things happen is given by v^2/r, where v is the velocity of the spinning and r the equatorial radius of the Earth. The centrifugal force that one of us of mass m would experience would be $F_c = mv^2/r$. For us to lift off at the equator, this centrifugal force would have to be greater than the countervailing tug of gravity pulling us back, whose force is $F_g = GMm/r^2$, where M is the mass of the whole Earth not wanting to let us go. The value of the gravitational force on us is thus always being challenged and offset by the centrifugal force of the Earth's rotation. Fortunately, there is not enough spin to worry us. Assuming a uniform, spherical Earth, the difference of forces at the equator is,

$$\Delta F = \frac{GMm}{r^2} - \frac{mv^2}{r}$$

---

[4] We use the term "centrifugal force" although, as explained in Chapter 1 and elsewhere, the perceived force on an object is really its simple resistance (inertia) to being compelled away from its Newtonian straight-line motion. We use the term extensively in this chapter because of its intuitive appeal in communicating the effects of the Earth's spin.

Or, in terms of the net gravitational acceleration,

$$\Delta g = g - f_c$$

where the subscripts represent gravitational and centrifugal acceleration, respectively. In our imaginary scenario, the spin rate could theoretically be adjusted to just barely offset the pull of gravity, so $f_c \approx f_g$, and create a kind of near-weightlessness on the surface of the Earth. We know Earth is not spinning nearly that fast because (among other obvious indicators) we are all firmly on the ground.

Observed g and True g

It is apparent from the above discussion that the observed force acting on a mass at the equator is the actual gravitational force of the Earth's mass less the centrifugal force pulling in the opposite direction. In terms of accelerations, the "true g" of a theoretically non-rotating Earth is diminished or diluted by the centrifugal acceleration of the actually rotating Earth:

$$g_{observed} = g_{true} - f_c$$

The observed g (or $g_{observed}$) is the value of g that is measured by instruments, such as pendulums, since the centrifugal force acts on those instruments and causes them to measure less apparent gravitational acceleration than they would if the Earth were not going around. If one wanted to find the "true g" (or g_{true}) in a given location that is independent of the centrifugal force, one would have to add the centrifugal acceleration to the observed, artificially high g:

$$g_{true} = g_{observed} + f_c$$

In other words, acceleration readings must be corrected for the centrifugal acceleration in order to find the true measure of g. True g of course can be calculated for any spherical body using Newton's Law of Gravitation. All that is needed is knowledge of the radius of the sphere and the Earth's mass. The result is the g for a non-rotating sphere, and should theoretically (for a perfect sphere of uniform density) be the same for all parts of the sphere. One could again also take measurements of the value of g, say at the equator of the sphere (to find the observed value of g there) then add to that measure the computed centrifugal acceleration at that location induced by the sphere's rotation to find true g there.

Newton's Analysis of the Effect of the Earth's Spin on Its Gravitational Pull

Isaac Newton addressed these issues in the *Principia*. He calculated the effect of the Earth's spin on observed g in Propositions XIX and XX in Book III, considering first that the Earth is a sphere of uniform density.[5] His ultimate purpose was to determine the oblateness or flattening of the Earth, which he did by a complex path which we shall not venture upon here; but his first steps are instructive in showing his method of working with net accelerations and finding true g. We will take a walking tour here through the early part of Proposition XIX here, and show how he turned his powerful insights on the motions of the heavens to the analysis of the Earth and its shape.

Newton's method turned to account the same principles of orbital motion that he had earlier applied to explain how the Moon and planets stay in their orbits. There are two countervailing forces. There is the mass on the Earth's rotating surface that tends to go in a straight line by virtue of Newton's First Law, but the force of gravity holds it to the surface and prevents it from going in a straight line. It forces the mass into an arc, just like the bug on the wheel discussed in Chap. 5. The mass resists that deflection from rectilinear motion (Newton's Second Law) and that resistance is perceived as centrifugal force in the direction away from the gravitational pull; that is, toward a perpendicular from the surface of rotation (Newton's Third Law).

Step 1: The observed value of g at the equator: The data Newton employed to ascertain the pull of the Earth's gravity (at the location of Paris, France) was identical to that which he put to work in his Moon test which we saw in Chap. 6. The fall distance of an object in one second (which is directly a function of *g*) had already been accurately measured by Huygens in Paris with a pendulum:

> In the latitude of *Paris* a heavy body falling in a second of time describes 15 *Paris* feet, 1 inch, 1 7/9 lines, as above, that is, 2173 lines 7/9. The weight of the body is diminished by the weight of ambient air. Let us suppose the weight lost thereby to be 1/11000 part of the whole weight; then that heavy body falling in a vacuum will describe a height of 2174 lines in one second of time.[6]

2,174 lines is equal to 4.904 m.[7] Remember that by the Galilean equation, $s = 1/2\ at^2$, the decline of a body in the first second is equal to half the acceleration acting on it. The fall value cited by Newton thus translates to a measured *g* in Paris of 9.81 m/s^2.

---

[5] Newton sought the ration of the axis of a planet to its diameter perpendicular to the axis, in proposition XIX. The next proposition dealt with comparing the weights of bodies in different regions of the Earth.

[6] Florian Cajori [1], (Short title) *Principia*.

[7] Since a Paris foot is .32484 m, and a line is 1/144 of a Paris foot (1/12th of an inch), 2,174 lines is 4.9042 m. The value of *g* would be twice that, or about 9.81 meters per second squared. This fall distance of 2,174 lines in one second on Earth is the same value Newton used for his Moon test discussed in Chap. 6.

172   9 Gravity on This and Other Worlds

Step 2: Centrifugal acceleration at the Earth's equator: The next step was to learn the effect of the Earth's spin on this gravitational field at the equator. Thus Newton first determined the Earth's radius from the best data on its circumference available to him. From Proposition XIX:

> And from these measures we conclude that the circumference of the earth is 123249600, and its semidiameter 19615800 *Paris* feet, upon the supposition that the earth is of a spherical figure.[8]

His value for the mean radius of the Earth was quite accurate.[9] With this data Newton could compute centrifugal acceleration at the equator just as he did with the Moon in its orbit, as we saw in Chap. 6:

> A body in every sidereal day of 23^h. 56^m. 4^s. uniformly revolving in a circle at the distance of 19615800 feet from the centre, in one second of time describes an arc of 1433.46 feet; the versed sign of which is 0.05236561 feet, or 7.54064 lines.[10]

Dividing the circumference of the Earth by the total seconds in a sidereal day[11] yields the velocity of a point on at the equator, which Newton found to be 1,433.46 Paris feet per second.[12] He then sought the versed sine from the calculated velocity. (For versed sine, think again of the fall distance, or amount of deflection from a straight line in an arc, in one second.) Since the circular acceleration is given by v^2/r and the versed sine is half this, $s = v^2/2r$, the result is a versed sine of about .05237 Paris feet or 7.54064 lines.

$$f_{equator} = \frac{v_e^2}{r_e}$$

$$f_{equator} = \frac{(1433.46)^2}{19,615,800}$$

$$f_{equator} = 0.10475 \text{ Paris feet/sec}^2$$

$$f_{equator} = 7.541 \text{ Paris lines in first second}$$

---

[8] *Principa*, Book III, Proposition XIX, Problem III, 425.

[9] Newton's value for the Earth's radius in Paris feet converts to 6,372 km, which is about one kilometer larger than the modern value of the Earth's mean radius of 6,371 km.

[10] *Principa*, Book III, Proposition XIX, Problem III, 425.

[11] The Earth's rotation using the Sun as the reference is called the solar day, consisting of 86,400 s. The true measure of the Earth's rotation is to use the fixed stars as the reference. This day is called the sidereal day and is slightly shorter than the solar day due to the movement of the Earth in its orbit. Newton uses the correct number of seconds in a sidereal day, 86,164.

[12] Using Newton's numbers for circumference and time, our calculations show the velocity value to be 1,430.4 Paris feet per second, about 3 feet per second less than Newton's value. It is unknown why there is this small discrepancy against Newton's value.

We now work with units of acceleration in meters per second per second, rather than in terms of versed sines; the centrifugal acceleration in modern units comes out to .034 m/s² or about 3½ cm/s in the first second, "upon the supposition that the earth is of a spherical figure."[13] The fall in one second in Paris caused by gravitation (2,174 lines) is thus compared with the versed sine deflection in one second at the equator from otherwise straight-line, rectilinear motion (7.54064 lines):

> And therefore the force with which bodies descend in the latitude of *Paris* is to the centrifugal force of bodies in the equator arising from the diurnal motion of the earth as 2,174 to 7.54064.[14]

Step 3: Centrifugal force at the latitude of Paris. The rotating radius of the Earth is less at the latitude of Paris than that of the equator. As shown in the accompanying figure, the new radius is the equatorial radius of the Earth times the cosine of the latitude. The component of the new centrifugal force perpendicular to the rotating Earth's surface at that latitude (pointing to its center, if the Earth is a sphere) is also the new acceleration times the cosine of the latitude. Hence the new acceleration is the old acceleration times the cosine *squared* of the latitude:

$$f_{paris} = f_{equator} \cos^2 \theta$$

where θ is the latitude, in this case Paris, France:

> The centrifugal force of bodies in the equator is to the centrifugal force with which bodies recede directly from the earth in the latitude of *Paris* 48° 50' 10", as the square of the ratio of the radius to the cosine of the latitude, that is, as 7.54064 to 3.267.[15]

In our calculations, the latitude in Paris in radians is $\pi/180$ times the latitude in degrees, or .852351 rad. Hence, using Newton's units and comparing the versed sines (again as surrogates for the acceleration values),

$$f_{paris} = (7.54064)\cos^2(.852351)$$

$$f_{paris} = 3.267 \text{ Paris lines in the first second}$$

The ratio of centrifugal forces mentioned by Newton is thus,

$$\frac{f_{paris}}{f_{equator}} = \frac{3.267}{7.54064}$$

---

[13] Recall that the term, versed sine, which is odd to our ears, is really just the deflection from an otherwise straight line that a particle of mass on the Earth's surface is forced to take (and by Newton's First Law tends to take) by reason of the rotation of the Earth forcing it into a curved path as it goes round. The resistance to going in a curve is as we noted perceived as centrifugal force. It is an example of Newton creatively using the same principles he developed governing orbital motion to analyze the forces of curved motion on the Earth's surface.

[14] *Principa,* Book III, Proposition XIX, Problem III, 425.

[15] Ibid.

Step 4: True g at the latitude of Paris. Given the gravitational and centrifugal accelerations at the latitude of Paris, it is now easy to compute the true g at that latitude by adding them:

> Add this [centrifugal] force to the force with which bodies descend by their weight in the latitude of *Paris*, and a body, in the latitude of *Paris*, falling by its whole undiminished force of gravity, in the time of one second, will describe 2177.267 lines, or 15 *Paris* feet, 1 inch, and 5.267 lines.[16]

This invokes the relation for finding true g mentioned above:

$$g_{paris \ (true)} = g_{paris \ (observed)} + f_{paris}$$

$$g_{paris \ (true)} = 2174 + 3.267$$

$$g_{paris \ (true)} = 2177.267$$

In modern units this translates to a true g acceleration (on the assumption of a spherical Earth) of about 9.823 m/s^2. Newton compares the true g at Paris with the centrifugal force at the equator:

> And the total force of gravity in that latitude will be to the centrifugal force of bodies in the equator of the earth as 2177.267 to 7.54064, or as 289 to 1.[17]

In our notation, Newton has found,

$$\frac{g_{paris \ (true)}}{f_{equator}} = \frac{2177.267}{7.54064}$$

$$\frac{g_{paris \ (true)}}{f_{equator}} = \frac{1}{289}$$

---

[16] Ibid.
[17] Ibid.

The Earth Is Not a Perfect Sphere

The accompanying illustration may help visualize the relationships:

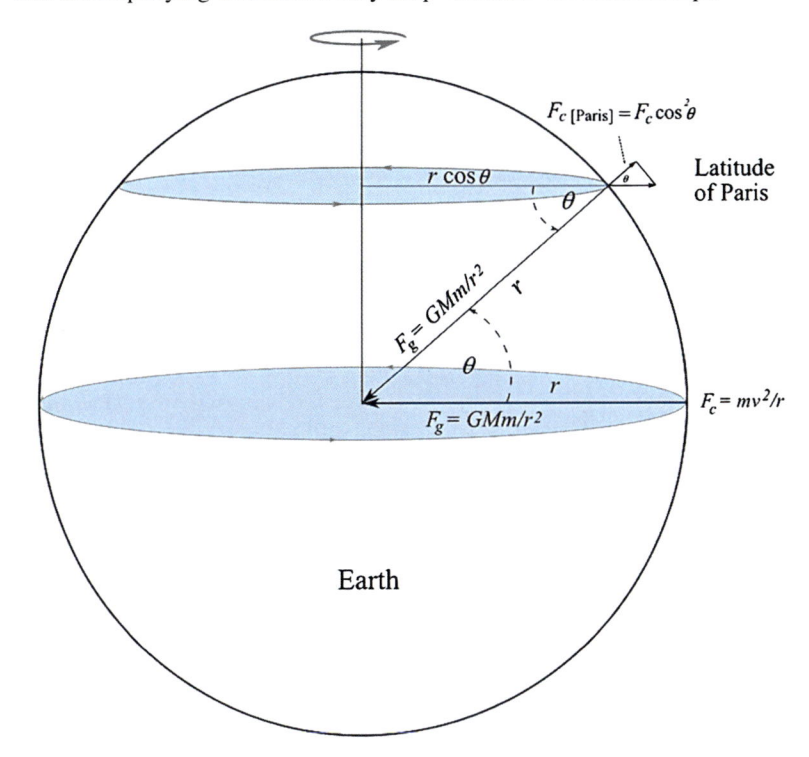

(Figure: Diagram of Earth showing rotation axis, latitude of Paris, with labels $F_{c\,[Paris]} = F_c \cos^2\theta$, $r\cos\theta$, $F_g = GMm/r^2$, $F_c = mv^2/r$, and angle θ.)

For our goal of understanding the concept of net acceleration, we can stop here. Recall that Newton's aim was to determine the oblateness or flattening of the Earth, its deviation from a perfect sphere of uniform radius and density. His determining of the "true g" at the latitude of Paris was the first task. If the Earth were a perfectly uniform sphere the true g (undiminished by rotational acceleration) should be the same all over the Earth. If the true g differed at different places, it would be proof that the Earth has a non-spherical shape. To determine the true gs on Earth, he would have to take the net g measured by instruments at different places on Earth and add in calculated centrifugal accelerations for the observers' latitudes. This would in theory give a measure of the extent to which the Earth is, or is not, a true sphere.

The Earth Is Not a Perfect Sphere

It is plain that that every point on the surface of the Earth is not equidistant from its center. The Earth is not a perfect geometrical figure. There are mountains, valleys, and also areas of uneven density within the Earth. All these affect the value of *g* where it is measured. But the much larger effect than this on *g* discovered by

Newton, is that the Earth is not a really not a geometrical sphere at all. It is an oblate spheroid: it has a *flatness* which makes its slightly swollen equatorial diameter larger than its squatter polar diameter. The Earth is an oblate spheroid generated by the rotation of an ellipse around its minor axis which passes through the poles. As such, the distance from the equator to the center of the Earth is longer than the distance from the poles to the center of the Earth. The result is that the gravitational acceleration at the two places on the surface are different, *less* at the equator and *greater* at the poles. This effect further compounds the effect of the more rapid velocity at the equator which also reduces the Earth's gravitational g there.[18] Newton analyzed measurements of the value of g gathered from various parts of the world, which together confirmed his theoretical calculations, with many qualifications regarding the accuracy of some of the data.[19] Given the fact that the value of g was determined in various parts of the world by pendulums, and his imperfect knowledge of the size of the Earth, it is amazing that Newton came relatively close to determining the actual value of the Earth's oblateness which is a little more than 13 miles.[20]

---

[18] Newton found that the reduction of gravitational g in Paris – of the weight of every part of a mass there – caused the centrifugal acceleration of the Earth's spin, when compared to the centrifugal force at the equator, is about one part in 289. But this ratio did not square with his earlier calculations that, considering the Earth as a sphere, showed the ratio to be 4 to 505 or about one part in 126. *Principa,* Book III, Proposition XIX, Problem III, 427. In other words, the spin seemed somehow to have a lesser effect on g than he originally thought. Newton devised a mathematical model of the Earth in which he supposed that a narrow canal filled with water penetrated it from one poles to the center and from the center to the equator. Assuming the water-filled canals in a rotating Earth must be in gravitational equilibrium, Newton calculated after considerable labor that the Earth must be about 17 miles shorter at the poles that at the equator:

> ... and therefore the diameter of the earth at the equator is to the diameter from pole to pole as 230 to 229. And since the mean semidiameter of the earth according to *Picard's* mensuration, is 19615800 *Paris* feet, or 3923.16 feet (reckoning 5000 feet to a mile), earth will be higher at the equator than at the poles by 85472 feet, or 17 1/10 miles And its height at the equator will be about 19658600 feet, and at the poles 19573000 feet. Ibid.

[19] A thoroughgoing critique and analysis of Newton's work on this subject is given in I. Todhunter [2]

[20] The propositions in question are rather obscure on the question of the shape of the Earth. Studies by Colin Maclaurin (1698–1746), Alexis Clairaut (1713–1765), and others treated the Earth's shape with considerable mathematical rigor. There is controversy about whether Newton arrived at approximate oblateness of the Earth more or less accidentally, using an incomplete method or whether Newton withheld much of his reasoning and simply didn't state his entire analysis. *See,* e.g., S. Chandrasekhar [3].

The Earth Is Not a Perfect Sphere

Finding the True Value of g at the Equator

Problem From the information given below, determine the components of gravitational and centrifugal accelerations acting on a 1kg mass at the equator.

Given

$F_g = \frac{GMm}{r^2}$	Newton's Law of Gravitation
$F_c = mv^2/r$	Centrifugal force on a mass m at velocity v at radius r
G	The universal gravitational constant: 6.674×10^{-11} N·m²/kg
M	The mass of the Earth: 5.9721×10^{24} kg
r_e	The mean radius of the Earth: $6{,}371 \times 10^3$ m

Assumptions We'll assume the Earth is a sphere of uniform density of the given mean radius, and the accuracy of the given data. We assume there are no other accelerations acting on the mass m other than the acceleration due to gravity.

Method Here we can use Newton's Law of Gravitation to find the acceleration due to gravity at the Earth's surface, whose distance in meters is the Earth's mean radius from its center. The number derived will be the gravitational acceleration of a perfect, non-rotating sphere. To find the net or expected observed acceleration acting on a mass at the equator, it is necessary to subtract the centrifugal acceleration from the gravitational acceleration. For this we determine the velocity at the equator and use v^2/r for the determination of this value. The result will be net g.

Calculations We know that the force acting on a mass at the surface of the Earth is given by Newton's Law of Gravitation:

$$F_e = \frac{GMm}{r_e^2}$$

It is given that m is a unit mass, so all our force numbers are accelerations:

$$g_e = \frac{GM}{r_e^2}$$

Inserting the given values on the right side of the equation, and solving for gravitational acceleration at the equator, g_e,

$$g_e = \frac{(6.674 \times 10^{-11})(5.9721 \times 10^{24})}{(6.371 \times 10^6)^2}$$

$$g_{e\,(\text{true})} = 9.82 \text{ m/s}^2$$

This value assumes the Earth is not rotating. It was derived just from mass and mean radius with no consideration given the spin of the Earth at the equator. It might be regarded as the "true equatorial g" for a spherical, uniformly dense Earth. (If the

Earth were a perfect sphere, this value should in theory match the observed gravitational acceleration at the equator.) We determine the countervailing centrifugal force component from the equation,

$$F_c = mv^2/r$$

which will offset the gravitational g_g. Again, thinking in terms of accelerations,

$$f_c = v^2/r$$

We can determine the rotational velocity of the Earth in the usual way, where P is the rotational period of the planet:

$$v = \frac{2\pi r_e}{P}$$

Inserting this expression for velocity into the acceleration equation, we get:

$$f_c = \frac{4\pi^2 r_e}{P^2}$$

The Earth's *sidereal* rotation period encompasses one revolution in 86,164 s.[21] Working through the numbers, we get:

$$f_c = \frac{4(3.14..)^2(6371 \times 10^3)}{86164^2}$$

$$f_c = .03388 \text{ m/s}^2$$

This is about three and a third centimeters per second per second of centrifugal acceleration. We now have two accelerations, one gravitational the other centrifugal. To find the net acceleration acting on a mass at the equator, we subtract on from the other:

$$g_{net} = 9.82 - .034$$

$$g_{net} = 9.786$$

Observations

1. The velocity of the spinning Earth at the assumed-spherical Earth at the equator is about 465 meters per second. Yet not withstanding this velocity the

---

[21] See note 11 for an explanation of this.

gravitational strength is about 9.82/.034 or 289 times that value, more than adequate to keep us from flying off into space!

2. Again assuming a spherical Earth, what would be the values of g and f and net g at the latitude of Paris using the above methods? If the Earth is a sphere, the true value of g should be the same, about 9.82 m/s². The rotational radius is less though, by the cosine of the latitude, which (using Newton's latitude) in radians is $\pi/180$ the latitude in degrees, or .852351. r_e cos (.852351) is 4,193.5 km. The velocity in this case has dropped to about 306 km/s and the centrifugal acceleration is now only .022 m/s². The net g at Paris by this reasoning is thus:

$$g_{net} = 9.82 - .022$$

$$g_{net} = 9.80 \text{ m/s}^2$$

This value is a higher g than at the equator, due to the reduced effect of centrifugal acceleration. What do you think the effect of the latter is as we approach the poles?

Using g to Derive Velocities in Circular Orbits

Consider the Earth's gravity at its surface as "one g" with an average value of about 9.8 meters per second per second acceleration.[22] It is instructive to compare the value of g as we leave the Earth's surface and go into space, even to the Moon, to see how it diminishes. Imagine that we have a "gravity meter" which would be just an accelerometer, able to measure accelerative forces acting on it. With the aid of our gravity meter (and a little mathematics) in our outward-bound travelling vehicle, we can observe how g drops dramatically as the inverse square law takes its toll. Also apparent is the corresponding decrease in orbital velocity and increase in period of satellites at increasing distances from Earth. Since surface gravity depends on mass and how far the surface is from the center of the mass, one can also calculate the "g" for the surface of the Moon or Sun or any other body whose mass is known. These values can be readily compared to the terrestrial g. The value of the lunar g will tell what velocity is needed to orbit the Moon. The value of the solar g will inform us why Earth orbits at the speed it does, 150 million km distant from the surface solar g.

By working with g as our unit, we can readily see the proportionate change in the pull of gravity as we leave Earth, by virtue of the inverse square law, and how that change affects velocity and the orbital period.

---

[22] By Newton's Second Law, force equals mass times acceleration. Acceleration is, as noted earlier, the same as force per unit mass, and this is how gravitational force is usually expressed, as acceleration, where units are typically m/s².

Problem Derive an equation for finding the velocity of an object in a circular Earth orbit in terms of g, the value of the Earth's gravitational acceleration, and distance from the center of the Earth.

Given

$f_c = v^2/r$	Centripetal acceleration f_c in a circular orbit of radius r at velocity v

Assumptions We assume for the moment that the orbiting body has negligible mass, and that it moves in a circle. We also assume that the Earth is a spherical object with uniform density, such that its mass can be represented by a point at its center of mass.

Method The velocity is straightforwardly derived from the centripetal acceleration equation. The gravitational force draws a mass toward the center of the Earth, and this must be balanced by the tendency, described by Newton's First Law, for objects in motion to continue to move in a straight line. This is why one feels a strong, centrifugal tug on a cord attached to a whirling object, except your resisting pull on the cord takes the place of gravity in the analogy. Here we will substitute g_r (note the subscript r) the Earth's gravitational acceleration at radius r, for f_c which is the assumed centripetal acceleration acting at the same distance. The radius of the circle of motion is the distance from the object to the center of the Earth.

Calculations We begin with the centripetal acceleration equation. Let g_r be the acceleration due to gravity g acting at distance r from the center of the Earth, and so expressed as g_r

$$g_r = \frac{v^2}{r}$$

Solving for velocity,

$$v = \sqrt{r g_r}$$

The circular orbital velocity is thus the square root of the object's distance from the center of the primary mass times the gravitational acceleration of the primary mass, at distance r.

Observation If the value of g_r is unknown, this equation requires first calculating g_r for an object's distance from the Earth before its velocity can be determined. Recall that $g_r = GM_E/r^2$ (where M_E here is the mass of the Earth) and that value, when substituted into the above velocity equation, simplifies to the form we will frequently see:

$$v = \sqrt{\frac{\mu}{r}}$$

where μ is the product of the constants GM_E, with the value $\mu = 3.99 \times 10^{14}$ m^3/s^2.

Using g to Derive Periods in Circular Orbits

Here we show that the tools used to find g on the surface of the Earth can be used to find the periods of bodies orbiting above it.

Problem Derive an equation for finding the period of an object in a circular orbit in terms of g_r, the value of the Earth's gravitational acceleration at radius r.

Given

$f_c = v^2/r$	Centripetal acceleration f_c on a particle in a circular orbit of radius r at velocity v
$v = 2\pi r/P$	Circular velocity of a particle in terms of the circle's radius r and period P

Assumptions We make the same assumptions about circular orbits around uniformly dense spherical bodies as we noted above.

Method Simply substitute the right-hand side of the expression for circular velocity for velocity in the first equation for centripetal acceleration. This will yield acceleration in terms period.

Calculations The right-hand terms of the equation $v = 2\pi r/P$ can be substituted for velocity in the centripetal acceleration equation:

$$f_c = \frac{v^2}{r}$$

$$f_c = \frac{4\pi^2 r}{P^2}$$

This is the relation we saw in the previous problem, but now applied at any distance from Earth. Rearrange the equation to solve for the orbital period. The term f_c is going to be equivalent to gravitational acceleration at distance r. Again denoting the value of g at radius r by g_r:

$$P^2 = 4\pi^2 \frac{r}{g_r}$$

Taking the square root of both sides we obtain

$$P = 2\pi \sqrt{\frac{r}{g_r}}$$

This is the period of an orbiting object at distance r from the center of the Earth, where at such distance the gravitational acceleration of the primary mass at that distance is g_r. More generally, for any centripetal acceleration,

$$P = 2\pi \sqrt{\frac{r}{f}}$$

which recalls the relation from Newton's corollaries to Principia Proposition IV of Book I:

$$f \propto \frac{r}{P^2}$$

Observations

1. These calculations may also be cleanly done using radial notation, where $\omega = 2\pi/P$ (circular angular velocity in radians), and $\omega r = 2\pi r/P = v$, or circular velocity v along the arc of radius r.[23] The acceleration due to gravity g_r acting at distance r from the center of the Earth, is

$$g_r = \omega^2 r$$

Since $\omega^2 = 4\pi^2/P^2$, substitution of the ω^2 term and solving for P yields the same result.

2. The equation for the period of an orbiting object is, interestingly, the same as the basic formula for finding the period of a simple pendulum (swinging in a small arc) which we saw in Chap. 3:

$$P = 2\pi\sqrt{\frac{l}{g}}$$

where l is the length of the string attaching the bob to the pivot point and g is the gravitational constant at the altitude of the pendulum. More generally, for any centripetal acceleration f

3. Again, the gravitational acceleration is itself determined by $g_r = \mu/r^2$. Substituting this into the period/pendulum equation simplifies to Kepler's Third Law:

$$P = 2\pi\sqrt{\frac{r^3}{\mu}}$$

where again (for the Earth as the orbited mass) μ has the value given above, again with the assumption that the secondary mass is small enough (as is the case with Earth satellites) to be discounted. Working directly with g is intuitive and has the advantage of allowing us to visualize the effects on both velocity and period as g changes with distance from Earth. This is illustrated with the examples below.

---

[23] As noted elsewhere, an object moving in a circle covers 2π *radians* ("radiuses") every revolution, or an actual distance (in mks units) of $2\pi r$ meters. The velocity v is the time it takes the object to accomplish one such revolution, called the period of the object, denoted by P. Therefore $v = 2\pi r/P$. But if the velocity is expressed just in radians per second (think of radii per second) we use the symbol ω for velocity. This is then called the *angular velocity*, or *angular frequency*, where $\omega = 2\pi r/P$.

Comparing Velocities and Periods at Increasing Distances from Earth

Problem Approximate the velocity of a spacecraft orbiting the Earth at the altitudes of (1) Mt. Everest, 8,848 km; (2) 500 km above the Earth's surface; and (3) 384,400 km away from the center of the Earth (the Moon's mean distance). Use g, the value of the Earth's gravitational acceleration, adjusted in each case by the inverse square law for the distance of the spacecraft from the center of the Earth.

Given

$v = \sqrt{rg_r}$	Velocity of an object in circular orbit of radius r
$P = 2\pi\sqrt{\frac{r}{g_r}}$	Period of an object in circular orbit of radius r
g	9.8 m/s², Earth's approximate surface g
r_E	6.38×10^6 m, the Earth's equatorial radius

Assumptions We assume the Earth is of uniform density and we'll completely ignore the effects of atmospheric friction, especially at the altitude of Mt. Everest!

Method We'll adjust the value of g to various r from the center of Earth, realizing that its value diminishes with the square of that distance (the inverse square law). Then we will use the new g in each case to find velocity and period using the given equations.

Calculations Here is the sequence of calculations:

Location of hypothetical satellite	Distance from center of Earth to satellite (m)	Number of Earth radii to satellite	Square of Earth radii to satellite	Inverse of square of number of Earth radii	Times g_E at Earth's surface	Velocity (km/s)	Period
Earth's surface r_E	6.38×10^6	1	1	1	9.8	7.91	84.5 min
Mt. Everest, 8,848 m	6.39×10^6	1.001567	1.003137	.996873	9.7694	7.9	84.7 min
500 km above the earth's surface	6.88×10^6	1.07837	1.162882	.85993	8.4273	7.61	94.6 min
384,400 km away from Earth's center	3.844×10^8	60.25	3,630.2	.000275	.0027	1.02	27.4 days

Observations

1. The fifth column from the left shows the fractional number of g compared to the Earth's surface – the relative g – and the sixth column from the left is actual the gravitational acceleration experienced at the given distance. We can see clearly how the velocity slows and the period increases with increasing distance from the

Earth's center. Even at the altitude of Mt. Everest, there is an appreciable diminution of Earth's gravitational force, expressed as a reduction in g, to 99.7% of its surface value. It is surprising to realize that a mere 500 km above the surface of the Earth, a little over 300 miles, where some satellites orbit, the value of g has dropped to about 86% of its value. At the Moon, our closest celestial neighbor, being virtually in our backyard in solar system terms, the value of g is less than 3 one-hundredth's of 1% of its value at the surface of the Earth. Yet it is eternally held there in its graceful orbit, continuously pulled back from straight-line motion by the less than 3 millimeter per second, per second acceleration exerted from Earth.

2. Note that the above equations using g as the parameter and correcting for distance by the inverse square law yield the same results as the conventional equations using GM (or μ). The values determined by each method are the same, as we would expect:

Location	Distance from Earth's center $r_{\oplus+h}$	Velocity $v = \sqrt{\frac{\mu}{r}}$ km/s	Period $P = 2\pi\sqrt{\frac{r^3}{\mu}}$
Earth's surface (radius = r_E)	6.38×10^6 m	7.91	84.5 min
Mt. Everest, 8,848 m	6.39×10^6 m	7.9	84.7 min
500 km above the Earth's surface	6.88×10^6 m	7.61	94.6 min
384,000 km away from Earth's center (the Moon's mean distance)	3.84×10^8 m	1.02	27.4 days

3. A geosynchronous satellite is one whose orbit that exactly matches the Earth's rotation with reference to the stars, so there is one revolution in 24 h 56 m 4 s, or 86,164.1 s. To do this it must be fairly high up. How high depends on solving Kepler's Third Law to find the radial distance:

$$P = 2\pi \sqrt{\frac{r^3}{\mu}}$$

$$r = \sqrt[3]{\frac{P^2 \mu}{4\pi^2}}$$

$$r = \sqrt[3]{(8.61641 \times 10^4)^2 \left(\frac{3.986 \times 10^{14}}{39.478}\right)}$$

$$r = 42,164 \text{ km}$$

This distance is from the center of the Earth.[24] Subtracting the mean equatorial radius of the Earth of 6,378.14 km from that number gives 35,786 km as the distance from Earth's surface.

---

[24] Note that we used a slightly more refined value of μ here or 3.986×10^{14}.

4. Since 42,164 km is 6.61 Earth radii, the value of the Earth's gravitational acceleration g at the surface of the Earth must be diminished $1/6.61^2$ times, or to about 2.3% of its surface value. The acceleration it experiences (g_r) is therefore .2243 m/s^2. The orbital velocity at this distance is

$$v = \sqrt{rg_r}$$

$$v = \sqrt{(4.2164 \times 10^7)(.2243)}$$

$$v \simeq 3.1 \text{ km/s}$$

How much slower this is than the seven and a half kilometers per second velocities of lower-flying Earth satellites!

5. These values for g_r and velocity make intuitive sense? To check the results with our intuition, we can think about it this way. Since the distance to the satellite's orbit (about 6.6 radii away) is about 1/9th the distance to the Moon (which is about 60 Earth radii away) the value of g_r at lunar distance should be roughly 1/81th the value for the geostationary satellite, and it approximately is (.0027 for the moon and .2243 for the satellite orbit). The Moon's velocity in orbit is about 1/3 that of the satellite's orbital velocity. This all makes sense because, while the gravitational force, and hence acceleration, is inversely proportional to the square of the distance,

$$g \propto \frac{1}{r^2}$$

the velocity is inversely proportional (from $v = \sqrt{\mu r}$) to the square root of the distance,

$$v \propto \frac{1}{\sqrt{r}}$$

Calculating the Lunar g: Gravitational Acceleration on the Surface of the Moon

We have so far focused upon the terrestrial g, and how it diminishes by virtue of the inverse square law as we venture into space. Let us examine g on other bodies. These values can be straightforwardly calculated if we know the mass of the body. (Later we will see how to calculate that mass.)

Problem Approximate the value of the acceleration g_m due to the gravity of the Moon, measured at the surface of the Moon.

Given

$g = GM_m/r_m^2$	Moon's gravitational acceleration g at distance r_m
G	The gravitational constant: 6.674×10^{-11} N·m²/kg
M_m	Approximate mass of the Moon: 7.35×10^{22} kg
r_m	Approximate mean radius of the Moon: 1.738×10^6 m

Assumptions We'll assume the Moon is a sphere of uniform density of the given mean radius.

Method We will find the value of g_m at the Moon's surface, so again considering the Moon's mass as located at a point at its center, we find the value of gravitational acceleration at the distance from the center to the surface, which is the radius of the Moon. Using the given values, we'll solve for g_m.

Calculations Begin with the acceleration equation:

$$g_m = \frac{GM_m}{r_m^2}$$

Inserting the given values for the constants on the right side of the equation, we have:

$$g_m = \frac{(6.674 \times 10^{-11})(7.35 \times 10^{22})}{(1.738 \times 10^6)^2}$$

$$g_m = 1.62 \text{ m/s}^2$$

Observations

1. The value of g for the Earth's surface is 9.8 m/s² which is 9.8/1.62 or about six times the g_m of the moon. Expressed another way, the Moon's gravity is about 1/6th that of the Earth, and its mass is $(7.35 \times 10^{22})/(5.98 \times 10^{24}) = .012$ the mass of the Earth (roughly 1/81th).
2. The Moon's gravitational parameter, GM_m or μ is about 4.9×10^{12} m³/s².
3. The Moon spins on its axis too slowly to experience the effects of reduced apparent g at its equator that the Earth experiences.

Calculating the Solar g : Gravitational Acceleration of the Sun

Problem Determine the value of the acceleration due to the gravity g_{sun} of the Sun at the Sun's surface and the gravitationally induced acceleration g_{1AU} at the distance of one astronomical unit.

Given

$g_{sun} = GM_s/R_s^2$	Equation for Sun's gravitational acceleration, g at the surface of the Sun, at one solar radius R_s
G	Gravitational constant: 6.674×10^{-11} N·m²/kg
M_s	The mass of the Sun: 1.9891×10^{30} kg
R_s	Radius of the sun at photosphere: 6.96×10^8 m
$g_{1AU} = GM_s/r^2$	Equation for Sun's gravitational acceleration, g at one astronomical unit (AU) from the Sun, r
r	Mean distance of the Earth from the Sun, 1.496×10^{11} m

Method We will first substitute the given values for the constants on the right side of the acceleration equation, and solve for g_{sun}. For the second part of the problem, we will use r instead of the solar radius, and see how the solar g_{sun} is diminished to g_{1AU} by the inverse square law at the Earth's distance from the sun.

Assumptions Though the Sun is a gas body, the radius is given to the photosphere of the Sun. This is the part that emits radiation (at about 5,800 K) and is the part that we can see. It is rather thin (.1% of the solar radius) and so appears to be a well-defined, sharp edge.

Calculations First to find the value of g_{sun} at the Sun's photosphere, one solar radius, and insert the appropriate values. We will use the convenient shorthand $\mu_{sun} = GM_s$ for the "Sun's gravitational parameter" here and for future calculations where the Sun is the gravitational center, just as we used $\mu = GM_{earth}$ for the geocentric gravitational parameter. The value of μ_{sun} will be 1.3275×10^{20}:

$$g_{sun} = \frac{\mu_{sun}}{R_s^2}$$

$$g_{sun} = \frac{1.3275 \times 10^{20}}{(6.96 \times 10^8)^2}$$

$$g_{sun} = 274 \text{ m/s}^2$$

This acceleration is almost 28 times the gravity of the Earth at its surface – person of average weight would weigh about two tons on the Sun. Now to find how this powerful gravity is attenuated by distance, we make the radius distance equal to 1 AU.

$$g_{1AU} = \frac{\mu_{sun}}{r^2}$$

$$g_{1AU} = \frac{1.3275 \times 10^{20}}{(1.496 \times 10^{11})^2}$$

$$g_{1AU} = .00593 \text{ m/s}^2$$

Observations

1. The force of gravity of 28 times Earth's g on the Sun's surface becomes 1/1,652 terrestrial g at the distance of the Earth, as the Sun's gravitational acceleration has dropped from 274 m s^{-2} to 5.9 mm s^{-2}.
2. We can check this result by scaling g_{sun} out to 1 AU and watch it diminish by the inverse square of the distance as we leave the Sun. How many solar radii are there to the Earth's orbit? It is the distance to the Sun divided by the radius of the Sun:

$$\frac{1.496 \times 10^{11}}{6.96 \times 10^8} \simeq 214.94 \text{ solar radii.}$$

We would thus expect the value of g_{sun} at the Sun's surface to be diminished, at a distance of 214.94 solar radii, by 1/214.94^2. This attenuation factor is .0000216. Since we found that at the photosphere of the Sun $g_{sun} = 274$ m/s^2, we deduce the value of the Sun's gravitational acceleration at 1 AU to be 274 × .0000216 or .00593, m/s^2, the same almost 6 mm value we found above.
3. What is the *velocity* of the Earth in its orbit? Using this equation readily yields the answer:

$$v = \sqrt{r g_{1AU}}$$

$$v = \sqrt{(1.496 \times 10^{11})(5.93 \times 10^{-3})}$$

$$v \simeq 29.8 \text{ km/s}$$

4. We can of course calculate the Earth's mean orbital velocity by using the equation with the heliocentric gravitational parameter,

$$v = \sqrt{\frac{\mu_{sun}}{r}}$$

$$v = \sqrt{\frac{(1.3275 \times 10^{20})}{(1.496 \times 10^{11})}}$$

$$v = 29.78 \text{ km/s}$$

The Earth's orbit is of course elliptical, and the mean velocity, which implies a circular average, differs from the actual velocities it experiences at the apsides of its orbit.

Exercises: How Gravity Is Just Right The eminent Astronomer Royal, Martin Rees, wrote a well-known book titled *Just Six Numbers*.[25] In it he marvels that the key constants of the universe are as they are, and speculates how different things would be if they were even a little larger or a little smaller. He begins his chapter about gravity by discussing its vast pervasiveness on large scales:

> If we were establishing a discourse with intelligent beings on another planet, it would be natural to start with gravity. This force grips planets in their orbits and holds the stars together. On a still larger scale, entire galaxies—swarms of billions of stars—are governed by gravity. No substance, no kind of particle, not to even light itself escapes its grasp. It controls the expansion of the entire universe, and perhaps it's eventual fate.[26]

Rees goes on to argue that gravity is actually quite a weak force, but its power lies in its steady accumulation on large, larger and colossal scales, not cancelling its effects as electric charges do. He questions what our universe would be like it gravity were not so weak. If gravity were even slightly stronger, he argues, then the world as we know it and the life forms which surround us could not exist: "In an imaginary strong-gravity world even insects would need thick legs to support them, and no animals could get much larger. Gravity would crush anything as large as ourselves."[27] He makes the interesting case that in such a universe, galaxies and stars would form quickly, would be smaller and more densely packed, and there would not be sufficient time to permit complex evolutionary processes:

> ... [I]n this hypothetical strong gravity world, stellar lifetimes would be a million times shorter. Instead of living for ten billion years, the typical star would live for about 10,000 years. A mini-Sun would burn faster, and would have exhausted its energy before even the first steps in organic evolution had got underway.[28]

This argument is consistent with the so-called anthropic principle, one version of which holds that if the world were any different we would not have evolved to observe and comment on it. We are fortunate indeed to live in a Goldilocks world!

---

[25] Martin Rees [2].
[26] Ibid., 27.
[27] Ibid., 34.
[28] Ibid.

Problems

1. Given the assumptions of this chapter, how much more would a 200,000 kg ship weigh at the poles than at the equator of an assumed spherical Earth?
2. Design an equation to answer to the following question: what would the Earth's rotational period have to be in order to make us nearly weightless at the equator? Does the equation look familiar? If so, explain why?
3. Use the above equation to calculate the Earth's period, in minutes, that would make us weightless at the equator. In this case, use the Earth's equatorial radius of 6,378.14 km. Its mass, as before, is 5.97219×10^{24} kg. What would the rotational velocity be in km/s?
4. If the Earth's density is increased by 10%, what would be the change of the gravitational field at its surface?
5. Assuming the lunar surface g is 1.624 m/s^2, and its radius is 1,737.5 km, what is the Moon's mean density?
6. Mars is a nearly spherical planet with a mass of 6.41693×10^{23} kg, an equatorial radius of 3,396.16 km, and a rotational period of 1.02595676 days. Assuming a uniform distribution of mass within Mars, what is the approximate true and apparent Martian g at the equator?
7. Suppose the Earth, whose mean radius is 6,371 km, were shrunk to the size of the Moon, whose mean radius is 1,737.5 km. Find the proportional change of the Earth's density and surface g.
8. Four "if's" to (try to) answer in your head first and then write down: (a) If both the mass and radius of the Earth were increased by five times, what would be the proportional change to its true surface gravitational acceleration? (b) If the Moon were half its current distance from Earth, what would be the proportional change to its period and velocity? (c) If Earth's g were three-fourths its current value, what would be the proportional change to the Moon's period? (d) If the mass of the Earth were suddenly made half its current value, what would be the proportional change to the velocity of the Moon? Express your answers in integers, fractions, and radicals, as applicable.
9. You are given the following facts: (a) Neptune's mass is 1.0241×10^{26} kg and its mean radius is 24,622 km. (b) Triton is Neptune's largest moon, in a zero eccentricity orbit. Its period is only 5.877 days. Find the value of the true surface g of Neptune, and its g at Triton's distance. Confirm Triton's period using the new g you found.
10. The period of Jupiter's small moon Amalthea (discussed in Problem 11.2), orbiting 181,200 km from Jupiter's center, goes around the giant planet once every .498179 days. A small circularly-orbiting satellite of Earth at that same distance would be 28.41 radii away, less than half the distance to our Moon. (a) Find the g of Earth acting on such a satellite, and from that, its period in days. (b) Just using proportions, determine the ratio of the mass of Jupiter to the mass of the Earth.

References

1. Cajori F (1949) Principia in modern English: Sir Isaac Newton's mathematical principles of natural philosophy. Book III, Proposition XIX, Problem III (Motte's Translation Revised). University of California Press, Berkeley, p 425
2. Todhunter I (1962) A history of the mathematical theories of attraction and the figure of the earth. Dover, New York, first ed. 1873
3. Chandrasekhar S (1995) Newton's principia for the common reader. Clarendon Press, Oxford, pp 384–397
4. Rees M (2000) Just six numbers. Basic Books, New York

Chapter 10
A Binary System Close to Home: How the Moon and Earth Orbit Each Other

The everyday notion of the planets revolving around the Sun, and the Moon revolving about the Earth – the concept of one large, fixed, and static body in the center and the satellite bodies revolving around it – is actually somewhat misleading. It depends upon one's frame of reference. Surely if we could stand on the Sun, that would be our impression as to the planets, even though we know that all of them, and Saturn and Jupiter especially, exert their own smaller gravitational pull on the Sun and cause it to wiggle around a little irregular orbit of its own. And from our fixed reference here on Earth, the Moon naturally appears to revolve around it. Yet it too pulls on the Earth, and they both in fact orbit, albeit in vastly different arcs. Relative motion is an idea that took mankind a long time to understand and appreciate. Did it not seem natural for the Sun to be orbiting the Earth, when clearly the Sun's motion was all that was detectable? Kepler wrestled with relative motion in trying to determine the motion of Mars from the Sun's and not the Earth's frame of reference. In fact all motions appear relative to the observer's frame of reference.[1] If one observes a system of two or more orbiting bodies from any frame of reference external to them, their orbits dance around a common *center of mass*. This point is usually called the *barycenter* of the system. The velocities of the masses in such a system depend upon the frame of reference from which they are measured. The velocity of one mass from the perspective of the center of mass will be quite different from the velocity measured from the other as the point of reference. Think about when you were on a train and another train on a parallel track inched by you in the same direction, or flew by in the opposite direction. You saw the other train's velocity relative to you. In fact, the actual velocity of the other train, from the perspective of an observer external to both, was neither inching nor flying.

---

[1] Einstein famously followed up on the implications of the idea of relative motion, beginning in 1905 with the notion of the absolute standard of the speed of light, ultimately developing theories which profoundly altered our conceptions of the fundamental nature of mass, energy, space, time and gravitation.

Imagine a double star system where each star has the same mass. From your point of view outside the system, looking down on it, they each appear to be moving around an invisible point midway between them, their center of mass. It is their balance point. The stars always seem to face each other, like children whirling around each other with clasped hands. They move in synch and their orbital periods are the same. Now shift your position to a vantage point on one of the stars (such things are possible in the imagination!). Now you are at rest watching the other star orbiting in the sky around you. And you would notice another thing: it is moving at twice the velocity that it appeared to move when you were out in space observing it. Its *relative* velocity is much greater. But from either frame of reference, their period is the same, determined by Kepler's Third Law which depends only upon their masses and distance between them.

Now we can change the picture to help us focus on the concept of center of mass and learn more about what it means. If we vary our scenario and make one mass larger than the other, the center of mass will be closer to the large mass. This seems intuitive, as the center of balance, so to speak, is shifted more toward the greater one, just like the adjustment two boys of unequal weight must make on a see-saw, where the heavier boy has to sit closer in to the center. As to the stars, each will still orbit around their center of mass, and again, their periods will be identical, still determined by Kepler's Third Law.

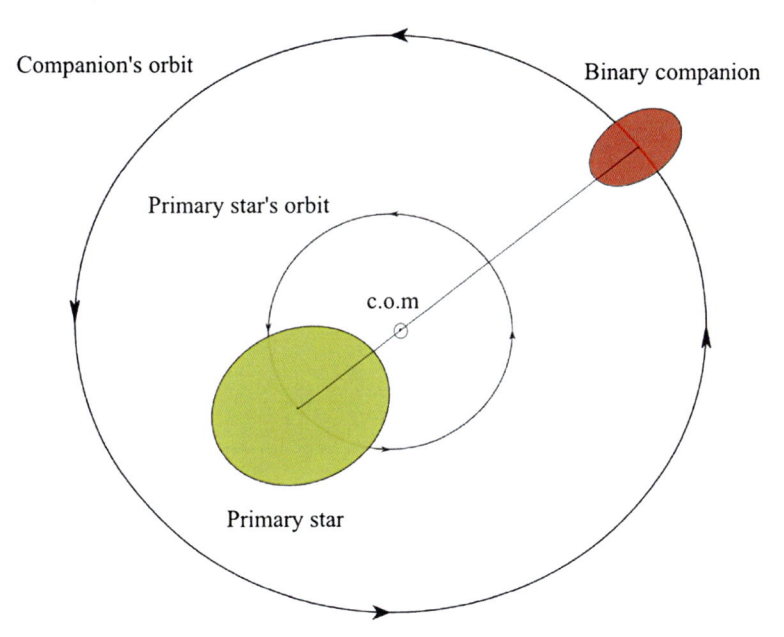

Mutual Revolution of Two Stars Around Common Center of Mass

10 A Binary System Close to Home: How the Moon and Earth Orbit Each Other

As we imagine the larger mass to become even larger still, the center of mass will be even closer to the larger mass, again what one would intuitively expect. If the larger mass is very much bigger than the smaller mass, the center of mass can be *inside* the larger mass. It is as if a very large boy on the see-saw had to position himself practically over the center pivot to allow balance with the much lighter boy. This is the case with the Earth – Moon system: the center of mass is inside the Earth. While the center of mass is inside the Earth, the Earth still orbits around that center of mass, like a frying pan being moved around in small circles on a stove. Remember that the mass of the Earth (or any other body) acts on other masses as if its entire mass were concentrated at a single point at its center. This was one of Newton's great breakthroughs in understanding gravity. This point, mathematically speaking, orbits the center of mass. That the Earth–Moon center of mass is inside the Earth doesn't affect the gravitational result: as with the double stars, the Moon and the Earth are like two whirling skaters holding hands, the pull on their arms being the gravitational force between them. Their period is determined again by their masses and the distance from their centers (Kepler's Third Law). And it is the barycenter of this system that is in orbit around the Sun.

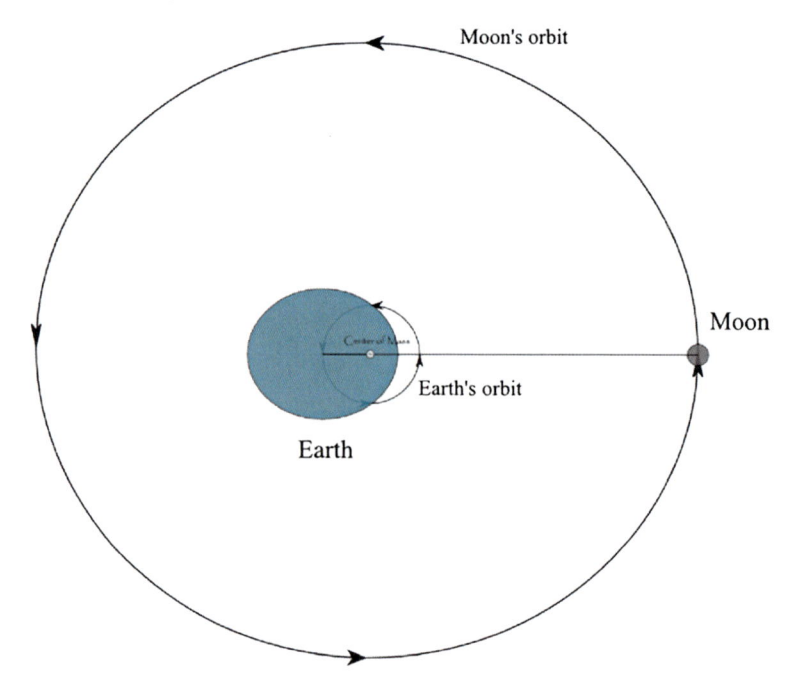

**Mutual Revolution of Earth and Moon
Around Common Center of Mass**

Locating the Center of Mass

If we wish to find the center of mass, it is easy to do. Intuitively it is where we would expect it to be. Imagine that you have mass m_1 and are standing on a balance scale. The center of mass is the fulcrum or pivot of the scale. To make the scale balance and find your weight, you slide the smaller weight, which we will call m_2 farther out along the little graduated scale, away from the pivot, making r_1 greater, so the weights are just balanced, where you then read off your weight. The heavier you are, the father out along the scale you must slide the weight. Hence, the greater is mass m_1, the farther from the center of mass must m_2 be slid. Of course you may not want to slide m_2 so far out; you may instead substitute a heavier weight m_2 to make it balance. Intuitively, you might say that the masses are inversely proportional to the distances to the center of mass, so $m_1:m_2 :: r_2 : r_1$. Let's see how this works out in the context of orbiting things.

We saw that the equations representing the gravitational and centripetal accelerations of each of two masses in orbit around their center of mass are:

$$\frac{Gm_1}{r^2} = \omega^2 r_2$$

$$\frac{Gm_2}{r^2} = \omega^2 r_1$$

Dividing one set of equations by the other to find the ratio of their accelerations, we have,

$$\frac{m_1}{m_2} = \frac{r_2}{r_1}$$

This is as we expected, and,

$$m_1 r_1 = m_2 r_2$$

This is a definition of the center of mass. Again, r_1 is the distance of m_1 from the center of mass and r_2 is the distance of m_2 from the center of mass. The separation distance is here denoted R, and is the sum of each of the individual distances to the center of mass:

$$R = r_1 + r_2$$

From the above two equations we can derive expressions for the distances r_1 and r_2 from the center of mass in terms of the masses m_1/m_2 and their distant r apart. From the first equation, it is apparent that $r_2 = r_1 (m_1/m_2)$. Substituting that result for r_2 in the second equation and solving for r_1 we have,

$$r_1 = R - \frac{m_1}{m_2} r_1$$

Locating the Center of Mass

which, after rearranging can be written as,

$$r_1 = R\left(\frac{m_1}{m_1 + m_2}\right)$$
$$\underbrace{\phantom{r_1 = R\left(\frac{m_1}{m_1 + m_2}\right)}}_{\text{Distance of } m_1 \text{ from the center of mass}}$$

The corresponding expression for r_2, obtained by the same reasoning, is:

$$r_2 = R\left(\frac{m_1}{m_1 + m_2}\right)$$
$$\underbrace{\phantom{r_2 = R\left(\frac{m_1}{m_1 + m_2}\right)}}_{\text{Distance of } m_2 \text{ from the center of mass}}$$

These equations tell us the respective distances of the two masses from their common center of mass. Approaching the matter intuitively, let us look at the first of the final two equations and roughly estimate the Earth's distance from the system's center of mass. Since the Earth (call it m_1) is over 80 times heavier than the Moon (m_2), and thus rather dwarfs it in mass, we would expect that the distance of m_1 from its center to the barycenter (r_1) would be a little less than one eightieth the distance to the Moon from the center of the Earth. Since the Moon is about 60 Earth radii from Earth, the barycenter should be a little less than 6/8th or ¾ of an Earth radii from the Earth's center. We will see below if this is roughly correct.

Finding the Earth–Moon Barycenter

Problem Find the distance from the Earth's center to the barycenter of the Earth–Moon system.

Given

m_1	The mass of the Earth: 5.9736×10^{24} kg
m_2	The mass of the Moon: 7.349×10^{22} kg
r_1	The radius of the Earth's orbit around the center of mass
r_2	The radius of the Moon's orbit around the center of mass
R	The mean distance between the centers of Earth and Moon: 3.844×10^8 m
$r_1 = R\left(\dfrac{m_2}{m_1 + m_2}\right)$	Equation for locating distance r_1 from m_1 to the center of mass

Assumptions We assume the Earth and Moon are spheres of uniform density in circular orbits, which is approximate enough for our purposes. We are also using the mean distance between Earth and Moon, and assuming a two-body system, ignoring for the moment the (significant) perturbing effects of the Sun and also the planets which render calculations of the Moon's actual movement around the Earth more complex.

198   10 A Binary System Close to Home: How the Moon and Earth Orbit Each Other

Method We need to find the radius r_1 of the Earth's orbit around the center of mass. We can do this by solving the given equation for r_1. This tells us what fraction of the distance between Earth and Moon equals the ratio of the Moon's mass to the mass of both bodies. From that and the equation $R = r_1 + r_2$ we can determine r_1 as well. Here again is a diagram of the system, with radii shown:

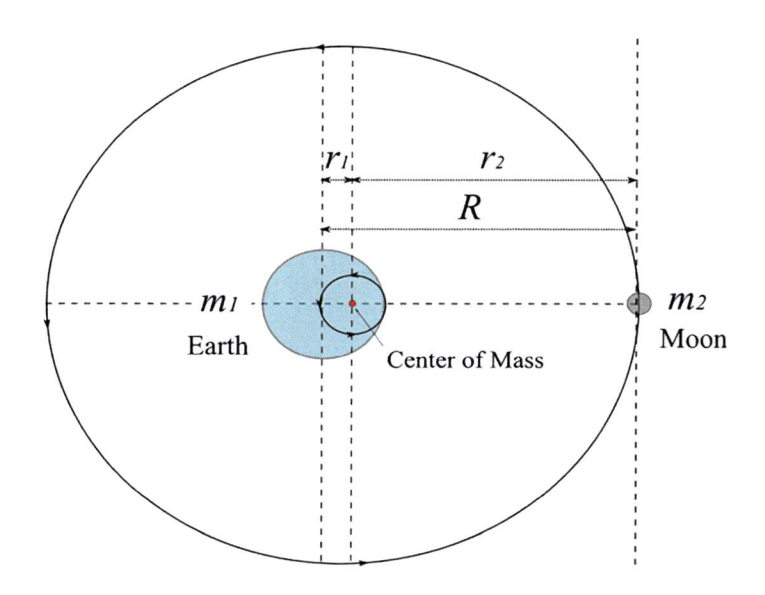

Earth and Moon Orbital Geometry

Calculations Substitute the given values for the masses and mean distance between them:

$$r_1 = R\left(\frac{m_2}{m_1 + m_2}\right)$$

$$r_1 = 3.844 \times 10^8 \left(\frac{7.349 \times 10^{22}}{5.9736 \times 10^{24} + 7.349 \times 10^{22}}\right)$$

$$r_1 = 4.6716 \times 10^6 \text{ meters}$$

4,671.6 kilometers

Observations
1. Since the radius of the Earth is 6,378 km, the distance from the center of the Earth to the barycenter is indeed a little less than 3/4ths of an Earth radius toward the Moon, from mid-Earth. To find the distance from the center of the Moon to the barycenter of the system, we could also do this:

$$r_2 = R\left(\frac{m_1}{m_1 + m_2}\right)$$

The result is about 379,728 km. This is about 98 % of the way to the Moon.

2. The barycenter of the Jupiter – Sun system, or any other system, can be found in the same way. The radius of r_1 (the Sun's small orbit) is to the distance between their centers, R, as the mass of the Sun is to their combined masses. The mass of Jupiter is only about .0009536 the combined mass of the system. (Jupiter's mass is 1.8986×10^{27} kg and the Sun's is 1.9891×10^{30} kg.) So we would expect the barycenter of the Jupiter–Sun system to be about that portion of the mean distance between the Sun and Jupiter, of about 778,570,000 km, or about 742,438 km from the center of the Sun. We could have estimated it pretty well just by taking the ratio of the Jupiter's mass to that of the Sun, or 1/1,048, since even Jupiter's mass is relatively tiny compared to the star. The result is 743,000 km. Since the Sun's radius is 696,000 km, the Jupiter-Sun barycenter lies just a little outside the Sun's surface (photosphere).

3. The Earth hardly budges the mighty Sun. The Earth-Sun mass ratio is about 1:333,000. Since their mean distance apart (semi-major axis) is 149.6 million kilometers, the barycenter of the Earth-Sun system is a mere 449 km from the center of the Sun. Again, when comparing masses of enormous disparity, we can effectively ignore the smaller mass, and estimate by using the ratio of their masses multiplied by the separation.

4. The foregoing discussion concerned only two-body systems. The situation for multi-body (or n-body) barycenters is somewhat more complex. But center of mass determinations at any moment follow the same procedure. One need only calculate one pair's barycenter, then the third from that, as if the barycenter of the first two were the second body. The motion of the barycenter in a multibody system can be interesting. Because of the movements of all the planets around the Sun, each with their own orbital periods, and each with their own perturbing effects, the solar system barycenter moves about the Sun in a complex path. Binary systems may and often do follow other more complicated paths around their barycenters. In some cases their elliptical orbits around the barycenter may even overlap, but their shapes will always be similar.

Deriving Equations for Velocities Around the Center of Mass in Circular Binary Orbits

There are various ways of determining the velocities in circular orbits around the center of mass. Here we demonstrate two, and it is illuminating to compare them. For each, we must ask, what is the common link between the two masses in their orbits? What physical laws or principles relate them to each other or apply to each which would give us the mathematical tools to solve the problem? Let us assume their masses (m_1 and m_2) significantly differ, and that they therefore have different orbital radii (r_1 and r_2) and different velocities (v_1 and v_2) (which is our task to determine).

Problem Derive equations for finding the velocities of two masses in circular orbits around their common center of mass, in two ways: (1) first, in terms of their masses and their relative velocity only; (2) second, using Newton's laws and the equilibrium of accelerations of each. Then, (3) combine the equations and solve for relative velocity.

Given

For the first part of the problem
$V = 2\pi r/P$	General equation for finding circular velocity, where P is the period
$v = v_1 - v_2$	Relative velocity of objects moving in the same direction; where two bodies are orbiting around center of mass, they are going in opposite directions, and we may write relative velocity here as $v = v_1 - (-v_2)$ or $v = v_1 + v_2$

For the second part of the problem
Gm_1/R	Gravitational acceleration acting on m_2 from m_1, where the masses are R distance apart
v^2/r_2	Acceleration of m_2 as it orbits the center of mass at r_2

For the third part of the problem
$r_1 = R\left(\frac{m_2}{m_1+m_2}\right)$	Equation for locating distance r_1 from m_1 to the center of mass

Assumptions Same assumptions as in previous problem.

Method For the first part of the problem, if the distances to the center of mass are known, then the velocities may be determined for each orbit around the center of mass using the notion that their periods are the same and that the relative velocity v is the difference of the individual velocities, v_1 and $-v_2$ (the negative indicates that v_2 moves in a direction opposite v_1). For the second part of the problem, we can use Newton's laws to equate gravitational and centripetal accelerations of each mass, and solve for their respective velocities. Finally, we can use the center of mass equation to find relative velocity, by substituting the value of radius for r_1.

Calculations *First part of the problem*: We know the periods of each mass are the same, so,

$$P = \frac{2\pi r_1}{v_1} = \frac{2\pi r_2}{v_2}$$

$$\frac{v_1}{v_2} = \frac{r_1}{r_2}$$

and, since the definition of the center of mass is,

$$m_1 r_1 = m_2 r_2$$

we deduce that

$$m_1 v_1 = m_2 v_2$$

Now we wish to solve for v_1, but this equation has two variables for velocity, so we need another equation in order to solve for one of them. We can use the equation for the relative velocity of the two masses moving in opposite directions, $v = v_1 - (-v_2)$ which becomes $v = v_1 + v_2$. First we isolate v_1:

$$v_1 = v_2 \left(\frac{m_2}{m_1}\right)$$

Now, we substitute this equation for v_1 into the equation $v = v_1 + v_2$:

$$v = v_2 \left(\frac{m_2}{m_1}\right) + v_2$$

When we work through the algebra, we arrive at the result for the velocity of m_1 about the center of mass:

$$v_1 = v \left(\frac{m_2}{m_1 + m_2}\right)$$

By the same analysis, the velocity of m_2 around the center of mass is,

$$v_2 = v \left(\frac{m_1}{m_1 + m_2}\right)$$

where again v is the *relative* velocity of the two bodies.

Second part of the problem: Here we use a different method, and consider the gravitational and centripetal accelerations of the two masses. The gravitational acceleration imparted by the mass m_1 on m_2 causes the equivalent centripetal acceleration of m_2 as it bends it to its orbit, so we may put the expressions for each equal to each other. Notice the radius of the orbit for the centripetal acceleration is the smaller orbit r_2, but the gravitational acceleration is, as always, determined by the distance between them.

$$\underbrace{\frac{v_2^2}{r_2}}_{\substack{\text{Centripetal acceleration of} \\ m_2 \text{ as it orbits the center} \\ \text{of mass at distance } r_2}} = \underbrace{\frac{Gm_1}{R^2}}_{\substack{\text{Gravitational acceleration} \\ \text{of } m_2 \text{ imparted by the} \\ \text{gravitational pull of } m_1}}$$

We want the left side to be just velocity, so we multiply each side by r_2:

$$v_2^2 = \frac{Gm_1}{R^2} r_2$$

Taking the square root of each side yields the circular orbital velocity of m_2 about the center of mass:

$$v_2 = \frac{\sqrt{Gm_1 r_2}}{R} \quad \left.\right\} \text{Velocity of } m_2 \text{ around center of mass}$$

By the same reasoning, the velocity of m_1 is,

$$v_1 = \frac{\sqrt{Gm_2 r_1}}{R} \quad \left.\right\} \text{Velocity of } m_1 \text{ around center of mass}$$

Third part of the problem: The above equations appear quite disparate in form and origins, but they are easily reconciled. Since they each represent velocities around the center of mass, let us pick one set and equate them to see what falls out. The two equations we found for v_2 are:

$$v_2 = v\left(\frac{m_1}{m_1 + m_2}\right) \qquad v_2 = \frac{\sqrt{Gm_1 r_2}}{R}$$

Equating them, we have,

$$v\left(\frac{m_1}{m_1 + m_2}\right) = \frac{\sqrt{Gm_1 r_2}}{R}$$

Squaring each side, cancelling common terms and rearranging yields this:

$$v^2 \left(\frac{m_1}{m_1 + m_2}\right) = \frac{G(m_1 + m_2) r_2}{R^2}$$

A nice simplification follows from the identity we discussed earlier, $r_2 = R(m_1/(m_1 + m_2))$, and results in an equation for the *relative* velocity of either mass around the other, at distance R apart:

$$v^2 = \frac{G(m_1 + m_2)}{R}$$

$$v = \sqrt{\frac{G(m_1 + m_2)}{R}}$$

Deriving Equations for Velocities Around the Center of Mass in Circular...

Observations

1. This last equation is the relative velocity of one mass orbiting another in a circular orbit. It is given with reference to one mass, as if we were on one of the masses, observing from at that fixed point. It is an important equation, and should be committed to memory. We will use it as a starting point later when we discuss velocities in elliptical orbits.

2. One frequently encounters problems involving binary systems where the calculations would be much simpler if we could change our frame of reference from the center of mass to the primary mass of the system. Instead of dealing with individual, changing distances of each mass to the center of mass, we would deal with one distance parameter, which is just the distance between them. A clever technique of calculating the *reduced mass* of the secondary does just this: it simply changes our frame of reference to one of the orbiting masses, as if it were standing still, and from there we can examine the motion of the other mass orbiting around it.

    Recall that according to Newton's Second Law, $F = mf$, where the force per unit mass is the acceleration. According to Newton's Third Law, each force effects the other equally and in the opposite direction, so $m_1 f_1 = -m_2 f_2$. Since acceleration is force per unit mass, this relation can be written this way:

$$f = \frac{F}{m_1} + \frac{F}{m_2}$$

$$f = F\left(\frac{1}{m_1} + \frac{1}{m_2}\right)$$

$$f = F\left(\frac{m_1 + m_2}{m_1 m_2}\right)$$

Or, rearranging to express Newton's Second Law where force is a function of mass and acceleration,

$$F = \left(\frac{m_1 m_2}{m_1 + m_2}\right) f$$

We can interpret this to mean that the system acts as if mass has been shifted to a stationary primary, the secondary having a *reduced mass* of $m_1 m_2/(m_1 + m_2)$. For the sake of shorthand, we will call the reduced mass μ':

$$\mu' = \frac{m_1 m_2}{m_1 + m_2}$$

3. We can relate reduced mass to relative velocity. If v is again relative velocity, and μ' is the reduced mass of the secondary, then

$$v = v_1\left(\frac{m_1}{\mu'}\right) \quad v = v_2\left(\frac{m_2}{\mu'}\right)$$

Calculating the Orbital Velocities of the Earth and Moon Around Their Center of Mass

Above we derived equations in two ways to find the velocities of two masses around their center of mass. We will now use one of the methods to determine the orbital velocities of the Earth and Moon around their center of mass.

Problem Using the second method given above, calculate the individual, non-relative, orbital velocities of the Earth and Moon around their center of mass.

Given

$v_2 = \sqrt{\frac{GM_1 r_2}{R}}$	Center of mass velocity equation for the Moon, m_2
$v_1 = \sqrt{\frac{Gm_2 r_1}{R}}$	Center of mass velocity equation for the Earth, m_1
m_1	Mass of the Earth: 5.9736×10^{24} kg
m_2	Mass of the Moon: 7.349×10^{22} kg
r_1	Radius of the Earth's orbit around the center of mass: $r_1 = 4.6716 \times 10^6$ m
r_2	Radius of the Moon's orbit around the center of mass: $r_2 = 3.7973 \times 10^8$ m
R	The mean distance between the centers of Earth and Moon: 3.844×10^8 meters
G	Gravitational constant: $6.674 \times 10^{-11} \text{N} \cdot \text{m}^2/\text{kg}$

Assumptions We assume the same things as in the previous problem: that the Earth and Moon are spheres of uniform density in circular orbits, in an ideal two-body system, unaffected by the gravitational perturbations of the Sun or planets.

Method Simply substitute the appropriate values into the equations and solve for velocity.

Calculations The velocity of the Moon around the CM is:

$$v_2 = \sqrt{\frac{Gm_1 r_2}{R}}$$

$$v_2 = \frac{\sqrt{(6.674 \times 10^{-11})(5.9736 \times 10^{24})(3.7973 \times 10^8)}}{3.844 \times 10^8}$$

$$v_2 = 1,012.2 \text{ m/s}$$

The velocity of the Earth around the CM is:

$$v_1 = \sqrt{\frac{Gm_2 r_1}{R}}$$

$$v_1 = \frac{\sqrt{(6.674 \times 10^{-11})(7.349 \times 10^{22})(4.6716 \times 10^6)}}{3.844 \times 10^8}$$

$$v_1 = 12.45 \text{ m/s}$$

Observation It is interesting to see that, while the Moon travels in its large orbit at the speed of a rifle bullet, the Earth lumbers along at a little less than 12 ½ meters per second in its small orbit, not even 28 mph, about the speed of a car along a residential street.

The Relations Between Masses, Accelerations and Period in Circular Binary Orbits

Two bodies coupled in an orbit around their common center of mass will naturally have the same period. It makes sense, but it can be instructive to show why that is so. In the case of the Earth and Moon, which we are idealizing as consisting of circular orbits, we saw that each one's orbit is a different size: the Moon revolves around its big orbit of radius r_2 as the Earth moves around its little orbit of radius r_1. Let us get "inside the math" a little and demonstrate by independent means that their respective periods are in fact the same, and show how their gravitational accelerations can be combined to derive the same period again. The latter statement is consistent with what we have been saying: we may treat the Earth as fixed, mathematically, with the Moon revolving around the Earth, and get the same result as if we used the absolute motions of each around the barycenter. In other words, if we shift our frame of reference to Earth and consider just the Moon's motion relative to a fixed Earth, the period will be unchanged. For variety of mathematical experience we will show this using angular notation, as we did in Chap. 8.[2] Recall that for circular orbits, and here referring to the Earth and Moon, respectively,

$$\frac{Gm_2}{R^2} = \omega^2 r_1 \qquad \frac{Gm_1}{R^2} = \omega^2 r_2$$

$$\underbrace{f_1 = \omega^2 r_1}_{} \qquad \underbrace{f_2 = \omega^2 r_2}_{}$$

Gravitational and centripetal accelerations of the Earth in its orbit around the center of mass

Gravitational and centripetal accelerations of the Moon in its orbit around the center of mass

Let us now add the accelerations such that their mutual acceleration is written as, $f_r = f_1 + f_2$. Given this, we will substitute the above identities for those accelerations, and factor the result:

$$f_r = \omega^2 (r_1 + r_2)$$

---

[2] Recall that we often use g to represent the gravitational acceleration at various distances from the mass in question. Here we return to the f notation for the gravitational acceleration. The Earth's gravitational acceleration imparted to the Moon at the lunar distance (caused by the gravitational pull of the Earth) is here denoted by f_2. This is the same as the Earth's g at lunar distance. The gravitational acceleration imparted to the Earth (from the Moon's pull of gravity) is denoted by f_1. This is the same as the Moon's g at Earth distance.

Since $\omega = 2\pi/P$, the above equation can be solved for period P,

$$P = 2\pi\sqrt{\frac{(r_1 + r_2)}{f_r}}$$

Because $R = r_1 + r_2$, the period of the orbiting masses does in fact reduce to a function of the mutual acceleration and center-to-center distance between them:

$$P = 2\pi\sqrt{\frac{R}{f_r}}$$

This is the period of both masses orbiting around the center of mass. The acceleration term is,

$$f_r = \frac{G(m_1 + m_2)}{R^2}$$

Substituting that term in the period equation yields Kepler's Third Law. It is mathematically equivalent to the period of one object of negligible mass revolving around the other where the total mass of the primary $m_1 + m_2$.

Problem Using the equations below, and the individual gravitational accelerations of the bodies induced by the mass of each upon the other at distance R apart, show *numerically* that: (1) the periods of the Moon orbiting at radius r_2 from the barycenter, and the Earth orbiting at radius r_1 from the barycenter are equal; and (2) the same result follows when using their mutual gravitational accelerations and total distance r between them.[3] For this problem we will again denote the Earth's and Moon's masses as m_1 and m_2, respectively. Gravitational accelerations will again be denoted by f (force per unit mass) with an appropriate subscript.

Given

$P = 2\pi(r_{cm}/f_c)^{1/2}$	Equation for finding the period of an object in circular orbit around a center of mass from the centripetal acceleration f_c and the radius of the center of mass orbit, r_{cm}. Below for f_c we will use the centripetal accelerations of the Earth f_1 and Moon f_2
$f_1 = \frac{Gm_2}{R^2}$	Equation for gravitational acceleration experienced by the Earth
$f_2 = \frac{Gm_1}{R^2}$	Equation for gravitational acceleration experienced by the Moon
m_1	Mass of the Earth: 5.9736×10^{24} kg
m_2	Mass of the Moon: 7.349×10^{22} kg
r_1	Radius of the Earth's orbit around the center of mass: $r_1 = 4.6716 \times 10^6$ m
r_2	Radius of the Moon's orbit around the center of mass: $r_2 = 3.7973 \times 10^8$ m
R	Mean distance between the centers of Earth and Moon: $R = 3.844 \times 10^8$ m
G	Gravitational constant: 6.673×10^{-11} N m²/kg

---

[3] For the illustrative purposes of the problem, we will not use Kepler's Third Law to find the period, but rather use the two-step process of first finding the individual gravitational accelerations of the Earth and Moon, then finding the period.

The Relations Between Masses, Accelerations and Period in Circular Binary Orbits

Assumptions Same as the previous problem.

Method For the first part of the problem, we determine the accelerations imparted to each body due to the gravitational pull of the other. These are each calculated at their mean distance R from each other. (It is important to realize that we use the center-to-center distance R to calculate their respective accelerations, f_1 and f_2). For calculating separate periods, we will use their respective orbital radii from the center of mass, r_1 and r_2. Then we insert the applicable accelerations of each body into the period equation and compute the period of the Moon in its large orbit about the barycenter, at distance r_2, and the period of the Earth about its smaller orbit around the barycenter at distance r_1. For the second part of the problem, we sum the accelerations to find the mutual acceleration of each toward each other.

Calculations *First part of the problem*: We find the values of the accelerations imparted to the Earth f_1 and the Moon f_2, caused by the gravitational pulls of the Moon and Earth, respectively.

Earth

$$f_1 = \frac{Gm_2}{R^2}$$

$$f_1 = \frac{(6.673 \times 10^{-11})(7.349 \times 10^{22})}{(3.844 \times 10^8)^2}$$

$$\underbrace{f_1 = 3.3188 \times 10^{-5}}$$

Acceleration of the Earth due to the Moon's gravity

Moon

$$f_2 = \frac{Gm_1}{R^2}$$

$$f_2 = \frac{(6.673 \times 10^{-11})(5.9736 \times 10^{24})}{(3.844 \times 10^8)^2}$$

$$\underbrace{f_2 = 2.6976 \times 10^{-3}}$$

Acceleration of the Moon due to the Earth's gravity

With these accelerations, we can calculate their periods around the center of mass:

Earth (small orbit)

$$P = 2\pi \sqrt{\frac{r_1}{f_1}}$$

$$P = 2\pi \sqrt{\frac{4.6716 \times 10^6}{3.3188 \times 10^{-5}}}$$

$$P = 2.357 \times 10^6 \text{ seconds}$$

$$P = 27.28 \text{ days}$$

Moon (large orbit)

$$P = 2\pi \sqrt{\frac{r_2}{f_2}}$$

$$P = 2\pi \sqrt{\frac{3.7973 \times 10^8}{2.6976 \times 10^{-3}}}$$

$$P = 2.357 \times 10^6 \text{ seconds}$$

$$P = 27.28 \text{ days}$$

The period of the Earth in its small orbit around the center of mass is precisely the same as the period of the Moon around its much larger orbit, which is the lunar *sidereal* month (measured against the fixed stars). The stronger pull of the Earth holds the Moon in a greater orbit. The lesser pull of the Moon holds the Earth in a relatively tiny orbit. And the periods of each are the same, as we have assumed.

Second part of the problem: Here we will ascertain numerically if the period is the same when we add accelerations and use the mean distance between the bodies rather than their respective barycenter distances:

$$f_r = f_1 + f_2$$

$$f_r = 3.3188 \times 10^{-5} + 2.6976 \times 10^{-3}$$

$$\underbrace{f_r = 2.7308 \times 10^{-3}}_{\text{Mutual gravitational acceleration of Moon and Earth}}$$

Insert this in the period equation, using the center-to-center radius R rather than distances to barycenter, and we arrive at the result we obtained above:

$$P = 2\pi\sqrt{\frac{R}{f_r}}$$

$$P = 2\pi\sqrt{\frac{3.844 \times 10^8}{2.7308 \times 10^{-3}}}$$

$$P = 2.35736 \times 10^6 \text{ seconds}$$

$$P = 27.2843 \text{ days}$$

Observations

1. The above exercise shows numerically that the combined accelerations over the total distance R between the masses yields the same period as the individual accelerations calculated with respect the shorter distances of each to the center of mass.
2. A quick comparison of the acceleration numbers on a calculator will show that the Earth's pull on the Moon is a little more than 81 times the Moon's pull on the Earth, which corresponds exactly to the ratio of their masses. Likewise the distances to their common center of mass: the center of the Moon is a little more than 81 times farther from the barycenter than the center of the Earth is. Thus the masses, barycenter distances and accelerations all bear the same ratio of about 81:1.

   Generalizing the relations of mass, velocity, radius and acceleration in circular orbits, in simple terms, for any binary pair of masses m_1 and m_2,

$$\frac{v_1}{v_2} = \frac{r_1}{r_2} = \frac{f_1}{f_2} = \frac{m_2}{m_1}$$

Thus, if we assume m_2 is the smaller mass, its velocity is greater, its orbital radius is bigger, and the gravitational acceleration *acting on it* is stronger, than the larger mass, whose velocity, radius and acceleration upon it are smaller.[4] These relations work for binaries in circular orbits. We will see later how they are modified for elliptical orbits. One would not, however, use them to compare, say, the orbits of Earth and Mars, since those bodies are not orbiting each other.

---

[4] Be careful of notation. The terms f_1 and f_2 are here the accelerations of m_1 and m_2. They are not here the forces induced by the masses m_1 and m_2.

3. The acceleration of the Earth induced by the Moon's mass can also be checked by looking at its orbital velocity, using the equation for centripetal acceleration discussed earlier: $f = 4\pi r/P^2$. Given that the lunar sidereal period is 2.537×10^6 s, and the radius of the Earth's orbit around the center of mass is 4.6716×10^6 m, the acceleration of the Earth induced by the Moon resulting from application of that equation is 3.32×10^{-5} m/s^2 as we saw before. The reader should feel free to check whether using this equation with respect to the Moon's longer orbit around the center of mass yields the correct greater lunar acceleration induced by the Earth.

4. Now that we have carefully calculated the orbital periods from the known masses of the Earth and the Moon, we can admit that the published orbital period of the Moon and Earth about their center of mass is slightly different from this. The published orbital period is 27.3217 days, which is slightly longer than the period of 27.2843 days we derived above. How could this be? What could effectively slow the Moon so? Whenever an orbit is off from a pure Keplerian period, or velocity, eccentricity or other orbital characteristic – that is, in a quantity that differs from the result one would expect from the straightforward application of Kepler's Third Law, one must suspect an external influence or disturbance. We are not concerned about friction, as in the case of a satellite in low-Earth orbit experiencing atmospheric drag. Nor can we credit the oblateness, or flattening of the Earth around the equator. The main reason is the Sun and to a much lesser degree, the other bodies in the solar system. The gravitational pull of the Sun on the Earth–Moon system creates *perturbations* in the lunar orbit. There are in fact a variety of different gravitational pulls of the Sun on the Moon, depending on where it is in its orbit around the Earth. The differences in these pulls in comparison to the Sun's pull on the Earth, called *tidal* influences, create disturbing effects on the lunar orbit which make the orbit somewhat "non-Keplerian."

Exercises: The Fascinating Orbits of Pluto and Charon Our knowledge of Pluto is comparatively recent, and illustrates how the rapid growth of technology has changed astronomy. The discovery of Pluto in 1930 began, like the discovery of Neptune, from attempts to understand anomalies in the orbits of Uranus and Neptune. Percival Lowell, the founder of Lowell Observatory in Flagstaff, Arizona (and unfortunately more widely known for his insistence on the presence of canals on Mars) had initiated the search for "Planet X" about 15 years earlier, and had derived hypothetical elements for the orbit. Clyde Tombaugh, a diligent Midwestern farm boy hired on by the observatory to help analyze photographic plates, spent many hours at a machine called a blink comparator, peering at plates of star fields, searching for the mysterious planet. Identical fields of stars were photographed on different days and examined with the comparator. The comparator rapidly illuminates one plate then another (though the eye sees only one continuous image). Changes in position of an object on the plate become apparent as it would appear to jump back and forth from one position to the other while the background stars remain fixed. Tombaugh thought that pictures taken about 6 days apart would be sufficient to show any movement of a distant, slow-moving planet. In January of that year, one faint speck of light was seen to have moved amid the multitude of stars in a particular region of Gemini. It was confirmed that that spot was the culprit, moving slowly almost 40 AU from the sun. Finding Pluto was the first significant discovery in the solar system since Neptune was found (through analysis of irregularities in the orbit of Uranus) in 1846. Beyond the basic facts of its orbit (its period is about two and a half centuries), and rough estimates of its small size, almost nothing else was known about this trans-Neptunian member of our solar system. But then, in 1978, James Christy at the U.S. Naval Observatory announced that deformed images of Pluto on photographic plates going back to the 1960s appeared to him to have a pattern, as if they moved periodically around it, indicating the possibility of a moon. Mutual eclipses of Pluto and this moon later confirmed its existence, and in the early 1990s, the Hubble Space Telescope finally resolved its relatively large satellite, named Charon. With adaptive optics, Charon was subsequently detected by ground-based telescopes. Two smaller moons, named Nix and Hydra were discovered in 2005, each orbiting the system's barycenter. Two other small moons of pluto have been discovered since, bringing the total to five moons. Over the years, and with the discovery of several other small bodies, it had become clear that Pluto (now called a "dwarf planet" by the IAU) was just one of innumerable primordial objects in the so-called Kuiper Belt of trans-Neptunian objects, whose character and origins could be key to understanding the formation of the solar system. In 2006, NASA launched the New Horizon's Mission to explore Pluto and Charon and the rest of the plutonian family and the Kuiper Belt. New Horizons received a gravity assist from Jupiter in 2007 and arrives at Pluto in mid-2015. NASA says, "The [Pluto–Charon] pair form a binary planet, whose gravitational balance point is between the two bodies. Although binary planets are thought to be common in the galaxy, as are binary stars, no spacecraft has yet explored one. New Horizons will be the first mission to a binary object of any type."[5] The following exercise will explore the dynamics of the fascinating Pluto–Charon binary system.[6]

Problems

1. Careful research of photographic images has indicated that the Charon-Pluto mass ratio is .1165 ($\pm$ .0055). With this information, ignoring any effects of Nix and Hydra, and using the notation developed in the chapter, what is the ratio of their distances from the barycenter of the system?
2. The combined mass of the Pluto-Charon system is estimated to be 1.457 10^{22} kg ($\pm$ .0009). What are the individual masses?
3. The semi-major axis of the Pluto-Charon system is only 19,571.4 km. What are the distances of Pluto and Charon from their barycenter, in kilometers? What is their ratio? Draw a sketch on any convenient scale you chose of the respective masses orbiting at the correct distances to the barycenter.

---

[5] See generally, http://pluto.jhuapl.edu/.
[6] The data on the Pluto–Charon system for these problems is taken from Buie et al. [1].

4. Pluto's estimated radius is 1,153 km. How many plutonian radii are there: (a) to the center of its orbit; (b) to Charon.
5. What is the ratio of the velocity of Charon to the velocity of Pluto in their orbits around the barycenter?
6. Calculate the respective velocities of Pluto and Charon in kilometers per second in their respective orbits around the barycenter. Do the ratios of velocities you found accord with your previous answer? Interpret your results.
7. What is the relative velocity of the two bodies? To check your work, find the relative velocity in two different ways.
8. What is Charon's reduced mass?
9. Does Charon's reduced mass times the relative velocity of Pluto and Charon yield the same value as Charon's mass times its velocity around the barycenter?
10. What are the values of the gravitational accelerations acting on the Pluto and on Charon, caused by the Charon and Pluto, respectively? Using these values, (a) calculate the periods of Pluto and Charon around their center of mass, in days; (b) confirm the period of the bodies, in days, using the combined acceleration and semi-major axis distance; and (c) confirm the period of the system using Kepler's Third Law.

Reference

1. Buie MW, Grundy WM, Young EF, Young LA, Stern AS (2006) Orbits and photometry of Pluto's satellites: Charon, S/2005 P1, and S/2005 P2. Astronom J 132:290–298, http://iopscience.iop.org/1538-3881/132/1/290/pdf/205133.web.pdf

Chapter 11
Using Kepler's Third Law to Find the Masses of Stars and Planets

How do we know the mass of a planet, asteroid, or of a distant star? If we are told that Saturn is light enough to float on water, that a tablespoon of neutron star stuff would weigh more than the Earth, or that there is a black hole at the center of our galaxy, how are those facts known? If you say, "You need to somehow use gravity as a probe to determine mass" you will be right. There are two ways to do this, one difficult and often uncertain; the other quite precise, depending on the quality of the observational data. In the first method, the mass of a body can be inferred by its perturbing effects on other bodies. This means there must be well defined orbits of other bodies that have passed close enough to the mystery mass to figure it out. Of course, the perturbed body need not be a another orbiting planet: in the rare case that a comet (or even a space vehicle) passes close by, the encounter may allow measurement of how much the body bends the visitor's path. The second, vastly preferable method employs Kepler's Third Law. It requires a body orbiting the mass we want to measure. That is, far and away, the most precise measuring tool of a body's mass. Without another body whirling around the star, planet or asteroid, one must fall back on other, imperfect means.

The Once Unknown Mass of Pluto

Pluto, discovered in 1930, illustrates the question. Before the discovery of Pluto's moon Charon, astronomers initially conjectured Pluto's size from its perturbing effects on Neptune, which were quite uncertain, and by guessing its size from its magnitude (once its distance was known) and supposed albedo (reflectivity). But reckoning mass from size is tricky business for a small, very distant body (even in the cases where a tiny disk may be shown in a telescope), since it is hard to know whether one is looking at a large, dusky sphere or a small, ice-covered, brilliant one. Fortuitous events can sometimes help discern size: a lucky line-up of satellite and planet (as in the case of Pluto and Charon in the late 1980s) can create eclipses and transits where the dip in total light from the system can be timed. Similarly, occasional, timed occultations with distant stars each can greatly aid in deducing size (as in the case with many asteroids and some Trans-Neptunian Objects, such as Eris). If size is constrained to within a plausible range, the mass of the object can sometimes be judged from its probable density and composition. This exercise can be interesting but is imprecise. If we don't know a body's composition, whether it is rocky, filled with the ice of some element, or layered with rock and ice or other substance, its mean density will be speculative.

In the case of Pluto, textbooks written even two decades after the planet's discovery were still hopelessly uncertain about its mass, guessing it to be at most about a tenth that of the Earth; off, it would later be found, by a factor of 40. Harold Spencer Jones, the Astronomer Royal from 1933 to 1955 at the Royal Greenwich Observatory, in England, wrote about Pluto in his 1951 book, *General Astronomy*:

> The mass of Pluto has been determined from the perturbations which it produces in the longitude and latitude of Neptune; the masses deduced from each co-ordinate were in close agreement, somewhat smaller than the mass of the Earth. This mass cannot reasonably be reconciled, however, with the diameter of $0''.23$ measured by Kuiper in 1950 under good conditions with the 200-inch reflector on Mount Palomar. The albedo, inferred from this diameter and the photovisual magnitude, is 0.17. The diameter is .46 times the Earth's, a value midway between Mars and Mercury. The volume is one-tenth that of the Earth; the mass derived from the perturbations of Neptune would therefore require a mean density for Pluto of 50, which is physically impossible. The mass of Pluto is not likely to be greater than one-tenth the Earth's mass. The dynamical determination depends to a considerable extent upon two observations of Lalande in 1795, which may be affected by large errors.[1]

Another decade later, the problem still had not been solved. Pluto's mass in one popular text was now said to be "nearly equal to the mass of the Earth," a ten-fold increase in the previous error.[2] The need to improve the mass estimate was acknowledged: "New observations of the perturbations of Neptune by Pluto, now in progress, may eventually resolve this problem."[3] Once Charon was discovered in 1978 and its close, rapid orbit eventually traced on photographs taken by excellent telescopes, the Pluto–Charon combined mass was ultimately determined to be less than 1/400th the mass of the Earth.[4]

---

[1] Harold Spencer Jones [1].

[2] Blanco and McCuskey [2].

[3] Ibid.

[4] The mass calculation of the Pluto + Charon is taken from Buie et al. [3]. Pluto appears to be about .002 mass of the Earth, or close to 1/460th of it.

Using Kepler's Third Law to Find Mass

With reliable data, even the amateur at his desk can use gravity as a probe to determine mass of planets, asteroids or other objects with admirable precision. Kepler's Third Law as modified by Newton can be reconfigured to solve for mass. Begin with the full statement of Kepler's Third Law:

$$P = \frac{2\pi}{\sqrt{G(m_1 + m_2)}} a^{3/2}$$

Now isolate the mass terms, first squaring each side:

$$P^2 = \frac{4\pi^2}{G(m_1 + m_2)} a^3$$

$$m_1 + m_2 = \left(\frac{4\pi^2}{G}\right) \frac{a^3}{P^2}$$

The sum of the masses equals a constant, $4\pi^2/G$, times the cube of the semi-major axis divided by the square of the orbital period, in units of meters, kilograms and seconds (the SI system). That's all there is to it. The real challenge is in obtaining accurate values for the variables of period and semi-major axis distance.

In the case of a star, the situation is no different. Again, the mass of a star of known spectral type and luminosity can be *approximated* from its position on the Hertzsprung-Russell diagram, or from the famous mass-luminosity relationship developed in the 1920s by A. E. Eddington, or by other means. But again, these methods are not exact, and it is only with binary star systems that astronomers obtain a precise picture of the combined masses of the system. It may take years of observations, however, to peg the period and semi-major axis of a double star system. The mass equation will yield the aggregate mass; the individual masses can often be teased out of the photographic or spectroscopic data on their motions if we can ascertain their center of mass, or their individual velocities, as in some of the problems below. Where the primary mass is overwhelmingly greater than its little satellite or planet, then the sum of the masses will be a good enough approximation of the mass of the primary alone.

If stellar spectroscopic observations reveal *velocity*, and we have an idea of the separation between the orbiting components, then one may be in luck and find mass that way. How is this so? Whether for a star or other body, the retooled Keplerian equation can easily give us mass if the velocity is known. Recall the equation for relative circular orbital velocity from the last chapter:

$$v = \sqrt{\frac{G(m_1 + m_2)}{R}}$$

Where R is the distance between the masses. This we solve for the sum of the masses, and we arrive at the useful relation,

$$m_1 + m_2 = \frac{v^2 R}{G}$$

And, where $m_1 \gg m_2$, then

$$m_1 \simeq \frac{v_2^2 R}{G}$$

where v_2 is the velocity of the small secondary mass. These variations on Kepler's Third Law are useful equations! Just by clocking the speed of a tiny whirling mass around its center of mass we can get at the total mass of the system. It can be applied to an earth satellite, a planetary probe orbiting another planet, an asteroid, a star, and even our entire galaxy. The physical laws stretch across the cosmos to tell us vital information regarding mass, if only we can spot one body orbiting another. The breathtaking power of Newton's gravitational theory, applicable in all reaches of the universe, is demonstrated time and again with great success.

Ratios of Ratios, All Is Ratios: The Historical Use of Kepler's Third Law to Determine Mass by Proportions

In the days before the gravitational constant was known, before the SI system was in wide use, or instruments existed that allowed measurements with any exactitude, evaluation of mass by ratios and proportions, in terms of solar masses (or Earth masses) was the coin of the realm. There is something refreshingly simple and intuitive about reasoning by proportions and approximations. Though texts nowadays rarely use proportional reasoning, facility with it enables understanding of the history behind the science and often gives greater insight into physical relationships. Take Kepler's Third Law expressed just in terms of its variables:

$$m_1 + m_2 \propto \frac{a^3}{P^2}$$

Historical investigations were limited to the known planets and satellites of our solar system. The primary mass of the Sun is overwhelmingly more massive that each of its planets; each planet orders of magnitude more massive than each satellite. In either case $m_1 + m_2 \approx m_1$ was (and is) a fair approximation and,

$$m \propto \frac{a^3}{P^2}$$

As we saw in Chap. 4, if one determines the Keplerian ratio of distance-cubed over period-squared for any planet (which we'll call its "Keplerian mass ratio") the result, assuming we use consistent units of measurement, is a *nearly* identical fraction, a constant, for each planet (discounting the small differences induced by the individual masses): this is the empirical relationship Kepler noticed that led to his law. This numerical fraction is unique to the solar system, because the primary mass is the Sun. Imagine a small particle of insignificant mass at any orbital distance from any body, be it the Sun, a planet, or distant star. The particle acts as a tracer of the mass of the primary: the ratio of its distance cubed to period squared will always be the same *for a given primary mass*. Another solar system with a different star of different mass would demonstrate another, unique, Keplerian mass ratio – another a^3/P^2 number – which would be the same for all the (presumed small) planets of that solar system.[5]

The special Keplerian mass ratio in a planetary system (where each $m_2 \ll m_1$), then, is a function of the primary star's mass. Similarly, all the satellites of a planet, be it one moon or many, will have their own, common, Keplerian mass ratio, as we saw, for example, with Jupiter's Galilean moons. This ratio will have one value for the moons of one planet, and another value for the moons of another planet. The Keplerian mass ratio will be individual to the particular satellite system, and purely a function of mass of the system. Again, since the moons of planets are only a small fraction of the mass of the primary (with our Moon and Charon being notable exceptions) the Keplerian mass ratios of planetary systems will in most cases be closely a function of the planet's mass alone.

Given these gravitationally determined relationships, one can calculate the relation of a planet's mass in our solar system to the mass of the Sun: *it is the ratio of the two Keplerian mass ratios*. In other words, the Keplerian mass ratio of the satellite system, determined by the *planet's* mass, can be compared to the Keplerian mass ratio of the solar system, determined by the *Sun's* mass; the ratio of the two yields the planet's mass in relation to the Sun. To illustrate, where a is the semi-major axis (or radius in circular orbits) and P the period, of a planet and its satellite:

This ratio:

$$\frac{a^3_{planet}}{P^2_{planet}} = \text{Keplerian mass ratio, determined by the mass of Sun}$$

---

[5] As we noted earlier, Kepler's Third Law may never have been discovered if the solar system had planets of very great mass (say, well in excess that of Jupiter), for then the clear a^3/P^2 relationship would not have been so apparent. The irregularities in the relationship, as we investigated in Chap. 8, would have been noticeable even in the crude instruments that Tycho Brahe used, and Kepler's Third Law might have been long delayed (and likely had another name). We were (and are!) fortunate that our system does not resemble one of the exoplanet systems where giant planets have been detected orbiting close to the parent stars. Of course, in that case, we probably would not be here to think about it.

divided by this ratio:

$$\frac{a^3_{satellite}}{P^2_{satellite}} = \text{Keplerian mass ratio, determimed by the mass of the orbited planet}$$

yields the mass of the planet in relation to the Sun. Or, more succinctly:

$$m_{\odot planet} = \left(\frac{a_{planet}}{a_{satellite}}\right)^3 \left(\frac{P_{satellite}}{P_{planet}}\right)^2$$

This equation in one step gives us the relative mass of a planet to the Sun. Its reciprocal gives the amount by which the Sun is more massive than the planet. Its use is not limited to the solar system of course: gravity is universal. For example, to find the mass of a star in solar masses circled by a planet:

$$m_{star+planet} = \left(\frac{a_{solar\ planet}}{a_{star's\ planet}}\right)^3 \left(\frac{P_{star's\ planet}}{P_{solar\ planet}}\right)^2$$

This tells us the mass of the combined star plus planet system which will usually closely approximate the mass of the star. One can derive other variations on Kepler's Third Law to accomplish the same result. Newton, in fact, used a simple alternative of this method in Book III of the *Principia* to approximate the masses of Earth, Jupiter and Saturn, using the solar mass as the unit of measurement.

How Newton Estimated Planetary Masses

It must have appeared stunning to others that with basic data and his profoundly original insight, Newton could purport to find the mass of the planets: to discern the actual quantity of matter and even density of these distant, unreachable objects that had been curious mysteries to mankind since ancient times.

In Proposition VIII of Book III of the *Principia*, Newton uses the term "weight" to refer to the force acting on an object, which depends on the mass of both the attracted and attracting bodies. The proposition states that in two spheres of assumed uniform density, the "weight of either sphere towards the other will be inversely as the square of the distance between their centers." This restatement of the inverse square law prepares us for what follows. In the first two corollaries to the proposition, Newton describes his method of finding the masses of orbiting objects:

PROPOSITION VIII, COROLLARY I

Hence we may find and compare together the weights of bodies towards different planets; for the weights of bodies revolving in circles about planets are (by Cor. II, Prop. IV, Book I) directly as the diameters of the circles and inversely as the squares of their periodic times...

How Newton Estimated Planetary Masses

> [B]y computation I found that the weight of equal bodies, at equal distances from the centres of the sun, of Jupiter, of Saturn, and of the Earth, were to one another as 1, $^1/_{1067}$, $^1/_{3021}$, and $^1/_{169282}$ respectively.[6]

PROPOSITION VIII, COROLLARY II

> Hence likewise we discover the quantity of matter in the several planets; <u>for their quantities are as the forces of gravity at equal distances from their centres</u>; that is, in the sun, Jupiter, Saturn, and the earth, as 1, $^1/_{1067}$, $^1/_{3021}$, and $^1/_{169282}$ respectively.[7]

Let us see what this means. From the first corollary:

> [T]he weights of bodies revolving in circles about planets are...directly as the diameters of the circles and inversely as the squares of their periodic times...

Since being proportional to a diameter is being proportional to a radius, we may put Newton's statement this way: the accelerations of circularly orbiting masses are proportional to their radii divided by the squares of the periods:

$$f_1 : f_2 :: \frac{a_1}{P_1^2} : \frac{a_2}{P_2^2}$$

This we recognize from the Corollary 2 of Proposition IV in Book I of the *Principia* which we considered in Chap. 7.[8] This relationship can be employed for any planet orbiting the Sun whose mass we wish to know, and for a satellite orbiting the planet. Then to compare them, we must mathematically make them the same distance away from their primaries:

> Hence likewise we discover the quantity of matter in the several planets; for their quantities are as the forces of gravity at equal distances from their centres...

Here Newton tells us that if we put masses the *same distance* from the orbited bodies, "at equal distances from their centres," then compare their gravitational accelerations toward their parent body, the different accelerations will be proportional to their different masses ("the quantity of matter" in each). By taking their ratio, one arrives at the mass of the planet in relation to the Sun.

Newton used the orbit of Venus as his reference distance to find the masses of Earth, Jupiter and Saturn. Those three planets were the only ones then known to have satellites; Mars' little moons were not to be discovered for another 200 years, and humanity knew of no other planets beyond Jupiter.

---

[6] Florian Cajori [4]. (Short title) *Principia*. (Underlining added for emphasis).

[7] Ibid., Cor. II (underlining added for emphasis).

[8] This expression, again, can be readily derived. Since circular velocity ($v = 2\pi r/P$) is proportional to radius over period ($v: r/P$), that proportion can be substituted for velocity in the equation for centripetal acceleration ($f = v^2/r$), such that for any circle, $f: r/P^2$. Here we use the more general notation for a as semi-major axis, which for a circle is its radius.

Newton's Calculation of the Mass of Jupiter

Problem Given Newton's data below, use the Newton's method given in the *Principia, Book* III, Proposition VIII, to find Jupiter's mass.

Given

$224^d.16\,\tfrac{3}{4}\,^h$	Venus' period given by Newton. This is 5392.75 h, or 224.6979 days. We will label the decimal day value P_v
72,400	Mean distance of Venus from the Sun, in hundred-thousandths of the Earth's distance from the Sun, according to Newton's *Phenomenon IV* of Book III. This translates to .724 astronomical units (AU). We will label it a_v
$16^d.16\,^{8/15\,h}$	Period of the "utmost circumjovial satellite revolving about Jupiter" according to Newton, which moon is now known as Callisto, equivalent to 440.5 h or 16.6889 days. We will call the decimal day value of Callisto's period, P_c
520,000	Mean distance of Jupiter from the Sun, in hundred-thousandths of the Earth's distance from the Sun, according to Newton's *Phenomenon IV* of Book III. This is equivalent to 5.2 AU. This is only used with reference to the value next given below
$8'\,16''$	Callisto's mean elongation from Jupiter, according to Newton. At Jupiter's given distance, this corresponds to about 1,250 hundred-thousandths of the Earth's distance from the Sun, or .0125 AU.[9] It will be denoted a_c

Assumptions We will assume as Newton did that the bodies are spheres of uniform density and that that no other forces are at work, and use Newton's data as given.

Method First, using Venus as a reference as Newton did, it is necessary to find the accelerations acting on Venus from the Sun (denoted f_{venus}), and on Callisto from Jupiter (denoted $f_{callisto}$). This means finding value of a/P^2 for both Venus and Callisto. Then, mathematically move Callisto as far from Jupiter as Venus is from the Sun, to find the new acceleration acting on that moon (denoted $f_{callisto\,new}$). This then enables comparison of the accelerations, respectively from the Sun on Venus and Jupiter on Callisto "at equal distances from their centres" of revolution. At that point, the task is to find the change in gravitational acceleration acting on Callisto by this mathematical relocation of Callisto from its orbit out to the distance of Venus. For this it is required to apply the inverse square law, since any change of distance will have an inverse squared effect on acceleration. Accelerations will not be in SI units of meters per second squared, but in AU per day squared. We could have chosen other units. The choice may be dictated by the scale of the problem and the data at hand. We are using the data Newton had available to him. Since we will

---

[9] One may imagine a circle with the Sun at the center and a radius of 5.2 AU drawn to Jupiter. Another line from the Sun drawn to Callisto at maximum elongation (distance from Jupiter) forms a tiny angle θ only $8'\,16''$ in extent, according to the best measures available to Newton. The reader can confirm that this angle in radians times Jupiter's distance yields the separation given of .0125 AU.

How Newton Estimated Planetary Masses   221

be dealing with proportions, it does not matter at all what units we use so long as we are consistent.

Calculations To find the acceleration of Venus in our AU/days system of units:

$$f_{venus} = \frac{a_v}{P_v^2}$$

$$f_{venus} = \frac{.724}{(224.6979)^2}$$

$$f_{venus} = 1.4334 \times 10^{-5}$$

Note that if Jupiter, or any other planet, were at Venus' distance from the Sun, it would experience the same acceleration by virtue of being the same distance from the center of the Sun. Similarly, to find the acceleration of Callisto at its regular distance from Jupiter,

$$f_{callisto} = \frac{a_c}{P_c^2}$$

$$f_{callisto} = \frac{.0125}{(16.6889)^2}$$

$$f_{Venus} = 4.488 \times 10^{-5}$$

Now we need to true-up the distances. We mathematically move Callisto a long distance away from Jupiter, all the way to the distance Venus is from the Sun. The change in gravitational acceleration will be the ratio of the squares of the distances:

$$f_{change\ factor} = \frac{a_c^2}{a_v^2}$$

$$f_{change\ factor} = \frac{(.0125)^2}{(.724)^2}$$

$$f_{change\ factor} = 2.981 \times 10^{-4}$$

This should be the factor reflecting the change in gravitational acceleration experienced by Callisto when we move as far from Jupiter as .724 AU. Let's apply that change factor to Callisto's acceleration:

$$f_{callisto\ new} = f_{change\ factor} \times f_{callisto}$$

$$f_{callisto\ new} = 1.33782 \times 10^{-8}$$

This Venus-distance acceleration on Callisto is understandably greatly less than it experiences in its regular orbit. We can now finally compare the ratios of the gravitational accelerations (determined by the respective masses of the attracting body) of the Sun on Venus, at Venus' distance (.724 AU), and Jupiter on Callisto, at the same Venus distance. The result will be the ratio of the mass of the Sun to the mass of Jupiter:

$$\frac{M_\odot}{M_J} = \frac{f_{venus}}{f_{callisto\ new}}$$

$$\frac{M_\odot}{M_J} = \frac{4.488 \times 10^{-5}}{1.33782 \times 10^{-8}}$$

$$\frac{M_\odot}{M_J} = 1072$$

This means that, using the given data, Jupiter is 1/1,072 the mass of the Sun. Or, there are about 1,072 Jovian masses in one solar mass.

Observations

1. This value is slightly different from that found by Newton of $1/1{,}067$, but working with Newton's method, as Dana Densmore expressed it, has "served our purpose in participating in Newton's reasoning."[10]
2. Newton used Venus' orbit as the convenient reference for all three planets, but this was not the only way to do it. He could have used the orbit of the parent planet in each case. For example, Newton could have used the acceleration of Callisto measured with reference to Jupiter's orbit. The steps would be: find the a/P^2 value for Jupiter where it is; find the a/P^2 value for Callisto; use the inverse square law to relocate Callisto to 5.2 AU from Jupiter; calculate their ratios.
3. That the method Newton used is reducible to the "ratio of Keplerian mass ratios" is easily shown by algebraic means. Let us apply it using Jupiter's orbit (rather than Venus') as the reference orbit. We will use Newton's data again to see what Newton would have come up with. Newton gives Jupiter's orbital period as

---

[10] Densmore [5]. Densmore gives a far more detailed and quite beautifully explicated account of Newton's method in her book. She arrives at a similar result to what we found above ($1/1{,}067$). She states: "It is not clear why we and Newton should not obtain exactly the same results if we both start with the same data. Perhaps there is a rounding discrepancy, or perhaps he used data he arrived at on another occasion (it is known that he did for one value that went into these calculations)." It is possible the discrepancy may be accounted for if one assumes that Newton used Jupiter, rather than Venus for the reference orbit when he actually ran the calculations. While his narrative clearly suggests that he used Venus, one can, as noted above, arrive at Newton's value more exactly this way.

4332.514 days.[11] The other values are the same as before. Using the equation below and inserting the numerical values for Jupiter and Callisto, we have,

$$m_{\odot planet} = \left(\frac{a_{planet}}{a_{satellite}}\right)^3 \left(\frac{P_{satellite}}{P_{planet}}\right)^2$$

$$m_{\odot planet} = \left(\frac{5.2}{.0125}\right)^3 \left(\frac{16.6889}{4332.514}\right)^2$$

$$m_{\odot planet} = 1068$$

This is actually closer to what Newton found, in Jovian masses (1,067).

4. Since Newton's day, the distances and ratios of the planets and their satellites have been determined with great refinement. Using data from the *Astronomical Almanac*, the ratios become:

$$m_{\odot planet} = \left(\frac{a_{planet}}{a_{satellite}}\right)^3 \left(\frac{P_{satellite}}{P_{planet}}\right)^2$$

$$m_{\odot planet} = \left(\frac{5.2028870}{.01258707755}\right)^3 \left(\frac{16.689}{4332.514}\right)^2$$

$$m_{\odot planet} = 1048$$

Notice there has been virtually no change to the deduced periods since Newton's time (the relatively rapid motion of planets and satellites enabling high precision on this, even from early times); however, the semi-major axis of Callisto is slightly larger than Newton had it. Because the distances are cubed, errors there have profound effect. Jupiter has "become" slightly heavier.

"Weighing" Jupiter the Modern Way

We could pick any Jovian satellite we wish to ascertain Jupiter's mass. Its little satellites spinning day and night around its great bulk give us an ideal tool for determining its mass using SI units. We will use Amalthea, one of Jupiter's tiniest inner moons. It has a nearly circular orbit (eccentricity .003) though that need not be a condition to using it.[12]

---

[11] *Principia*, Book III, Phenomenon IV, 405.
[12] Satellite data is from *The Astronomical Almanac for the Year 2012*, F2 (Washington: U.S. Govt. Printing Office, 2102).

11 Using Kepler's Third Law to Find the Masses of Stars and Planets

Problem Using the orbital information about Amalthea given below, find the approximate mass of Jupiter in SI units.

Given

$m_1 + m_2 = \left(\frac{4\pi^2}{G}\right)\frac{a^3}{P^2}$	Kepler's Third Law
G	The gravitational constant, 6.674×10^{-11} N·m²/kg
a	The semi-major axis of Jupiter's satellite Amalthea, determined from observation: 181,200 km
P	Orbital period given for Amalthea: .498179 days

Assumptions We will assume the accuracy of our data, no perturbing effects on Amalthea's orbit, and a pure "Keplerian" two-body system. We will ignore Amalthea's own mass.

Method Since we are using SI mks units, we will need to convert Amalthea's semi-major axis distance into meters and its period into seconds. Then we solve for the sum of the masses in the equation given. Because Amalthea is so tiny in comparison to Jupiter, we can assume that that the sum of the masses is, to a high order of accuracy, Jupiter's mass, that is $m_1 + m_2 \approx m_1$.

Calculations First we must convert Amalthea's period into seconds: take .48179 days times 24 h times 60 s to yield 43,042.666 s. The semi-major axis distance of 181,200 km is equivalent to 181,200,000 m. Inserting these values into the mass equation given above, we have,

$$m_1 + m_2 = \left(\frac{4\pi^2}{G}\right)\frac{a^3}{P^2}$$

$$m_1 + m_2 = \left(\frac{4(3.1415..)^2}{6.674 \times 10^{-11}}\right)\frac{(1.812 \times 10^8)^3}{(4.304266 \times 10^4)^2}$$

$$m_1 + m_2 = 1.899 \times 10^{27} \text{ kg}$$

Since $m_1 \gg m_2$,

$$\underbrace{m_1 \simeq 1.899 \times 10^{27}}_{\text{Mass of Jupiter in kilograms}}$$

Observations

1. This value accords closely with the published value for Jupiter of 1.8981×10^{27} kg.[13] Would the assumptions we made account for the small difference in the result?

---
[13] *The Astronomical Almanac for the year 2012* E4.

"Weighing" Jupiter the Modern Way   225

2. The Sun's is about 1.989×10^{30} kg, which puts Jupiter's mass at 1/1,048 the solar mass, consistent with our last result above. The Sun is thus roughly a thousand times more massive than Jupiter. Recall Isaac Newton had determined the ratio to be 1/1,067, an extremely close match![14]
3. Almathea's mass (reciprocal mass ratio) is 1/909090909 that of Jupiter.[15] That is, Jupiter is not quite a billion times more massive. To put it in a more familiar context, Amalthea's mass is only (relatively speaking) 2.1×10^{18} kg, or about one thirty-five-thousandth the mass of our Moon. Hence, it goes without saying that the error in disregarding its mass in the computation of Jupiter's mass is slight in the extreme.
4. We could use Amalthea's velocity to find Jupiter's mass. Amalthea orbits Jupiter at a speedy 26.45 km/s. If we take the equation we derived above, $m_1 \approx v^2 r/G$ (where $m_1 \gg m_2$), and remember to use mks units, we get the same result:

$$m_1 \simeq \frac{v^2 R}{G}$$

$$m_1 \simeq \frac{(26.45 \times 10^3)^2 (1.812 \times 10^8)}{6.674 \times 10^{-11}}$$

$$m_1 \simeq 1.899 \times 10^{27} \text{ kg}$$

5. Once we know Jupiter's mass, other physical properties can be derived. For example, since we know the mass of Jupiter, and we have an idea of its size from observation, we can determine its density. The volume of a sphere is $V = 4\pi r^3/3$. Density (usually symbolized by ρ) is by definition mass per unit volume, $\rho = m/V$. Hence the density is $\rho = 3m/4\pi r^3$. From its distance and angular diameter, and from the eclipses and transits of its moons, astronomers know quite accurately that Jupiter's mean radius is 69,911 km (it is not a perfect sphere, and its equatorial radius is actually larger). Hence the density of the planet is about 1.3 kg/m^3, or about a quarter of Earth's density. From this, astronomers have long inferred its mainly gaseous composition.

Calculating the Combined Masses of Quaoar and Its Satellite Weywot

Kuiper Belt objects (KBOs) are icy members of a planetesimal population of our solar system beyond Neptune. One of those objects, Quaoar, has a satellite, Weywot, studied by Wesley Fraser and Michael Brown of Caltech. From images

---

[14] *Principia*, Book III, Prop. VIII, Theorem VIII, Cor. I, 416.
[15] Ibid., F3.

taken with the Wide Field Planetary Camera 2 aboard the *Hubble Space Telescope*, Fraser and Brown deduced a period for Weywot of 12.438 days, a semi-major axis of 1.45×10^4 km.

Problem Using the Kepler's Third Law, approximate mass of the Quaoar–Weywot system, in kilograms.

Given

$m_1 + m_2 = \left(\frac{4\pi^2}{G}\right)\frac{a^3}{P^2}$	Kepler's Third Law, rearranged to solve for mass
G	The gravitational constant, 6.674×10^{-11} N·m²/kg
a	Weywot's semi-major axis: 1.45×10^7 m
P	Orbital period given for Weywot: 12.438 days

Assumptions We will assume the accuracy of the given data and that again we are dealing with a pure "Keplerian" two-body system.[16]

Method We will need to convert Weywot's period into seconds (the distance conversion to meters is already done). Then we solve for the sum of the masses using Kepler's Third Law adapted for finding mass.

Calculations We first convert 12.438 days into seconds, by multiplying that number of days by 86,400 s in a day. The resulting period is 1.0746432×10^6 s. Inserting the appropriate values into the equation, we get this result:

$$m_1 + m_2 = \left(\frac{4\pi^2}{G}\right)\frac{a^3}{P^2}$$

$$m_1 + m_2 = \left(\frac{4(3.1415..)^2}{6.674 \times 10^{-11}}\right)\frac{(1.45 \times 10^7)^3}{(1.0746432 \times 10^6)^2}$$

$$\underbrace{m_1 + m_2 \simeq 1.6 \times 10^{21}}_{\text{Aggregate mass of Quaoar and Weywot in kilograms}}$$

Observations

1. This result is in accord with the result published in the investigators' article cited above. By estimates of the likely density and size of Quaoar, the authors deduced the likely mass of Quaoar to be about 1.55×10^{21} kg, or about 97% of the mass of the whole system. This puts Weywot at about 1.15×10^{19} kg, or about 24 times Amalthea's mass, and 1/1,460 the mass of our Moon.

---

[16] The additional assumptions in gathering and processing the data can be found in Wesley C. Fraser and Michael E. Brown's article [6].

2. This above calculations required that we know only a few things from observation about the planet, or dwarf planet, and their satellites. Unless you have software to lend a hand, the above math might have seemed a little laborious, even on a calculator, involving SI units and the value of G which are a rather tedious to use. There are of course shortcuts we can remember and use. For example, $4\pi^2/G \approx 5.9 \times 10^{11}$, but even so one would still need to accomplish the conversion of distance and period into SI units and work through the calculations. Below we will look at some ways in which the Kepler–Newton equation has been greatly simplified and its usefulness increased.

Manipulating Units to Simplify Equations

We can achieve startling simplifications of the Kepler's Third Law when we leave the SI system and use units such as solar masses, years, and AU as the units.

Problem Find a simple algebraic value for the constant G using Newton's modification to Kepler's Third Law, using units of solar masses, the astronomical unit and seconds, then years, where the secondary mass is insignificantly small in relation to the primary mass.

Given

$P = \dfrac{2\pi}{\sqrt{G(m_1+m_2)}} a^{3/2}$	Kepler's Third Law, in Newtonian form
P	Period, where units will be *years*
a	Distance from the primary to the secondary mass, where units will be the *astronomical unit*
m_1	Primary mass $m_1 \approx m_1 + m_2$ in *solar mass units*
G	The gravitational constant

Assumptions We assume the secondary mass is insignificant compared to the primary, and ignore all other potential physical effects.

Method First change the units so that a is one astronomical unit, and m is one solar mass. Then insert the given values, which are ones, and solve for G. Since the Keplerian relationships, once defined, should hold for any situation, and for planet in the solar system, it doesn't matter which planet we pick to start with. Because the Earth is 1 AU distant from the Sun, and because its period is 1 year, we know the units will be simple. We will therefore choose the Earth as the starting point.

Calculations We start with Kepler's Third Law, where we assume negligible secondary mass:

$$P = \frac{2\pi}{\sqrt{Gm_1}} a^{3/2}$$

and take the first step to isolate G:

$$\sqrt{G} = \frac{2\pi}{P\sqrt{m_1}} a^{3/2}$$

Our mass unit is given as "one solar mass," the solar mass being unity. Thus as we are referring to the Sun, $m_1 = 1$. Furthermore, the distance term a is defined in units of the distance between the Earth and Sun, which is 1 astronomical unit (AU). Earth revolves around the Sun at $a = 1$. (We'll leave the period units in seconds for now). Since the mass and radius terms are unity, the equation simplifies dramatically:

$$\sqrt{G} = \frac{2\pi}{P}$$

$$\underbrace{G = \frac{4\pi^2}{P^2}}$$

G where units are solar masses, AU, and seconds

Now if we use the *year* as our unit of time, such that for the Earth–Sun system, $P = 1$, then this final simplification makes life even easier:

$$\underbrace{G = 4\pi^2}$$

G where units are solar masses, AU, and years

The above holds for the Earth-Sun system of units (AU, one solar mass and 1 year), where we assume a particle of near-zero mass is orbiting the Sun. That is, we assume a negligible secondary mass. The results of such calculations will be in AU, solar masses and years, which astronomers are used to using.

Reducing Kepler's Third Law to Its Simplest Form

Imagine that you are a new young astronomer hired by an observatory that has been observing double stars for some time (this used to be a major occupation of some observatories). Over the years, diligent people before you gleaned extensive data on the distances and periods of various stars, but now someone needs to analyze this data and ascertain their masses in the simplest way.

Problem Use algebraic means to simplify the equation for Kepler's Third Law to find the collective mass of a binary system, using only units that every astronomer is familiar with: the astronomical unit, the year, and the solar mass.

Manipulating Units to Simplify Equations

Given

$P = \dfrac{2\pi}{\sqrt{G(m_1+m_2)}} a^{3/2}$	Kepler's Third Law, appropriate for mks units
$G = 4\pi^2$	The inertial constant derived above, when the units are solar masses, AU and years

Assumptions Assume motion in perfectly elliptical orbits of any eccentricity (including zero). Other than ignoring all potential physical effects, we need make no further assumptions.

Method We use the units given above, and simplify using the given inertial constant. Since the question asks us to find the collective masses of the stars, we'll isolate the mass terms on the left of the equation.

Calculations The first step is a modest rearrangement for clarity, grouping the inertial constant and π terms together:

$$P^2 = \left(\frac{4\pi^2}{G}\right)\frac{a^3}{m_1 + m_2}$$

Now, because in the case where the time units are years, $G = 4\pi^2$, the fraction cancels to unity. Solving for the sum of the masses, we are left with a most beautiful equation and one of the most practical in astronomy. It is the ultimate reduction of Kepler's Third Law, which in this form the name Kepler's Harmonic Law seems most appropriately applied:

$$m_1 + m_2 = \frac{a^3}{P^2}$$

Observations

1. Again, these units are *solar masses, astronomical units* and *years*. If the orbital period is given in days, instead of years, we would divide the period by 365.25 as part of the equation, to still be using these basic "astronomer's units."
2. Perfectly elliptical orbits are not really possible in the solar system because of the perturbations of other planets. But double star systems with enough separation such that tides play no role travel in almost perfect ellipses and are excellent illustrations of Kepler's Laws.[17] There too, the orbits of each around their mutual center of mass is often very obvious, unlike the case in our solar system where the Sun's mass so dominates the masses of any planet that the center of mass is virtually indistinguishable from the Sun itself.

---

[17] See a concise description of this in James Kaler [7].

3. Applying the above equation to a double star system requires finding the separation of the stars from each other and the period of revolution. This may take many years of observations. But once they are approximated, the sum of the masses can be derived. And from the star orbits, the center of mass can be approximated because as we saw from the "big mass, small orbit" rule of the previous chapter, the semi-major axes of each star are inversely proportional to the respective masses of the stars: $a_1/a_2 = m_2/m_1$.
4. Astronomers can determine the semi-major axes of visual double star orbits by measuring their *angular* separation, or *parallax* (denoted usually by ρ with units of *arcseconds*), of the star from its companion, as seen from Earth. By definition, a parallax separation of 1 s of arc (1 arcsecond) means the star is 1 *parsec* (3.26 light years) away. A useful thing about parallax for our purposes is that 1 AU is the separation of a star from its companion seen from a distance of 1 parsec – where the angular separation is 1 s of arc. If the angular separation is therefore divided by the parallax, the result is the separation in astronomical units. If we apply this to the equation for Kepler's Third Law, using arcseconds as the units for the semi-major axis, the equation becomes:

$$m_1 + m_2 = \frac{a^3}{P^2 \rho^3}$$

The next problem applies this variation on Kepler's Harmonic Law to a well-known double star system.

Finding the Combined Masses of Sirius and Its Companion

Problem With the data given below, find the combined mass of Sirius (Sirius A), in the constellation of Canis Major, and its companion (Sirius B).

Given

$m_1 + m_2 = \frac{a^3}{P^2 \rho^3}$	Kepler's Third Law, in units given below
a	Semi-major axis of Sirius A and B: 7.56″ (arcseconds)
ρ	Parallax of Sirius: .379″
P	Orbital period of Sirius A/B: 49.9 years

Assumptions While satellite instruments such as the European Space Agency's *Hipparcos* mission[18] have greatly improved the accuracy of stellar parallaxes, such measurements still have a range of error associated with them. We will ignore measurement error in this problem.

---

[18] See http://www.rssd.esa.int/index.php?project=HIPPARCOS.

Manipulating Units to Simplify Equations

Method Since the equation and data are given, simply substitute the appropriate values into the equation and solve for the sum of the masses.

Calculations We begin with Kepler's Harmonic Law, modified as discussed above to accommodate our choice of units:

$$m_1 + m_2 = \frac{a^3}{P^2 \rho^3}$$

$$m_1 + m_2 = \frac{(7.56)^3}{(49.9)^2 (.379)^3}$$

$$m_1 + m_2 = 3.19 \, M_\odot$$

Observations

1. The M with the subscript is the conventional notation for solar masses. The sum of the masses is a little more than three solar masses. Interestingly, it turns out that about one-third of the contribution to this total mass of the system (almost one solar mass) comes from Sirius B, yet it is only about the size of the Earth! Sirius B is in fact one of the earliest discovered of those brilliant, dense, small stars known as *white dwarfs*.
2. If the actual separation of the masses in astronomical units has been calculated and is known, we can again use this equation:

$$m_1 + m_2 = \frac{a^3}{P^2}$$

For example, Alpha Centauri is a visual double star system with a semi-major axis of 20.6 AU and a period of 68 years. The sum of the masses of the two stars is,

$$m_1 + m_2 = \frac{(20.6)^3}{(68)^2}$$

$$m_1 + m_2 = 1.9 \, M_\odot$$

Determining the Individual Masses in the Double Star System Alpha Centauri

Problem In the case of Alpha Centauri, discussed just above, astronomers have learned from spectroscopy the stellar velocities of the orbiting stars. It is known that one star (which we'll call a_1) orbits .7 times as far from the center of mass as its companion, a_2. Knowing the combined masses for Alpha Centauri and its companion from the above example, and given the information below, calculate the individual masses of the components of the system.

Given

$a_1/a_2 = m_2/m_1$	Relationship between masses and distances from the center of mass of two objects in orbit around their common center of mass
$a_1 = .7a_2$	The relationship of stellar distances from the center of mass in the α Centauri system

Assumptions Again we will take as given the observational data and initial conclusions as to distances, and understand that the result will be only as accurate as our input data. For example, the preciseness of the determination of distances from spectrally measured velocities will depend in part on the size and quality of the telescope, the sophistication of the spectrographic equipment, the quality of the atmosphere when the spectroscopic images were recorded, and many other variables.

Method The simple relationship among masses and distances is the key to finding individual masses. If we know the ratio of the distances, we know the ratio of the masses. From the ratio of masses, and knowing the sum of the masses, we have sufficient information to derive the individual masses.

Calculations Since from the center of mass equation, $a_1 m_1 = a_2 m_2$, and given that $a_1 = .7a_2$, we substitute the right-hand term for a_1:

$$.7a_2 m_1 = a_2 m_2$$

Cancelling the distance terms yields,

$$.7m_1 = m_2$$

consistent, of course, with the statement that the ratio of the masses is in inverse relation to their distances from the center of mass. But since we are given the sum of the masses,

$$m_1 + m_2 = 1.9 \, M_\odot$$

we can substitute for m_2 from the previous expression and obtain the first mass:

$$m_1 + .7m_1 = 1.9 \, M_\odot$$

$$m_1 = 1.1 \, M_\odot$$

Manipulating Units to Simplify Equations

And, since $.7\, m_1 = m_2$, then the second mass will be,

$$m_2 = .8\, M_\odot$$

Alpha Centauri A is therefore about 1.1 solar masses, and its companion, Alpha Centauri B, is about .8 solar masses.

Observations

1. The shortest period of any binary star known is HM Cancri, a twenty-first magnitude star and an x-ray source 21,000 light years away. A spectrum by the Keck I telescope showed the objects to be tightly revolving white dwarf stars. They have a period of 5.4 min. If we assume a separation of about 65,600 km (a little over five Earth diameters, center to center distance) this translates to about .0004385 AU. We can then find their combined masses. Using Kepler's Harmonic Law, we may make the substitutions for semi-major axis and period. But we first need to convert minutes to years, and that is .000010267 years. We can also mathematically treat the stars as if one star were fixed and the other is revolving around it, and call the distance given as equal to the semi-major axis. Hence, we have

$$m_1 + m_2 = \frac{a^3}{P^2}$$

$$m_1 + m_2 = \frac{(.0004385)^3}{(.000010267)^2}$$

$$m_1 + m_2 = .8\, M_\odot$$

The combined masses of these two white dwarfs are therefore about the mass of Alpha Centauri B.[19]

2. The spectral redshifts and blueshifts of HM Cancri show that one star is revolving with a velocity that is 2.2 times faster than the velocity of the other. From this information alone we can deduce their individual masses. The velocities are inversely proportional to their masses, so the fastest star is the lightest. You must therefore multiply the mass of the lighter star (which we'll call m_2) by 2.2 to get the heavier star's mass: $m_1 = 2.2\, m_2$. Hence the equation for the combined masses is $2.2\, m_2 + m_2 = .8$. Thus,

$$m_2 = .25\, M_\odot$$

---

[19] See "Fastest Known Binary Star," *Sky & Telescope*, June 2010, p. 15. The article reports that the companions are "only about three Earth diameters apart." We don't know from the article if that is a surface-to-surface or center-to-center distance, but we have in any event used five Earth diameters as the only plausible distance given the masses and the period stated in the article.

234   11   Using Kepler's Third Law to Find the Masses of Stars and Planets

or only a quarter of the Sun's mass. The heavier star is 2.2 times this,

$$m_1 = .55 \, M_\odot$$

This is quite a lightweight stellar system!

What Is the Mass of Our Galaxy? (And Observations on Dark Matter)

We have seen how simple applications of Kepler's Third Law enable us to weigh planets and stars. Why shouldn't we be able to weigh whole systems of stars, such as globular clusters and galaxies? The next problem approaches the determination of mass of our Milky Way galaxy. The key that makes this possible is the knowledge that our galaxy, in common with all other observed spiral galaxies, is rotating about its massive center. Our solar system is an outlier in the galaxy, and rides around this massive, flattened pinwheel about once every quarter *billion* years. To give a sense of how long this is, the last time we were at this location in the galaxy, our planet was at the beginning of the Mesozoic Era, and the whole long age of the dinosaurs lay ahead, after the great Permian Extinction. This great, rotating colossus of billions of stars and untold quantities of gas and dust surprisingly does afford us a simple means of finding its mass. Since it is rotating, we can apply the same principles as work for any other orbiting thing, and determine mass from rotational velocity.

Problem Given the Sun's velocity, determine the mass of the galaxy interior to the Sun's orbit around the galactic center.

Given

$m_1 \simeq \frac{v^2 R}{G}$		Equation for determining mass from velocity
R		The distance of the Sun from the galactic center. This is about 8 kiloparsecs (thousands of parsecs), or $8 \times 1000 \times 3.26$ light years, or about 2.5×10^{20} m.[a]
v		The velocity of the Sun in its orbit around the galaxy: 220 km/s
G		Gravitational constant: 6.674×10^{-11} N·m²/kg

[a] One light year is 9.46×10^{12} km

Assumptions The values of distance and velocity are approximate. The mass ascertainable by this method is the mass *interior* to the Sun. We will discuss below how astronomers have analyzed the remaining mass.

Method The first step is to be sure all units are in SI units, then insert the appropriate values into the given equation. The solar mass is utterly insignificant in relation to the galactic mass.

What Is the Mass of Our Galaxy? (And Observations on Dark Matter)

Calculations All we need to do is put the solar velocity in meters per second, then solve for mass:

$$m_{galaxy:partial} = \frac{v^2 R}{G}$$

$$m_{galaxy:partial} = \frac{(220000)^2 2.5 \times 10^{20}}{6.674 \times 10^{-11}}$$

$$m_{galaxy:partial} = 1.8 \times 10^{41} \text{ kg}$$

This is the enormous mass of the Milky Way galaxy interior to the Sun, including gas, dust and any other form of matter, including a black hole at our galaxy's center.

Observations

1. Since the Sun's mass is about 1.99×10^{30} kg, there are about 9×10^{10} solar masses from the center of our galaxy out to the Sun. That is 90 billion suns worth of matter!
2. We could of course do this calculation using the simple units we used before, of years, AU and solar masses. Given a period of 225 Ma and a distance in AU of 8,000 parsecs × 206,000 AU/parsec = 1.648×10^9 AU, we arrive at the same result:

$$m_{galaxy(partial)_{\odot}} = \frac{a^3}{P^2}$$

$$m_{galaxy(partial)_{\odot}} = \frac{(1.648 \times 10^9)^3}{(2.25 \times 10^8)^2}$$

$$m_{galaxy(partial)_{\odot}} \approx 9 \times 10^{10}$$

3. The visible mass of the galaxy appears concentrated in the core and inner regions. Far out from our Sun into the galactic fringes, the apparent density of stars, dust and gas is vastly lower. Yet herein lies a puzzle. It appears that particles at the outskirts of our galaxy rotate about as fast as our Sun. Why is the velocity the same? We know from earlier discussion that the velocity v should be proportional to the inverse square root of the radius: $1/\sqrt{R}$. If the mass interior to the Sun were substantially all of the mass in the galaxy, such that we were on its very edge, the expected velocity at twice the distance out would be $v/\sqrt{2}$, not the same v. We would in that case expect particles at $2R$ to be orbiting

at about 156 km/s. In other words, there would be a normal Keplerian drop-off of velocity. But even though there is in fact some galactic material beyond the Sun, which should increase speeds somewhat, the velocity curve is still relatively flat. Thus, at distance $2R$, particles orbit with the same velocity as our Sun, so the galactic mass must be about twice what it is within the solar system's orbital distance of R. (Do you see from the equation why this should be so?) The overarching question is, why is the galactic mass sufficient to create a velocity of 220 km/s twice as far out from the Sun, even though the mass density we see is clearly toward the center, interior of the Sun's orbit, and is significantly lower farther out? These findings in this and other galaxies have led to the conclusion that the mass we see is only part of the picture. There must be mass invisible to us – the so called *dark matter* which must account for these higher-than expected rotation speeds.

Exercises: What Oberon and Triton Tell Us About Uranus and Neptune Astronomers would have had to approximate the masses of Uranus and Neptune by perturbation effects were it not for the lucky fact that they each have satellites. Uranus, discovered by Herschel in 1781 was soon found to have two satellites, also discovered by him in 1787. They were named Oberon and Titania, and over time enabled computation of Uranus' mass. The same held for even remoter Neptune, discovered in 1846 by the German Galle in Berlin, from the predictions of the Frenchman Leverrier (and concurrently calculated by the Englishman Adams, but not then communicated to the world). Its large satellite, later named Triton, was found the following year by Lassell using the U. S. Naval Observatory's 26 in. refractor in Washington.

Even with the discovery of a satellite, the task of determining precise mass of a distant, orbited body, however, can still be difficult. Reliable determination of mass of a planet or smaller body depends on accurate measurement of the distance of the satellite from its primary, and its period. These measures are affected by the remoteness and magnitude of the body, the quality of the optics, the atmospheric seeing, and the experience of the observer, among other factors. In the last century and a half, the distance to the satellite was typically determined by micrometer and given in semi-diameters of the primary planet. Early determinations by visual means with the telescopes at hand were respectable, and quite close to modern values for period, but more in error for the difficult-to-determine distance of the faint moon from the glare of the planet.

The reported masses of both Uranus and Neptune have changed over the years as data on period and distance has improved. Notice in the accompanying chart the differing historical values of the stated masses of Uranus and Neptune. The masses are given in terms of fraction of the solar mass, usually called the reciprocal mass ratio.

What Is the Mass of Our Galaxy? (And Observations on Dark Matter) 237

Historical derivations of outer planet masses from satellite periods and distances[a]

Reporter	Date of data (approx.)	Calculated mass of planet to mass of sun	Satellite used for the computation	Period (days)	Planetary radii to satellite
Uranus					
John Herschel	1847	1/20,470	Oberon	13.46334	22.56
Simon Newcomb	1882	1/22,600	Oberon	13.463269	22.88
Spencer Jones	1951	1/22,869	Oberon	13.463194	23.5
NASA/JPL	2012	1/22,913	Oberon	13.46	22.83
Neptune					
John Herschel	1847	1/18,780?	Triton	5.876887	~12
Simon Newcomb	1882	1/19,380	Triton	5.87690	12.715
Spencer Jones	1951	1/22,869	Triton	5.876389	13.3
NASA/JPL	2012	1/19,423	Triton	5.877	14.3256

[a]Data for the table was drawn from these sources: John F.W. Herschel [8], which relied on the original data of Lassell; Simon Newcomb [9]; Harold Spencer Jones [10].

The following exercises will explore the methods and challenges of obtaining accurate masses and other planetary information based on available data.

Problems

1. Use Kepler's Third Law to find the Keplerian mass ratios (distance cubed over period squared) of Earth, Mars and Jupiter to the Sun, to seven significant digits. Use astronomical units as your distance units and days as your period units (with Earth's period being 365.25 days). If you did these calculations for all the planets, explain what would be the result. Are results for each planet dependent or independent of the mass of each planet? Explain.

2. The historically reported values of the reciprocal mass ratios are shown for Uranus and Neptune in the accompanying table. Compare the early reciprocal mass ratios for Uranus as reported by John Herschel against the modern values. What is the percentage error in Uranus' mass implied by these differences? In Neptune's?

3. The semi-major axis of Oberon is about 583,500 km, and its period is 13.46319444 days. Using the Keplerian mass ratio found in the first problem, find the mass of the planet Uranus in relation to the solar mass (i.e., the reciprocal mass ratio in relation to the mass of the Sun). Take the length of an astronomical unit to be 149,598,000 km. *Tip:* Be sure to use units consistent with Problem 1. Compare your result against the historical determinations of those ratios given in the table in this chapter.

4. The semi-major axis of Triton is about 354,759 km, and its period is 5.876886574 days. Using the Keplerian ratio mass ratio found in the first problem, find the mass of the planet Neptune in terms of solar masses. Again, be sure to use consistent units. Check your result against the historical determination of this ratio.

5. Find the Keplerian mass ratio of the Earth in units of days and kilometers, and apply that to the data given for the orbit of Triton to find the mass of Neptune in solar masses.

6. Find the mass of Uranus and Neptune directly in SI units by using the respective orbits of Oberon and Triton and Kepler's Third Law, and its solar reciprocal mass ratio. Take the mass of the Sun to be 1.9891×10^{30} kg. How do your results differ from the modern values for the ratios in the accompanying table? How do they compare with the answers you found by the methods in the problems above?
7. It is also common to find planetary masses given in terms of Earth masses. The Earth's mass is about 5.7219×10^{24} kg. Using the results of the previous problems, how many Earth masses is Uranus? How many is Neptune?
8. Our Earth has a density of about five and a half grams per cubic centimeter. The respective diameters of Uranus and Neptune are 51,118 and 49,528 km. Given the size and masses of Uranus and Neptune, calculate their respective densities, in gm/cm^3 rounded to the nearest tenth, assuming they are perfect spheres. If Uranus were the size of the Earth, but retained its same mass, how would Oberon's orbit be affected?
9. Triton's orbit is nearly circular, with an eccentricity of .000016. You are told that its orbital velocity is 4.39 km/s, more than quadruple the orbital speed of our Moon. Calculate the mass of Neptune using the velocity of Triton and its distance from the center of the planet, and compare your result with what you obtained in problem 6.
10. What happens to the period of a satellite if we imagine that the primary mass is increased? Triton's distance from Neptune is very close the Moon's distance from Earth when it is at perigee. If Neptune were as dense as the Earth, or about 5.51 g/cm^3, what would Neptune's new hypothetical mass be? What would Triton's hypothetical orbital period be, in days? (*Hint*: Think in terms of ratios).

References

1. Jones HS (1951) General astronomy, 3rd edn. Edward Arnold & Co., London, pp 262–263
2. Blanco VM, McCuskey SW (1961) Basic physics of the solar system. Addison-Wesley, Reading, 42
3. Buie MW, Grundy WM, Young EF, Young LA, Stern AS (2006) Orbits and photometry of Pluto's satellites: Charon, S/2005 P1, and S/2005 P2. Astronom J 132:290–298, http://iopscience.iop.org/1538-3881/132/1/290/pdf/205133.web.pdf
4. Cajori F (1949) Principia in modern English: Sir Isaac Newton's mathematical principles of natural philosophy, Book III, Prop. VIII, Theorem VIII, Cor. I (Motte's translation revised). University of California Press, Berkeley, p 416
5. Densmore D (1996) Newton's principia: the central argument. Green Lion Press, Santa Fe, 473
6. Fraser WC, Brown ME (2010) Quaoar: a rock in the Kuiper Belt. Astrophys J 14(2):1549, 10 May 2010
7. Kaler J (2006) The Cambridge encyclopedia of stars. Cambridge University Press, Cambridge, pp 145–147
8. Herschel JFW (1901) Outlines in astronomy, vol Part one, A library of universal literature. P.F. Collier & Son, New York, p 897, originally published 1849
9. Newcomb S (1882) Popular astronomy. Harper & Brothers, Washington, p 540
10. Jones HS (1951) General astronomy, 3rd edn. Edward Arnold & Co., London, p 257, first published in 1922

Chapter 12
Motion in Elliptical Orbits

We venture here beyond mainly circular orbits and introduce masses orbiting in either circular or elliptical orbits. Our attention will be focused on comparing the uniform motion of the circular orbit and the motion at two points on the elliptical orbit: the so-called apsides, where the mass is either closest or farthest from the center of mass of the system.

Velocity Along the Apsides of Elliptical Orbits

The law of conservation of angular momentum leads directly to Kepler's Law of Areas—that planets in our solar system sweep out equal areas in equal times. These ideas, as we saw in Chap. 4, are useful to the understanding of motion in elliptical orbits. From this we can learn something about the velocity of a mass in an elliptical orbit. Again, in this chapter we focus on the apsides of the ellipse.

Suppose we think of a small orbiting mass moving through a tiny angle in a very small amount of time Δt. This distance from one focus will be r, which is called the *radius vector*. The distance travelled along this small arc of the ellipse will be approximately $v\Delta t$. Here v is the velocity component perpendicular to the radius vector. In this short period of time, the mass will sweep out a small, triangular slice of the ellipse. This distance $v\Delta t$ may be regarded as the base of the little triangle drawn from one focus of the ellipse. The height of the triangle is r.

Recalling one of the first lessons of elementary geometry, the area of the triangle should be half the base times the height. But the base is actually a curve, and as we have seen in dealing with arcs, the chord distance in a given small angle is not precisely the length of the arc subtended by it. The area given by that formula may approximate, but will not equal, the actual area within the radius vector lines. But using the tried and true principle of limits, however, we can imagine ever shorter increments of time, where the angle is made as small as we please. As $\Delta t \to 0$, this difference ultimately vanishes, as Newton showed in his Lemma 7 of Book I of the *Principia*. As the chord length thus ultimately approaches the arc length, the

D.W. MacDougal, *Newton's Gravity: An Introductory Guide to the Mechanics of the Universe*, Undergraduate Lecture Notes in Physics, DOI 10.1007/978-1-4614-5444-1_12, © Springer Science+Business Media New York 2012

velocity vector v becomes perpendicular to the radius vector. Representing all this mathematically, the area of this little triangle ΔA—approximately one-half the base $v\Delta t$ times the height r—can be expressed this way:

$$\Delta A \approx \frac{v\Delta t}{2} r$$

The change in this area in each increment of time (dividing each side by Δt) will be:

$$\frac{\Delta A}{\Delta t} \approx \frac{vr}{2}$$

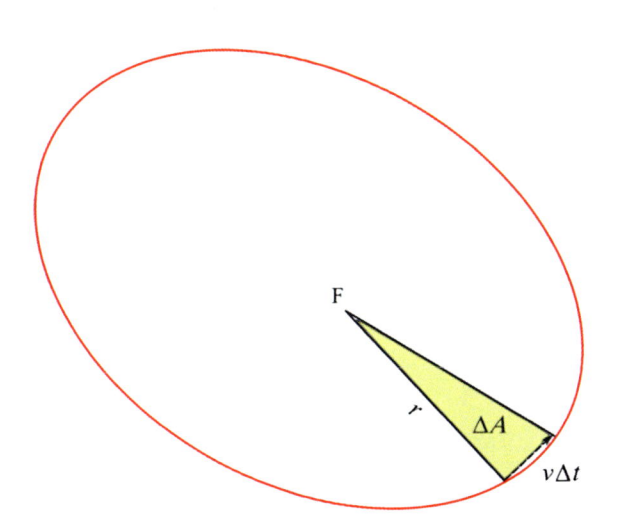

Small triangular area swept out by an orbiting mass
in a small increment of time

If enough of the small units of time when added up are sufficient for the mass to complete a full orbit, then we say that the radius vector has swept over the whole ellipse, covering its entire area A, in period P. That is, we imagine the sum of all the smallest little triangles around the ellipse as the total area, and the sum of all of the corresponding little increments of time as the period of the orbit.[1]

---

[1] In the language of calculus, we would integrate the expression $dA/dt = vr/2$.

Velocity Along the Apsides of Elliptical Orbits

Because the change of area with time is constant (this is another way of stating Kepler's Law of Areas), the term on the right must be constant. Having done this, the equation becomes:

$$\frac{A}{P} = \frac{vr}{2}$$

Now we can solve for velocity,

$$v = \frac{2A}{Pr}$$

Let us put the area term A and the radial term r in the familiar dimensions of the ellipse. Recall that the area of an ellipse is $\pi a b$, where a and b are again the semi-major and semi-minor axes of the ellipse, and for perihelion, $r_P = a(1-e)$, and for aphelion, $r_A = a(1+e)$. Let us arbitrarily select the perihelion distance and make the substitutions to find the velocity at perihelion in an elliptical orbit:

$$v_P = \frac{2\pi a b}{P[a(1-e)]}$$

We found before that the semi-minor axis is $b = a(1-e^2)^{1/2}$ so the further substitution gives:

$$v_P = \frac{2\pi a \left[a(1-e^2)^{1/2}\right]}{P[a(1-e)]}$$

The nice thing about these relationships is that they factor easily, because $1 - e^2 = (1-e)(1+e)$ The result for the velocity at perihelion is:

$$v_P = \frac{2\pi a}{P}\left(\frac{1+e}{1-e}\right)^{1/2} \quad \text{Velocity at perihelion}$$

By the same reasoning, the velocity at aphelion is:

$$v_A = \frac{2\pi a}{P}\left(\frac{1-e}{1+e}\right)^{1/2} \quad \text{Velocity at aphelion}$$

These are useful equations when we know only the period, the semi-major axis and eccentricity of an orbit. Such data is often discernable in the observations of double stars. Note that if eccentricity becomes zero, in the case of a circular orbit, then the velocities in each case are $v = 2\pi r/P$.

The Orbital Velocities of Earth Around the Sun

To illustrate how the above equations may be used, take the example of the Earth's modestly elliptical orbit around the Sun. All we need to know is Earth's semi-major axis distance, its eccentricity and its orbital period:

$$\text{Eccentricity: } e = .0167$$
$$\text{Semi-major axis: } a = 149,600,000 \text{ kilometers, or } 149.6 \times 10^9 \text{ meters}$$
$$\text{Period: One year or } 3.1558 \times 10^7 \text{ seconds}$$

Inserting these values into the applicable equations just given above, the velocities of Earth (in km/s) are:

$$\text{Velocity of Earth at perihelion: } 30.29 \text{ km/s}$$
$$\text{Velocity of Earth at aphelion: } 29.29 \text{ km/s}$$

As expected, the Earth is moving faster in its orbit when it is nearer the Sun (which happens to occur in January) than it is when it is farthest away (in June). These equations did not require us to know the mass of the Earth. The same equations are easily applied to satellites and other orbiting bodies. In the case of relatively close exoplanet or double star systems, with repeated observations astronomers can determine the period and eccentricity, and from their distance and apparent separation, deduce the semi-major axis. But the most precise determinations of velocity and period are deduced if we know (or can find out) the masses of the orbiting bodies.

Comparing Circular and Elliptical Orbits

The velocities with which we were concerned above were those just at the ends of the elliptical orbit. That is because at those special, apsidal, points the radius vector lies conveniently along the major axis of the ellipse. At those places, the distance r for perihelion or aphelion is easily translated into the semi-major distance a, by the equations $r_P = a(1 - e)$, and $r_A = a(1 + e)$.

Look again at the velocity equations. The first term on the right-hand side of each is $2\pi r/P$. This we recognize as the familiar velocity in a circular orbit, the circumference divided by the period: $v = 2\pi r/P$. But in an elliptical orbit, the circular velocity appears to be modified by a factor dependent upon the eccentricity of the ellipse. In the velocity equations, the circular velocity terms were simply altered for aphelion by the elliptical correction factor $[(1 - e)/(1 + e)]^{1/2}$ and for perihelion by $[(1 + e)/(1 - e)]^{1/2}$ where e is the eccentricity of the ellipse.

Comparing Circular and Elliptical Orbits

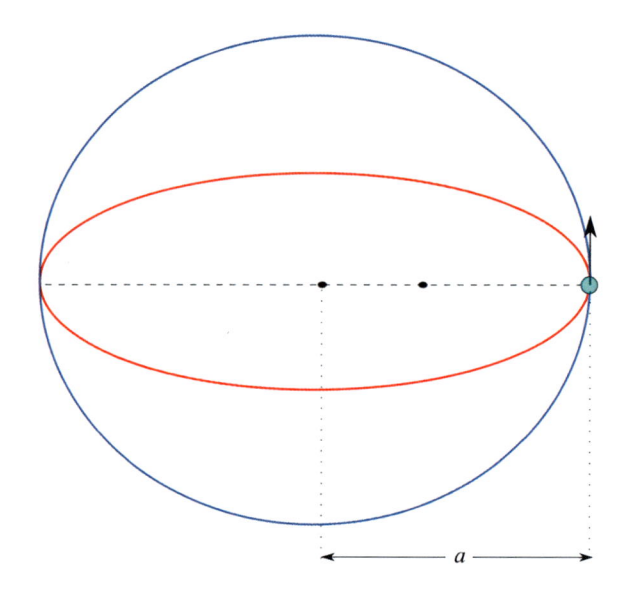

These equations suggest we might find further fertile ground for exploration if we compare circular to elliptical orbits. It is good where possible to picture the physical meaning of an equation, so let us imagine an ellipse that is circumscribed by a circle such that the circle's radius equals the semi-major axis of the ellipse, as in the accompanying diagram.

We will continue to focus only on the end points, the apsides, where the orbits touch. Consider a circle of radius a. Remember that circular angular velocity can be expressed in angular notation: $\omega_c = 2\pi/P$, where P again is the period. This is the angular motion around the circle in radians per unit time. Actual velocity along the rim of the circle is this angular velocity times radius, or $v = 2\pi a/P$. In angular notation this is written $v = \omega a$, and its square is $v^2 = \omega^2 a^2$. Recalling the above perihelion equation, the velocity (squared) is,

$$v_P^2 = \omega_c^2 a^2 \left(\frac{1+e}{1-e}\right)$$

Dividing each side by a yields,

$$\frac{v_P^2}{a} = \omega_c^2 a \left(\frac{1+e}{1-e}\right)$$

We would arrive at an equivalent expression, with signs in the fraction reversed, for aphelion velocity. Venturing an interpretation of this equation, we recognize the left-hand term as centripetal acceleration of a mass in a circular orbit of radius a. (The centripetal acceleration in a circular orbit is $v^2/a = \omega^2 a$.) The square of the velocity over the radius equals the inertial acceleration $\omega^2 a$ of the mass (here regarded as unit mass) in such a circular orbit. But on our inscribed ellipse of semi-major axis a it is multiplied by the ratio for aphelion of $(1 + e)/(1 - e)$. The velocity equations inform us that the uniform velocity of an object in a circular orbit needs correction when we are seeking the apsidal velocity along an elliptical orbit; they show that an orbiting object moves faster than circular at perihelion and slower than circular at aphelion. Similarly, the relative acceleration (in elliptical orbits over circular) increases at perihelion and decreases at aphelion. Note how this difference is magnified as we imagine the eccentricity to become greater (as it approaches one). If on the other had the eccentricity diminishes, the difference is less. If we make the eccentricity zero, where the two orbits in the diagram become congruent, the equation becomes,

$$\frac{v^2}{r} = \omega_c^2 r$$

Which matches the familiar expression for the equivalence of centripetal and inertial acceleration in a circular orbit (of radius r).

Angular Velocity in Circular and Elliptical Orbits

From the above discussion, we found the relation between circular orbital velocity (in an orbit of radius a) and the velocity along an elliptical orbit (of semi-major axis a) at perihelion to result in this equivalence of acceleration terms:

$$\frac{v_P^2}{a} = \omega_c^2 a \left(\frac{1+e}{1-e}\right)$$

These relations may be explored further if on each side we divide by a and take square roots. Remembering that angular velocity is $\omega = v/a$, the angular velocities for perihelion and aphelion of the ellipse (denoted respectively by ω_P and ω_A) become, in relation to circular angular velocity ω_c:

$$\omega_P = \omega_c \left(\frac{1+e}{1-e}\right)^{\frac{1}{2}}$$

$$\omega_A = \omega_c \left(\frac{1-e}{1+e}\right)^{\frac{1}{2}}$$

These equations clearly convey the differences between the angular velocities at the apsides of an elliptical orbit and the uniform angular velocity in a circular orbit in our comparison case, where the radius of the circular orbit equals the semi-major axis of the elliptical orbit.

To summarize, the uniform circular velocity when multiplied by the applicable correction factor $[(1 - e)/(1 + e)]^{1/2}$ or $[(1 + e)/(1 - e)]^{1/2}$ (depending upon whether the respective aphelion or perihelion angular velocity is being calculated) easily generates elliptical velocities along the respective apsides. The perihelion angular velocity in an elliptical orbit is faster than circular velocity, and aphelion velocity is slower, consistent with Kepler's Second Law.

Gravitation and Elliptical Orbits

We compared the uniform velocity in circular orbits with the apsidal velocities of inscribed elliptical orbits. Though we used Kepler's Law of Areas, the equations have been geometrically derived. We have considered the phenomena of motion in circles and ellipses, but did not directly refer to mass or gravity. Here we examine how Newton's law of gravitation is entirely consistent with what we have been discussing, and yields the same results.

We noted that in our orbit construction, where the radius of the circular orbit equals the semi-major axis of the inscribed elliptical orbit, this identity holds at perihelion:

$$\frac{v_P^2}{a} = \omega_c^2 a \left(\frac{1+e}{1-e}\right)$$

Since according to Newton's Third Law, the gravitational acceleration balances the inertial (or centrifugal) acceleration in a stably orbiting system, then, for a circular orbit of radius a, as we found in Chap. 8,

$$\omega_c^2 a = \frac{G(m_1 + m_2)}{a^2}$$

The right-hand term is the familiar Newtonian expression of gravitational acceleration between two masses separated by distance, which in this case is a. Now we may substitute the term for gravitational acceleration for the right-hand acceleration term in the velocity equation:

$$\frac{v_P^2}{a} = \frac{G(m_1 + m_2)}{a^2} \left(\frac{1+e}{1-e}\right)$$

Multiplying through by a, and using the same reasoning for the aphelion velocity, the result is:

$$v_P = \sqrt{\frac{G(m_1 + m_2)}{a} \left(\frac{1+e}{1-e}\right)} \quad \text{velocity at perihelion}$$

$$v_A = \sqrt{\frac{G(m_1 + m_2)}{a}\left(\frac{1-e}{1+e}\right)} \quad \text{velocity at aphelion}$$

These are again identical to the circular velocity equations except for the eccentricity corrections on each. We can also cast these expressions in terms of the perihelion and aphelion distances for each, often expressed in the literature in terms of q and Q, respectively. These are the radius vectors of the ellipse at the apsides, r_P and r_A. We substitute for a from the equations $r_P = a(1 - e)$, and $r_A = a(1 + e)$. This yields an even simpler pair of equations:

$$v_P = \sqrt{\frac{G(m_1 + m_2)}{q}(1+e)} \quad \text{velocity at perihelion}$$

$$v_A = \sqrt{\frac{G(m_1 + m_2)}{Q}(1-e)} \quad \text{velocity at aphelion}$$

If we call $\mu = G(m_1 + m_2)$, we make it prettier:

$$v_P = \left(\frac{\mu}{q}(1+e)\right)^{\frac{1}{2}}$$

$$v_A = \left(\frac{\mu}{Q}(1-e)\right)^{\frac{1}{2}}$$

The Orbital Velocities of Mars

Let's take an example using the apsidal distances of Mars. If we were to find the velocity of Mars at perihelion and aphelion, using the equations just mentioned, we would need to know the eccentricity e and either the semi-major axis distance of the Martian orbit or the distances at aphelion and perihelion. For the sake of variety we will use the Q (aphelion) and q (perihelion). The appropriate constants for the Martian orbit are,

$$\mu = 1.3275 \times 10^{20}$$
$$e = .0934$$
$$q = 206.62 \times 10^9 \text{ meters}$$
$$Q = 249.23 \times 10^9 \text{ meters}$$

Plugging those values into the above equations, the velocities of Mars are:

$$v_q = 26.5 \, \text{km/s} \qquad v_Q = 21.98 \, \text{km/s}$$

Given that the mass of the Sun is over three million times the mass of Mars, the planet's mass may safely be neglected in the above calculations, so that $\mu = Gm_1$. These values show greater variation than those of Earth, calculated earlier. This is because the eccentricity of Mars is greater than that of the Earth. Here is the comparison of the orbital velocities of Earth and Mars as they orbit our Sun in graphical form:

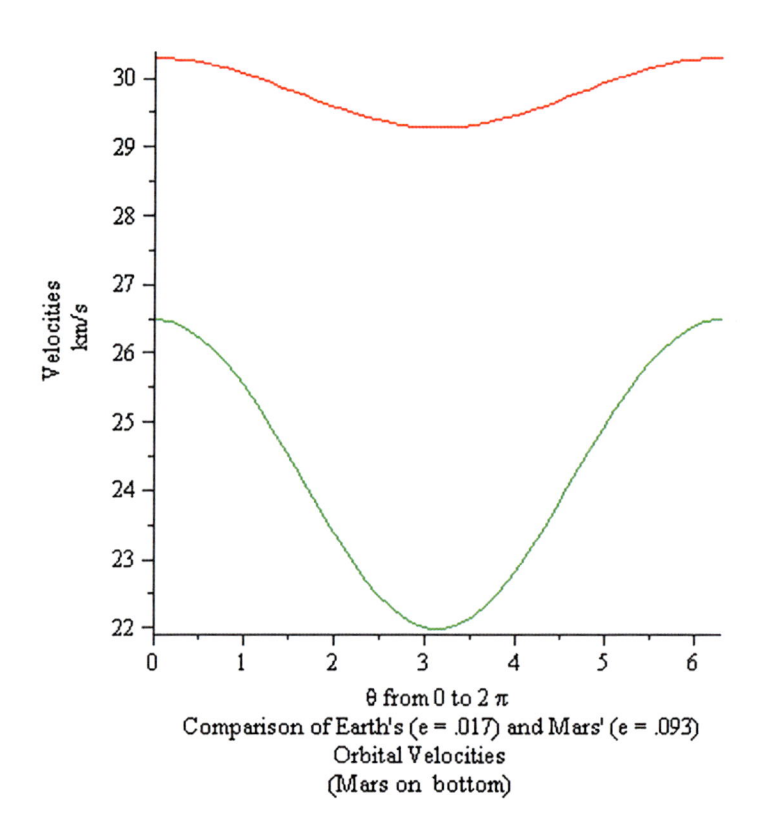

Comparison of Earth's (e = .017) and Mars' (e = .093)
Orbital Velocities
(Mars on bottom)

Kepler Revisited

We have thus considered several forms of the velocity equation, each applicable along the apsides of an elliptical orbit. Both require knowing eccentricity, although eccentricity is readily computed by a variety of means. One method is to compute the ratio of A_p and P_{er}.[2]

Let us compare two of the above equations. The left-hand equation below outputs velocity when the period and semi-major axes are known; the right-hand equation gives the velocity when the masses and semi-major axes are known:

Perihelion velocity: $\quad v_P = \dfrac{2\pi a}{P}\left(\dfrac{1+e}{1-e}\right)^{\frac{1}{2}} \quad\quad v_P = \left(\dfrac{G(m_1+m_2)}{a}\left(\dfrac{1+e}{1-e}\right)\right)^{\frac{1}{2}}$

Aphelion velocity: $\quad v_A = \dfrac{2\pi a}{P}\left(\dfrac{1-e}{1+e}\right)^{\frac{1}{2}} \quad\quad v_A = \left(\dfrac{G(m_1+m_2)}{a}\left(\dfrac{1-e}{1+e}\right)\right)^{\frac{1}{2}}$

Since both sets of equations equal the same thing (velocity) they present the opportunity to see what happens when we put them equal to each other:

Perihelion: $\quad \dfrac{2\pi a}{P}\left(\dfrac{1+e}{1-e}\right)^{\frac{1}{2}} = \left(\dfrac{G(m_1+m_2)}{a}\left(\dfrac{1+e}{1-e}\right)\right)^{\frac{1}{2}}$

Aphelion: $\quad \dfrac{2\pi a}{P}\left(\dfrac{1-e}{1+e}\right)^{\frac{1}{2}} = \left(\dfrac{G(m_1+m_2)}{a}\left(\dfrac{1-e}{1+e}\right)\right)^{\frac{1}{2}}$

Squaring each side and cancelling out the common eccentricity terms, we end up with one equation, which devolves into Kepler's Third Law:

$$\dfrac{4\pi^2 a^2}{P^2} = \dfrac{G(m_1+m_2)}{a}$$

$$P^2 = \dfrac{4\pi^2}{G(m_1+m_2)} a^3$$

This shows that the basic Keplerian relationship emerges when we independently construct velocity equations: the first from the period of an object in an elliptical orbit and the Law of equal areas in equal times (Kepler's Second Law); the second from the balance of gravitational and inertial accelerations maintained in the elliptical orbit. The result is the same Kepler's Third Law that we derived for the circular orbit. It is independent of eccentricity and thus applies to elliptical and circular orbits.

---

[2] This ratio can be sometimes be discerned in observations of the trace of the orbit, whether by radar or other determinations of the height of an orbiting spacecraft, or the maximum and minimum separations of a satellite from its planet or of a double star system.

Finding the Velocity of an Artificial Satellite in Earth Orbit When Just Its Perigee and Apogee Are Known

Problem An artificial satellite orbits the Earth in an elliptical orbit whose perigee is 160 km (100 miles), and whose apogee is 16,000 km, measured from the surface of the Earth. Find the eccentricity of the satellite orbit and the velocities of the satellite at perigee and apogee in kilometers per second.

Given

$v_P = \sqrt{\frac{Gm_E}{q}(1+e)}$	Equation for perigee velocity, where $m_2 \ll m_E$
$v_P = \sqrt{\frac{Gm_E}{Q}(1-e)}$	Equation for apogee velocity where $m_2 \ll m_E$
$e = \frac{Q-q}{Q+q}$	Equation for eccentricity e from perigee q and apogee Q
6.38×10^6	Mean radius of the Earth in meters
Gm_E	The product of the Earth's mass and the gravitational constant, often abbreviated (in the Earth-orbiting context) by the symbol μ, where $\mu = 3.99 \times 10^{14}$

Assumptions That the Earth is a spherical body of uniform density, so the path of the satellite follows in an idealized ellipse, without distortions in its velocity caused by gravitational anomalies of the Earth, and without influences of the Moon or other bodies.

Method First find the perigee and apogee distances by referring them to the center of the Earth, by converting to meters and adding them to the Earth's radius. Then ascertain the eccentricity of the orbit, using the equation given, and solve the applicable velocity equations.

Calculations Before we can compute the velocities at the apsides with the given equations, the eccentricity must be known. Remember that the values for perigee, 160 km, and apogee, 16,000 km, must be corrected for Earth's center, and converted to SI units of meters. Since Earth's radius is 6.38×10^6, this must be added to the distances given. Doing this, we find:

$$q = 6.54 \times 10^6 \text{ meters}$$
$$Q = 2.24 \times 10^7 \text{ meters}$$

Calculating eccentricity:

$$e = \frac{Q-q}{Q+q}$$
$$e = \frac{2.24 \times 10^7 - 6.54 \times 10^6}{2.24 \times 10^7 + 6.54 \times 10^6}$$
$$e = .548$$

This eccentricity shows that the orbit will be an elongated ellipse. Now we can determine velocities at the apsides, again using meters for units.

For perigee	For apogee
$v_P = \sqrt{\frac{\mu}{q}(1+e)}$	$v_A = \sqrt{\frac{\mu}{Q}(1-e)}$
$v_P = \sqrt{\frac{3.99 \times 10^{14}}{6.54 \times 10^6}(1+.548)}$	$v_A = \sqrt{\frac{3.99 \times 10^{14}}{2.24 \times 10^7}(1-.548)}$
$v_P = 9718$ m/s	$v_A = 2837$ m/s
$v_P = 9.72$ km/s	$v_A = 2.84$ km/s

Observation We can see what the above orbit looks like by graphing the equation for the conic:

$$r = \frac{p}{1+e\cos\theta}$$

where the parameter $p = a(1-e^2)$, eccentricity $e = .548$, and the semi-major axis a is the average of the apsidal distances: $a = (Q+q)/2$ or 14,470 km. The parameter (semi-latus rectum of the ellipse), is thus 10,125 km. Let π go for a complete revolution, from 0 to 2π. Graphing this with *Maple* software displays this figure:

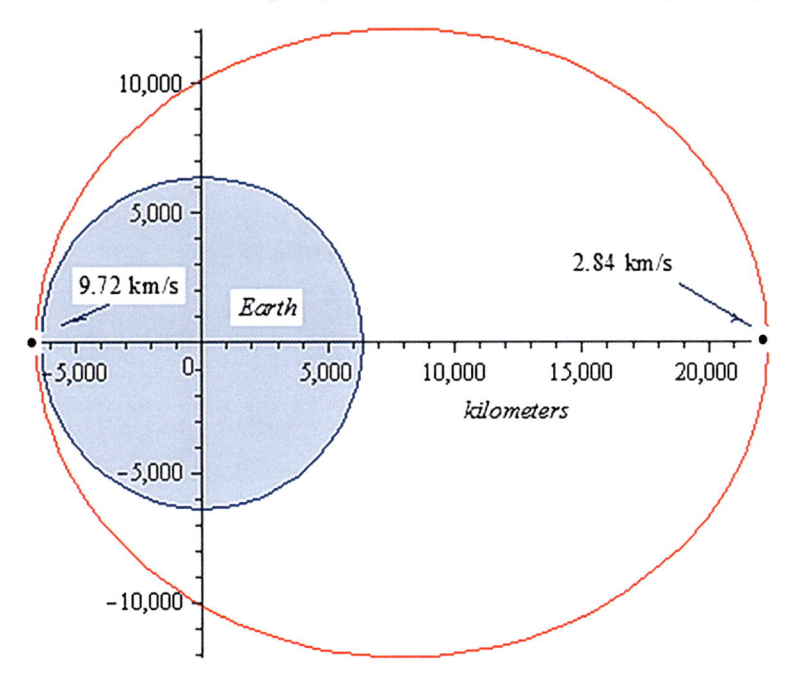

Orbit of Earth Satellite at e=.548

One can see the dimensions of the orbital ellipse clearly here, and the significant variation in velocity, consistent with Kepler's Second Law, between perigee and apogee.

Deriving Apsidal Velocities in an Elliptical Orbit from the Laws of Conservation of Energy and Momentum

The equations above for determining apsidal velocities were derived by using the geometry of an ellipse inscribed in a circle. Here we show how those same apsidal velocities arise naturally from the laws of conservation of mechanical energy and momentum. We had passing encounters with those laws in earlier chapters, and the subject is typically covered generally in elementary physics texts, but the energy laws are so vitally useful in the context of orbits that we will benefit here by examining them more closely. This will also serve as a brief, introductory energy course for the chapters that follow.

One of the foundation principles in our natural world is that energy is never lost, it is always conserved. Energy is not created out of nothing, either. What energy there is in the universe is what energy there will always be, so far as we know, though it comes in different forms, and one form can change to another. For example, mechanical energy can be converted to heat through friction or impact. The most important energy exchange in dealing with orbits and spacecraft is between kinetic and potential energy. In elementary physics courses these two forms of energy are usually written this way:

$$KE = \frac{1}{2}mv^2 \Big\} \text{Kinetic energy}$$

$$PE = -\frac{GMm}{r} \Big\} \text{Gravitational potential energy}$$

Look at these equations. The first depends on v, the velocity of a mass. The second is determined by r the distance between the mass and the primary mass attracting it. The first equation is, roughly speaking, the energy of motion; the other, the ability to make something have motion. The "potential" for motion exists when something is lifted away from something that attracts it: it "wants" to fall back. As something is lifted from the Earth, which is constantly pulling it back, the Earth's gravitational pull creates the potential of accelerating the object into motion toward the Earth. This potential is called the gravitational potential energy, often symbolized by the letter U. Notice the accompanying plot of the change in gravitational energy on a 1 kg mass as it is transported from the Earth to the Moon. A physicist would say that it takes work to move the mass away from Earth; that is, the exertion of a force over a distance. As the distance is greater, the gravitational pull lessens. The sum of all the incremental amounts of work to get something from the ground into space is the amount of gravitational potential energy that has been converted to kinetic energy of motion. This gravitational potential energy is conventionally expressed in the negative, moving toward zero as r becomes greater and greater. It becomes zero when the distance becomes infinite. At very great distances from the gravitational source, there is no more practical ability for the attracting mass to draw it back. That is why the potential energy approaches zero.

Gravitational Potential Energy Upward from Earth to Moon

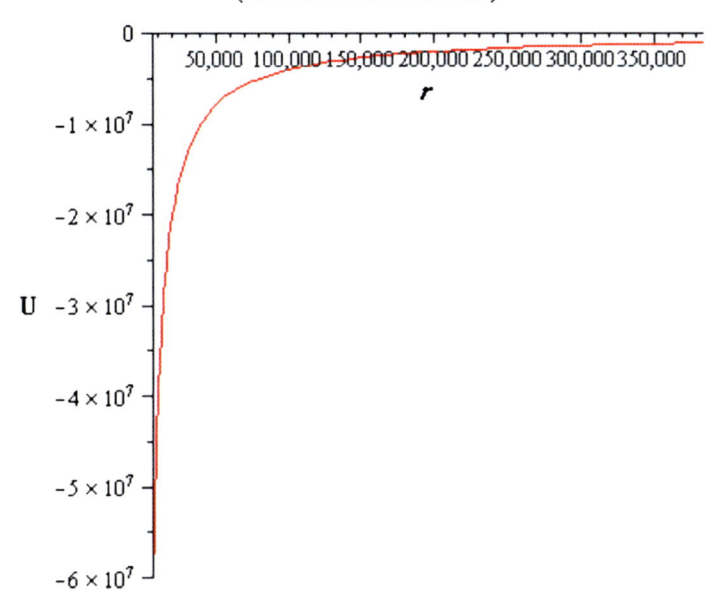

(Distance is in kilometers)

Similarly, when a meteor flies toward Earth, its kinetic energy increases as its potential energy becomes more and more negative: it becomes ever less as it descends. It is the conversion of the gravitational potential energy to kinetic energy that makes it speed up on the approach. In liftoff of a rocket, the opposite occurs, only here it is the chemical energy of the rocket giving thrust that does the work of converting gravitational potential energy to kinetic energy. The rocket gains distance (potential energy becomes closer and closer to zero) and it gains velocity (kinetic energy increases). The key thing to remember is that the sum of the two energies is always the same. One form converts to another, in either direction or along any path, but the sum is always the same.[3] Energy is conserved. Returning to

---

[3] We assume for the sake of our discussion that no energy is dissipated by friction as a meteor or spacecraft travels through the atmosphere, which would convert some of the kinetic energy to heat (and slow the object down). In fact heat through atmospheric friction is a significant source of energy loss in such cases. It is what melts and destroys most meteors before they hit the Earth.

the case of a circular orbit, its energy will not change unless something external, a force, changes it. Such a force could be friction, the influence of other perturbing masses, the thrust of a rocket, impacts or other influence that either augments or dissipates the orbital energy in a particular case.

Examining the graph we see that the gravitational potential energy (think of it as the "pull-it-back" energy) acting on a mass increases sharply as it leaves the Earth, but its *rate* of increase drops the farther away from Earth it goes, as the pull of the Earth's gravity weakens by the inverse square law, and the work at each succeeding interval of distance is less. The mighty effort of getting it up and away from the powerful, close-in hug of Earth's gravity is manifest in the steep slope near the Earth. The U graph is the record of the work required to extract the mass from the tight grip of our planet. Notice that as we get farther away from Earth, the curve runs almost horizontally along the x axis line, but does not touch it, and would theoretically go to infinity. There, U energy would be zero. If the graph went that far out, it would record all the work required to separate a unit mass from *all* influence of the Earth's gravity. But this graph just shows the work necessary to remove the mass from the immediate pull of Earth's gravity and hang it in space at the Moon's distance. Of course from there it would tend to fall back, converting all that gravitational potential energy into kinetic energy as it fell toward Earth. At every moment of the fall, its $K + U$ energies will be the same, a constant.

If we were to consider the graph with r at every point rotated completely around the vertical axis, the three-dimensional graph would look like funnel, or well. This "gravitational well" or "pit" exists around every mass, is determined by the quantity of the mass, and determines how much work is needed to "climb out of the well" and escape it. That is, the size of the gravitational well determines the amount of kinetic energy for a given mass that would be needed to overcome the gravitational potential energy pulling it back.

Let us go out in space and consider ideal circular orbits of a two-body system, where each mass moves in lock-step at a uniform speed around the center of mass. Throughout the course of their orbits, the K energy of a given mass, based as it is solely on velocity, is constant. Likewise the U energy is constant. It is based upon the distance of the masses from each other, and in circular orbits this U energy is unchanged. Constant distance, constant velocity means no exchanges of K and U energies: each remains steady in any circular, eccentricity–zero orbit. Where orbits are not circular but elliptical, however, the non-zero eccentricity means that the distance of the masses from each other changes as they orbit the center of mass: the U energy changes. This change in U requires that the K energy must change, since the sum of U and K must always be constant. As U increases (as work is employed to separate the masses), then the K energy (the velocity) needed for this work must decrease. As U decreases (as "negative" work brings the masses together), the K energy is the beneficiary of this decrease in U, and velocity must increase. The energy needed to perform the work of bringing the masses together and of separating them—for example as a planet approaches perihelion then half an orbit latter reaches perihelion—comes respectively from the U energy of gravity and the K energy of motion, each continually being exchanged for the other throughout the course of each orbit. This explains in energy terms the velocity curves of Earth and Mars as they swing through their orbit, moving closer, then farther from the Sun, perpetually converting one form or energy to another then back again.

We can put these powerful principles to use to test the results we obtained earlier regarding the velocity along the apsides of the ellipse. We will also refer in our proof to the conservation of angular momentum, which we encountered earlier. We will consider any two masses orbiting their common center of mass.

Problem Using the laws of conservation of energy and angular momentum, find the velocities in an elliptical orbit at the apsides, where two masses orbit their common center of mass, at distance R from each other.

Given

$K = \frac{1}{2}mv^2$	Kinetic energy of any mass at velocity v
$U = -\frac{Gm_1m_2}{R}$	Potential energy of revolving masses separated by distance R
mv_Ar_A	The angular momentum of a mass at aphelion or apogee, where v_A and r_A are the respective velocity and radius vector of the mass at that point
mv_Pr_P	The angular momentum of a mass at perihelion or perigee where v_P and r_P are the respective velocity and radius vector of the mass at that point

Assumptions We assume symmetrically spherical masses in idealized elliptical orbits in a two-body system with no perturbing or other external forces which alter affect it. Therefore, we assume no influences which either augment or dissipate its total energy.

Method The first step is to design an equation for the total energy of the orbiting system. Consider two masses m_1 and m_2 separated by distance R, with two velocities, v_1 and v_2:

$$E_T = \frac{1}{2}m_1v_1^2 + \frac{1}{2}m_2v_2^3 - \frac{Gm_1m_2}{R}$$

This represents the total energy as equal to the sum of the kinetic energies of each of the masses, at their respective velocities, and their potential energy, which is a function of their masses and distance apart. It doesn't matter where they are in their orbits, or whether the orbits have high or low eccentricity: the total energy will always be constant. But note that there are two velocity terms. How can we simplify this? By using the concept of relative velocity, it is mathematically possible to convert this two-body problem, where two masses orbit the center of mass, into a problem of one mass orbiting the other mass, with our rest-frame of reference being on one of the masses. In other words, rather than begin from the point of view of an observer outside the system, we will, by the technique described in Chap. 10, mathematically place ourselves on one mass. The relative velocity of the other mass is then the sum of the two velocities. We will call the rest-frame-of-reference mass m_1, and solve for relative velocity v. We simplify the energy calculations by evaluating the kinetic energy of the single orbiting mass, and the potential energy occasioned by the distance between them. Since energy in this conservative system is neither gained nor lost, the total mechanical energy of any orbiting mass will still be the same at all points in the orbit. The total energy E_T in any kind of closed orbit, circular or elliptical, should equal $K + U$ everywhere. This means that the total energy E_T is also constant at the apsides, where the calculations are easier. Hence, energy at perihelion equals energy at aphelion, $E_P = E_A$. We can therefore, equate the sum of K and U energies at aphelion and perihelion and set up two equations for the apsidal velocities: an equation is written for each of the apsides, one using r_1 for R and the other using r_2. Each of the radius vectors can be equated with its respective elliptical identity: $a(1 \pm e)$. Then, since we know the total energy at one of the apsidal points will equal that of the other, we set the two E_T equations equal to each other and solve for the applicable velocity at each point. Finally, the momentum conservation equations can be employed to obtain ratios of velocities and radius vectors. Because at the apsides we also know the velocity of the mass in terms of elliptical eccentricity and semi-major axis, we can make substitutions and solve for velocities.

Calculations In the initial model, the sum of the kinetic energy K and potential energy U is equal to the total energy of the binary system:

$$E_T = \frac{1}{2}m_1 v_1^2 + \frac{1}{2}m_2 v_2^2 - \frac{Gm_1 m_2}{R}$$

Now we translate the separate velocities into relative velocity, v. We make use of these identities for converting individual velocities into relative velocity:

$$v_1 = v\left(\frac{m_2}{m_1 + m_2}\right) \qquad v_2 = v\left(\frac{m_1}{m_1 + m_2}\right)$$

These separate velocities can now be plugged into the energy equation to covert the whole thing into a single equation with only one velocity term:

$$E_T = \frac{1}{2}m_1 \left[\frac{m_2 v}{m_1 + m_2}\right]^2 + \frac{1}{2}m_2 \left[\frac{m_1 v}{m_1 + m_2}\right]^2 - \frac{Gm_1 m_2}{R}$$

Simplifying,

$$E_T = m_1 m_2 \left[\frac{v^2}{2(m_1 + m_2)} - \frac{G}{R}\right]$$

This is the total energy for a given relative velocity in the binary orbit. From here we can create two equations. Again, an equation is written for each of the apsides, one using $R = r_1$ and the other using $R = r_2$. Then, each of the radius vectors can be equated with its respective elliptical identity: $r_P = a(1 - e)$, and $r_A = a(1 + e)$. Then, since we know the total energy at one of the apsidal points will equal that of the other, we set the two E_T equations equal to each other and solve for the velocity at each of the apsides.

$$\underbrace{m_1 m_2 \left[\frac{v_P^2}{2(m_1 + m_2)} - \frac{G}{a(1-e)} \right]}_{\text{Total energy at perihelion}} = \underbrace{m_1 m_2 \left[\frac{v_A^2}{2(m_1 + m_2)} - \frac{G}{a(1+e)} \right]}_{\text{Total energy at aphelion}}$$

We'll call $\mu = G(m_1 + m_2)$ and simplify by canceling common terms and collecting the velocity terms on one side:

$$\frac{1}{2}\left[v_P^2 - v_A^2\right] = \frac{\mu}{a}\left[\frac{1}{(1-e)} - \frac{1}{1+e}\right]$$

While this is promising, the apparent difficulty here is that we again have two velocities, and we are looking for one or the other. As it stands, we cannot find the velocity at one of the apsides without knowing the velocity at the other. We have two variables; hence we need a second equation to solve for one. Here we can use another conservation law, the law of conservation of angular momentum. We know that the angular momentum will be the same at every point on the ellipse, so we can derive a useful ratio:

$$m v_P r_P = m v_A r_A$$

$$\frac{v_P}{v_A} = \frac{r_A}{r_P}$$

Using the radius vector–semi-major axis translations can simplify this neatly:

$$\frac{v_P}{v_A} = \frac{1+e}{1-e}$$

This equation has interest in its own right, since it shows clearly the relationship of the velocities to each other at the apsides. Returning to the previous equation, we can set it up to use that elegant ratio, and choose first to solve for aphelion velocity:

Deriving Apsidal Velocities in an Elliptical Orbit from the Laws...

$$\frac{v_A^2}{2}\left[\frac{v_P^2}{v_A^2} - 1\right] = \frac{\mu}{a}\left[\frac{1}{(1-e)} - \frac{1}{(1+e)}\right]$$

$$\frac{v_A^2}{2}\left[\frac{(1+e)^2}{(1-e)^2} - 1\right] = \frac{\mu}{a(1+e)}\left[\frac{1+e}{(1+e)} - 1\right]$$

Working through the algebra and cancelling in the numerators yields,

$$\frac{v_A^2}{2}\left[\frac{4e}{(1-e)^2}\right] = \frac{\mu}{a(1+e)}\left[\frac{2e}{(1-e)}\right]$$

With a little more simplification we finally arrive at this result for aphelion velocity:

$$v_A^2 = \frac{\mu}{a}\left(\frac{1-e}{1+e}\right)$$

Solving for perihelion velocity similarly yields,

$$v_P^2 = \frac{\mu}{a}\left(\frac{1+e}{1-e}\right)$$

These of course are the same equations we found before, but derived from energy considerations. Unpacking the concise notation and summarizing, we have,

$$v_A = \underbrace{\sqrt{\frac{G(m_1+m_2)}{a}\left(\frac{1+e}{1-e}\right)}}_{\text{Perihelion, periapse, or perigee velocity}} \qquad v_P = \underbrace{\sqrt{\frac{G(m_1+m_2)}{a}\left(\frac{1-e}{1+e}\right)}}_{\text{Aphelion, apoapse, or aogee velocity}}$$

Observations

1. Note once again how, if the orbit is circular ($e = 0$), the equation simplifies to the common expression for velocity in a circular orbit, $v^2 = \mu/r$.
2. If the one of the masses is insignificantly small in relation to the other, $m_2 \ll m_1$, we again have our approximation that

$$v_A^2 = \frac{Gm_1}{a}\left(\frac{1+e}{1-e}\right) \qquad v_P^2 = \frac{Gm_1}{a}\left(\frac{1-e}{1+e}\right)$$

3. We used the relative velocity to provide a relative picture, with one mass as the frame of reference. If we used *reduced mass* instead, we would arrive at the same result. We leave that derivation for the reader to explore.

The Gaussian Constant in Celestial Mechanics

The asteroid Ceres had been discovered in 1803, but was soon lost as its path took it into the Sun's glare. The tools then available to astronomers were not so well developed as to determine the whole orbit of a thing from just a few observations. Then a young man named Karl Friedrich Gauss developed a method to determine such orbits. His method was used by astronomers to recover Ceres after its passage around the Sun. It was a stunning success and Gauss became instantly famous. Gauss evolved into one of the greatest mathematicians of all time, and contributed brilliantly to a broad range of problems. Of interest to us here is what became known as the *Gaussian constant*.

Gauss sought a constant for use with Kepler's Third Law that would be valid for all objects in any orbit around the Sun, and would make orbital calculations simpler. He derived the constant k that bears his name in his famous 1809 work, *Theory of Motion of the Heavenly Bodies Moving About the Sun in Conic Sections,* sometimes referred to by its shortened Latin title, *Theoria Motus*. Gauss used the Earth's orbit around the Sun as his reference. He took the distance between the Earth and Sun as the unit distance, and the Sun's mass as one unit.

Earlier we saw how when we changed units the equations became much simpler. The mass unit was one solar mass; the distance unit was one astronomical unit (AU). The secondary mass was assumed to be a particle of near-zero mass is orbiting the Sun. When the unit of time was in seconds, the inertial constant became this:

$$G = \frac{4\pi^2}{P^2}$$

G where units are solar masses, AU and seconds

When the year is the time unit (such that $P = 1$) the constant further simplified to this:

$$G = 4\pi^2$$

G where units are solar masses, AU and years

If, on the other hand, we use *days* as our unit of time, as Gauss did, then we have a constant that will tell us orbital periods of planets in days, which can also be very useful. In Gauss' day the year's length was not as precisely known as it is today, and with a slight adjustment for the mass of the Moon, he arrived at the value for $\sqrt{G}$[4]:

$$\sqrt{G} = .01720209895$$

---

[4] In Gauss' day the period of revolution of the earth was taken as 365.2563835 days. The combined mass of the Earth-Moon system was assumed by Gauss to be 1/354710 of the Sun's mass.

Gauss called this value k so,

$$k = .01720209895$$

This is the *Gaussian constant* or the *Gaussian gravitational constant*. Thus, with these units (solar masses, AU and days),

$$\underbrace{G = k^2}_{G \text{ where units are solar masses, AU and days}}$$

This value of k is used today in celestial mechanics for the solar system, where the mass and distance units are solar and AU, respectively, and where the time units desired are days.[5] It has fundamental importance as an astronomical standard constant.

Use of the Gaussian constant can simplify calculations in celestial mechanics. And even though the values for mass of the Earth-Moon system and Earth's orbital period have been refined over the years since the early 1800s, the Gaussian constant has been adopted as a fundamental constant in astronomy.

Using the Gaussian Constant to Find Heliocentric Periods

For bodies of negligible secondary mass orbiting the Sun, where the units are AU and days, Kepler's equation in terms of the Gaussian constant can be written as:

$$P = \frac{2\pi}{k} a^{3/2}$$

Using the Gaussian constant, we can as an example find the orbital period of Mars (neglecting its own mass), given that its semi-major axis is about 1.52 AU (or, more precisely, 1.52371034 AU)[6] from the Sun:

$$P = \frac{2\pi}{.01720209895} (1.52371034)^{3/2}$$

$$P = 686.99 \text{ days}$$

which is 1.881 years. This is an easy method to get quite accurate results in celestial mechanics calculations.

For bodies of more considerable mass, the equation reverts to this general form (where the 1 is the Sun's mass in solar mass units):

---

[5] The modern astronomical unit has been redefined (a little) to, in effect, take out the mass adjustment Gauss made in determining his constant.

[6] See http://ssd.jpl.nasa.gov

$$P = \frac{2\pi}{k\sqrt{1+m}} a^{3/2}$$

where m is the mass of the body in heliocentric orbit, in solar masses. The Gaussian constant was derived from bodies orbiting the Sun, and is commonly used for such bodies. It applies to *any* body: planet, comet, asteroid, or spacecraft in solar orbit. It was derived from the particular facts of the Earth-Sun system, because they were known, but the same constant could just as easily have been derived from another planet's data in the solar system.

Applying the Gaussian Constant to Find Heliocentric Orbital Velocities

The Gaussian constant is expressed in terms of solar masses, astronomical units and days. For finding the period of heliocentric orbits, this is very useful. Later we will be calculating velocities in heliocentric orbits. The usual result in such calculations is kilometers per second. So it would seem the AU/days units for the Gaussian constant would not be much of a time-saver. However, we can derive a conversion factor for such purposes. For the conversion we divide 1 AU in meters by the length of the day, which is 86,400 s. Hence the Gaussian AU/day units can become meters per second. To find its value, we divide

$$\underbrace{\frac{\text{AU}}{\text{Day}}}_{\text{Units for Gaussian constant k}} \rightarrow \underbrace{\frac{\text{AU in kilometers}}{\text{Day in seconds}}}_{\text{Conversion factor when seeking result in km/s}}$$

Inserting the correct values for the numerator and denominator,

$$\frac{149,597,871 \times 10^8}{86,400}$$

$$\underbrace{1731.46}_{\text{Conversion factor from AU/days to km/sec}}$$

---

[7] The precise result is 1731.45684 which can be used where more precision is desired.

The Gaussian Constant in Celestial Mechanics

This number is good to remember.[7] To see how this is useful, recall the circular velocity in a heliocentric orbit is given by

$$v = \sqrt{\frac{Gm_\odot}{a}}$$

Using the units for the Gaussian constant k (where mass units are solar, time units are days and distances are in AU) the equation becomes,

$$v = \frac{k}{\sqrt{a}}$$

To take the example of Mars again, its *mean* velocity should be,

$$v = \frac{.01720209895}{\sqrt{1.5237}}$$

$$v = .0139358 \, \text{AU/day}$$

These are awkward units! Converting to kilometers per second, multiply the result by the conversion factor derived above and we obtain, for a mean orbital velocity of Mars, the value,

$$1731.46 \times .0139358 = 24.1 \, \text{km/sec}$$

We used two constants, the Gaussian constant and a conversion factor. Combining these into one "modified" Gaussian constant for a heliocentric orbit, and calling the result the Greek letter kappa, we get

$$\kappa = 1731.46 k$$
$$\kappa = 29.785$$

We will use this number frequently in heliocentric velocity calculations.[8] The equation for mean velocity in kilometers per second thus becomes,

$$v = \frac{\kappa}{\sqrt{a}}$$

---

[8] The result for κ to more decimal places is 29.78469189. We will use the rounded value for the instructional purposes of this book.

Simple Computation of Apsidal Velocities of Objects in Heliocentric Orbits

These equations may be applied to generate results for heliocentric apsidal velocities in orbits that are in kilometers per second. Referring to the earlier-developed equations for such velocities, we can apply the shorthand Gaussian constant in this way (where units are AU and $\kappa = 29.785$):

$$v_P = \frac{\kappa}{\sqrt{a}} \left(\frac{1+e}{1-e} \right)^{\frac{1}{2}}$$

<u>Velocity at perihelion in km/s for helicentric elliptical orbits</u>

$$v_A = \frac{\kappa}{\sqrt{a}} \left(\frac{1-e}{1+e} \right)^{\frac{1}{2}}$$

<u>Velocity at aphelion in km/s for helicentric elliptical orbits</u>

Sometimes it is convenient to work with just the perihelion and aphelion distances in computation. Because aphelion $Q = a(1 + e)$ and perihelion $q = a(1 - e)$, the two equations above become:

$$v_P = \kappa \left(\frac{1+e}{q} \right)^{\frac{1}{2}}$$

<u>Velocity at perihelion in km/s for heliocentric orbits, using AU units</u>

$$v_A = \kappa \left(\frac{1-e}{Q} \right)^{\frac{1}{2}}$$

<u>Velocity at aphelion in km/s for heliocentric orbits, using AU units</u>

Where again, $\kappa = 29.785$

Looking back, it is instructive to note that all we have done, ultimately, is to replace $\sqrt{G(m_1 + m_2)}$ in the original velocity equations with a new proportionality constant κ which was designed (as Gauss did with his constant) to make the units come out the way we wanted them to. If you prefer other units, you may repeat the basic analysis and design your own constant.

Exploring Sedna's Orbit

Sedna is one of the more fascinating trans-Neptunian dwarf planets discovered in recent years. It is remote and very red. It is most interesting in respect to its unusual orbit, which will be explored below.

The Gaussian Constant in Celestial Mechanics

Problem Given the just the orbital elements below, determine the period of Sedna in days and years and its velocities at perihelion and aphelion.
Given

.85904862	Eccentricity, e, of Sedna's orbit
541.4295	Semi-major axis, a, of Sedna's orbit in AU
.01720209895	Gaussian constant, k, for finding heliocentric periods in units of days
29.785	Modified Gaussian constant, κ useful for computing orbital velocities in kilometers per second

Assumptions We will assume no perturbing influences on Sedna.
Method Using the equation with the Gaussian constant entails substituting the given parameters and solving for period. The period will be in days. Divide by 365.25 to yield years. Similarly, using the modified Gaussian constant enables computation in kilometers per second of the apsidal velocities of Sedna.
Calculations First, to find the period in days using the Gaussian constant, then years,

$$P = \frac{2\pi}{k} a^{3/2}$$

$$P = \frac{2\pi}{.01720209895} (541.4295)^{\frac{3}{2}}$$

$$P = 4,601,625.8 \text{ days}$$

$$P = 12,598 \text{ years}$$

In order to find the perihelion velocity, we use the equation,

$$v_P = \frac{\kappa}{\sqrt{a}} \left(\frac{1+e}{1-e} \right)^{\frac{1}{2}}$$

$$v_P = \frac{29.785}{\sqrt{541.4295}} \left(\frac{1+.85904862}{1-.85904862} \right)^{\frac{1}{2}}$$

$$v_P = 4.65 \text{ km/s}$$

By the same process, using the aphelion equation, we obtain for the aphelion velocity,

$$v_A = 0.352 \text{ km/s}$$

Observations

1. At aphelion, Sedna travels about a third as fast as our Moon does in its orbit. Even at perihelion, it moves more slowly that Neptune's mean velocity.
2. As can be seen from these results, this is a most unusual orbit for a solar system body! The last time Sedna was in its current position, the Northern hemisphere was emerging from the last cycle of glaciations. According to the JPL website, Sedna will reach perihelion on July 6, 2075.[9]
3. We can find perihelion distance q by remembering that $q = a(1 - e)$. Thus,

$$q = 541.4295(1 - .85904862)$$

$$q = 76.315 \text{ AU}$$

This *perihelion* distance is well beyond the orbit of Pluto. At aphelion, Q, given by the relation $Q = a(1 + e)$, Sedna is 1,006.54 AU from the Sun! The nature of this orbit raises interesting questions about its origin.

Exercises: The Many Strange Worlds Beyond Pluto The creative use of some of the world's most powerful optics has led to the discovery of a host of distant "dwarf" planets far beyond the orbit of Pluto. Pluto itself was demoted to the status of dwarf planet after the discovery of Eris by Michael Brown and his team at CalTech, who led the way to finding Eris and several more of these so-called Kupier Belt Objects (KBOs). Eris is more massive than Pluto, known to be so because of its tiny satellite, Dysnomia. Yet Eris is only one of the far-out objects discovered by Brown's group. They have found a menagerie of distant worlds in the Kuiper Belt which may shed light on the origins of our solar system. The variety of these KBOs is remarkable: enormous Eris (accompanied by its moon), rocky Quaoar (with a moon), snow-white Makemake (with no known moons), blimp-shaped Haumea (with two moons), and remote, red Sedna, so distant that it is in a class by itself, with an orbital period of about 12,600 years. Because Eris was the troublemaker that began the demote-Pluto movement, we begin the exercises below with investigations of its orbit.

Problems

1. The dwarf planet Eris has a semi-major axis $a = 68.047440$ AU, and an eccentricity $e = .4347542$. Using just that data, calculate its velocities in kilometers per second at perihelion and aphelion.
2. Given Eris' semi-major axis, calculate the period of Eris in days and years using the Gaussian constant.
3. Imagine a circle circumscribing Eris' orbit whose radius is equal to Eris' semi-major axis. (1) Calculate its mean velocity in that fictitious circle in kilometers per second. (2) Apply the elliptical correction factors for perihelion and

---
[9] http://ssd.jpl.nasa.gov/sbdb.cgi#top

aphelion to determine the velocities of Eris at the apsides in kilometers per second. (3) Compare the resulting apsidal velocities with the answers to Problem 1. Take the length of an astronomical unit to be 149,598,000 km.

4. Given the semi-major axis and eccentricity of Eris and the mass of the Sun of 1.9891×10^{30} kg, compute the velocity of Eris in SI units at perihelion and aphelion.

5. Using the SI system of units, calculate: (1) the velocity of Eris on the fictitious circle described in Problem 3; (2) the centripetal acceleration for Eris as it would be on this fictitious circle; and (3) Eris' centripetal acceleration on its actual elliptical orbit at the apsides.

6. As a check on your results in the previous problem, compute the gravitational acceleration imparted upon Eris by the Sun at Eris' perihelion and aphelion.

7. Eris at perihelion (q) is 38.4635 AU distant from the Sun. Its aphelion distance is 97.63 AU. Assuming a mass of Eris of 1.659584×10^{22} kg, determine the potential energy U and the kinetic energy K of Eris at both apsidal points. Compare the sum of U and K at each of those points. Explain your results. Use SI units.

8. The elongated, dwarf planet Makemake has a semi-major axis $a = 45.4363$ AU, and an eccentricity $e = .1625448088$. Using just that data, calculate its velocities in kilometers per second at perihelion and aphelion, and its period in days and years.

9. The dwarf planet Haumea has a semi-major axis $a = 42.9849$ AU, and an eccentricity $e = .197523$. Using just that data, calculate its velocities in kilometers per second at perihelion and aphelion, and its period in days and years.

10. Haumea's two small satellites, Hi'iaka and Namaka, were discovered in 2005 by Michael Brown and his team using the adaptive optics technology at Hawaii's W.M. Keck Observatory. The brightest moon, Hi'iaka, orbits its parent with a semi-major axis of 49,880 km once in 49.462 days with an orbit eccentricity of .0513. Find (1) the mass of Haumea, and (2) the velocities of Hi'iaka in meters per second at its apsides.

Chapter 13
The Energy and Geometry of Orbits

The velocities we found for the elliptical orbit were at the apsides, the places where the comet, or the orbiting planet, moon, star or spacecraft is nearest or farthest from one focus of the ellipse. These again are the perihelion and aphelion distances (when referring to the Sun), perigee and apogee (the Earth), periastron and apastron (stars), or more generally for any object, periapsis and apoapsis. These two points in the orbit are important to know, but it is usually essential as well to be able to calculate the precise orbital velocities at other points of the elliptical orbit. How fast will the comet be moving against the stars when it travels past the Earth? Is it on its predicted path, or has its speed been altered by the gravitational influences of another planet during its trip? In another context, it can be critically important to know at each moment the velocity of a vehicle destined for a rendezvous with a moon, planet, comet or asteroid. We want to compare actual with predicted velocity of the vehicle to see if it is on course for the interception. In the Apollo Moon program days, it was vital to know the speed of the returning spacecraft to predict whether it would have a successful "insertion" at just the right spot on reentry, or fly past the Earth and be lost in space. The many smaller probes sent into the corners of our solar system have relied on our accurate knowledge of velocity at any point in the orbit. Velocity corrections are commonplace for space vehicles, which often employ short bursts of rocket thrust to keep them on course.

How can this type of velocity calculation be done? The key is for us to refer to the total energy of the orbit, which is constant everywhere in the orbit. If we can calculate the energy of the orbit in one place, say at the apsides, then the energy at every other place must be the same, due to the law of conservation of energy. If we can then relate velocity to that energy, we'll have the key to unlock the mathematical puzzle, and create an equation for velocity at any point along the ellipse.

The law of conservation of energy offers rewards in the study of orbital dynamics. We have seen its use in deriving the basic velocity equations. More fundamentally, the energy of a mass in orbit will characterize the orbit itself. It will tell us what kind of conic section the orbit is: ellipse, parabola or hyperbola. Energy considerations invite remarkable clarity into the subject. To continue our inquiry into the energy of orbits, we must first examine the total energy in the circular orbit, then the more general case of the elliptical orbit. From there we will be positioned to evaluate the other conic sections in terms of energy.

The Total Energy in a Circular Orbit

Recalling the discussion on the conservation of energy in the last chapter, the total energy E_T in any orbit is the sum of kinetic and potential energies: $E_T = K + U$, which, in the absence of external forces, is constant everywhere in the orbit. As before, we can express the total energy in a circular binary orbit this way:

$$E_T = m_1 m_2 \left[\frac{v^2}{2(m_1 + m_2)} - \frac{G}{r} \right]$$

where r is the distance between the two masses. Remember this equation; we will use it again soon. Since the velocity in a circular orbit is this,

$$v^2 = \frac{G(m_1 + m_2)}{r}$$

we can substitute the right-hand side of this equation for velocity in the energy equation. The result, after simplifying, is total energy in a circular orbit of radius r:

$$E_T = -\frac{G m_1 m_2}{2r}$$

Examine this equation. It appears that the total energy in the case of, say, a particle of mass m_2 in a circular orbit around a primary mass m_1, is dependent solely upon the radius of the orbit. Does this make intuitive sense? Velocity is constant in a circular orbit, and, with an unchanging radius, there will be no variation in gravitational potential energy (which would change only with a change of distance from the center of mass). Hence there should be one and only one total energy for a mass orbiting a primary at a given radius. Moreover, the total energy is *negative*. This is again because potential energy is always negative, but gets less so with distance; hence it increases at a decreasing rate (moves closer to zero) as radius increases. At infinite distance it is zero.

A circular orbit has a finite radius, and is said to be *bound*. This is not always easy to understand intuitively, but think of the gravitational potential as being sufficiently great (negatively great!) to *bind* the mass into this orbit. An object may be thought of as circling fully within the gravitational well. If the total energy in a given case is negative, the gravitational pull factor (the gravitational potential energy, $-U$) in the equation is dominant over the velocity factor (the kinetic energy, K) and the orbit is regarded as bound.

The Total Energy in an Elliptical Orbit

The circular orbit may be regarded as the simplest case, a special form of the ellipse. What is the total energy of a mass in an elliptical orbit? For an elliptical orbit, we again begin with the equation for the total energy in an orbit, using r for the distance between the masses:

$$E_T = m_1 m_2 \left[\frac{v^2}{2(m_1 + m_2)} - \frac{G}{r} \right]$$

Now let us investigate the energy at one location of the orbit, for example at perihelion, whose velocity we know. Recall from the previous chapter that at perihelion, the velocity (squared) is,

$$v_P^2 = \frac{G(m_1 + m_2)}{a} \left(\frac{1+e}{1-e} \right)$$

We could just have easily picked the aphelion velocity; the energy will be the same at each point. We now substitute the right-hand side of this velocity equation for v in the energy equation, recalling that $r = a(1 - e)$ at perihelion:

$$E_T = G m_1 m_2 \left[\frac{(1+e)}{2a(1-e)} - \frac{1}{a(1-e)} \right]$$

Simplification of this expression conveniently cancels out the eccentricity terms and yields the total energy of the elliptical orbit:

$$E_T = -\frac{G m_1 m_2}{2a}$$

This expression is functionally equivalent to the circular orbit energy, except that instead of radius it refers to the semi-major axis a. For given masses, the total orbital energy depends only on the length of the semi-major axis. Eccentricity is irrelevant.

This equation refers to *total* energy in the orbit. Let's look again at the types of energy involved. Unlike the case of the circular orbit, the values of K and U in the elliptical orbit continuously change as the mass orbits the ellipse, their sum remaining constant. Increases or decreases in distance from the mass (and thus changes in potential energy U), will demand corresponding decreases or increases in kinetic energy K, and hence velocity, to keep E_T constant in accordance with the law of conservation of energy. As we saw before, when the distance from the Earth to an orbiting satellite is less in its elliptical path (perigee), its velocity will be at maximum, and where its distance is greatest (at apogee) its velocity is less, consistent with Kepler's Second Law. There is thus a constant balancing of the two energies at every point in the orbit such that the total energy between and at these extremes is *always* constant. Of course, other dissipative changes to the total energy of the system outside of this idealized hypothetical, such as atmospheric friction, would alter this result.

Considering the geometry of elliptical and circular orbits, and these conclusions make intuitive sense. Look again at the figure in the last chapter where the semi-major axis of the ellipse is equal to the radius of the circumscribing circle:

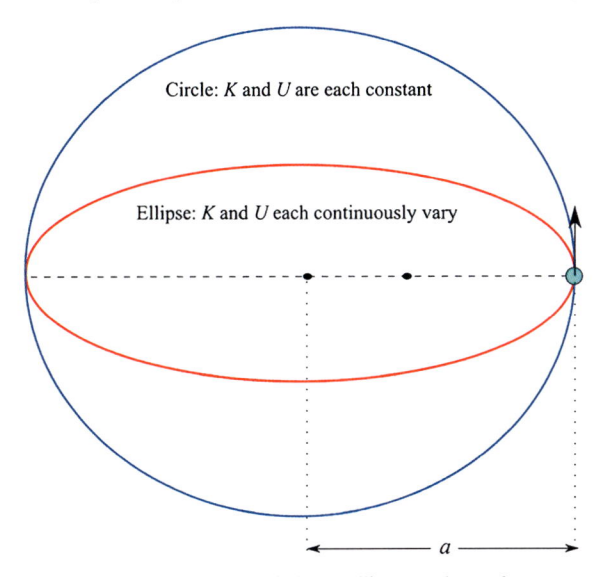

Circle: K and U are each constant

Ellipse: K and U each continuously vary

a

But *total energy* $K + U$ in circle and ellipse are always the same

In the ellipse, the K and U energies are always varying, but their sum in each case is constant. As we decrease the eccentricity of the ellipse, it becomes nearer to a circle and the variability of the K and U energies will diminish, but their sum will not change at all. When the variability ceases, at circularity, the sum is still what it always was, and is no different from that of a circle. The total energy of the circular orbit is the same as the total energy of the elliptical orbit in the case where the radius and semi-major axis are the same.

We have seen that the total energy in a bound orbit is negative. Is there ever a case where total orbital energy is zero or positive? We will see shortly that where velocity of the mass is sufficiently great, the kinetic energy component can *equal* the gravitational potential energy—such that the K and U energies are exactly equal to each other—and the total orbital energy is *zero*. In that very special case we have a *parabolic* orbit, where the mass has just the right energy to *unbind* itself from the gravitational pull and *escape*. If the velocity is still greater, the total energy is *positive* and we have a *hyperbolic* orbit. Here the K energy of velocity dominates the energy equation wholly, until at infinite distance (where U is zero), the *only* remaining component is the kinetic energy.

Velocity Anywhere Along the Elliptical Orbit

Returning to the equation for the conservation of energy, $E_T = K + U$, we found that the total energy of the elliptical orbit equals,

$$E_T = -\frac{Gm_1m_2}{2a}$$

This must be set equal to the expression for total orbital energy as a function of velocity, we found above:

$$m_1m_2\left[\frac{v^2}{2(m_1+m_2)} - \frac{G}{r}\right] = -\frac{Gm_1m_2}{2a}$$

Canceling, simplifying, and solving for velocity yields,

$$v^2 = G(m_1+m_2)\left(\frac{2}{r} - \frac{1}{a}\right)$$

Using the shorthand notation $\mu = G(m_1 + m_2)$, (or Gm_1 if $m_1 >> m_2$) the equation appears in its most often-cited form:

$$v^2 = \mu\left(\frac{2}{r} - \frac{1}{a}\right)$$

This equation, known as the energy equation or *vis viva* equation, is especially important in celestial mechanics. (It is commonly written in this velocity-squared form.) We derived it completely using the law of conservation of energy. The semimajor axis a being given, one can mathematically place one end of the radius vector r at the focus of the ellipse and point the other to any spot on the curve where the velocity is sought, and the equation will yield the velocity at that point on the ellipse. Demonstrate for yourself that directing the vector to either of the apsides generates the same apsidal equations for velocity we found before.

If the center of force is the Sun and it is convenient for the units to be in astronomical units and kilometers per second, the equation may be modified as before, where the constant $\kappa = 29.785$:

$$v^2 = \kappa\left(\frac{2}{r} - \frac{1}{a}\right)$$

Where the ellipse devolves into a circle, such that $a \to r$, the equation becomes the familiar equation for the velocity (squared) in a circular orbit,

$$v^2 = \frac{\mu}{r}$$

Energy-wise, the fundamentals of the elliptical and the circular orbits are the same: both are bound and the total energies of each are independent of eccentricity, functions solely of their respective semi-major axis and radius. Imagine moving an orbiting particle to a place infinitely far away from its orbit. In that case, its potential energy would have increased to become zero (since the distance to the center of force has become infinite), and its kinetic energy (and so its velocity) would have decreased to become zero, their sum always remaining the same. This change would require work, and the amount of that work exactly equals the total energy of its original orbit.

The Very Particular Parabolic Orbit and the Velocity of Escape

Suppose there is a body in an elliptical orbit with the Sun at one focus, and you are somehow able to reach down into the solar system and stretch it along its major axis, such that the value of a grows and grows. The Sun remains at one focus as we stretch it past the orbit of Pluto, past the Kuiper belt, all the way perhaps to the Oort Cloud. The secondary focus gradually becomes very remote from the Sun. The eccentricity of this orbit increases as we gradually flatten it, becoming closer and closer to unity. At .999999 eccentricity it is still an ellipse, and closed, but suppose we push the aphelion distance even farther out, in fact infinitely far out, such that our elongated orbit finally breaks. In such an orbit, the semi-major axis would become infinitely great, the eccentricity would be exactly 1 and the ellipse would become a parabola. What this means geometrically can be shown by examining the equation of the conic:

$$r = \frac{p}{1 + e \cos \theta}$$

Recall from the discussion in Chap. 4 that when $e = 1$ (parabola) and $\theta = 0$, then $r = \frac{1}{2} p$. When $\theta = \pi/2$ (which is 90°), then $r = p$. But when $\theta = \pi$ (180°), $\cos \theta = -1$ and r goes to infinity. Graphing the conic section when eccentricity is one yields this:

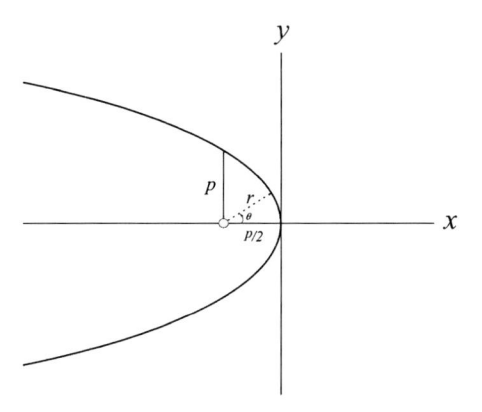

The Very Particular Parabolic Orbit and the Velocity of Escape

The parameter is the vertical line from the focus, which is here depicted as the Sun and the center of force. Notice that the perihelion distance is just half the parameter. What can we say about the velocity of a particle of mass in this parabolic orbit? Here is the *vis viva* equation derived above:

$$v^2 = \mu\left(\frac{2}{r} - \frac{1}{a}\right)$$

As a in the ellipse becomes larger and larger, the second term gets smaller and smaller. When it "opens" and we have a parabola, at $a \to \infty$, the second fraction becomes zero, and the equation reduces to this form:

$$v = \sqrt{\frac{2\mu}{r}}$$

Compare this with the equation for circular velocity. It differs only by the square root of 2. If the velocity of a comet is $\sqrt{2}$ times what its circular orbital velocity would be at a given distance from the Sun, we know it cannot be in a circular orbit; it will be *unbound* and escape. It will have climbed fully out of the Sun's gravitational well. If a spacecraft is in any orbit around the Sun, and it is given a thrust kick to propel it at 1.414 (the square root of 2) times its circular orbital velocity at that distance, it will escape from the solar system (unless other gravitational influences of, say, encounters with other planets pull it back).

From the energy perspective, we can describe what happened in this stretching. The total energy of a particle of mass in an elliptical orbit is $E_T = -Gm_1m_2/2a$, but because a is infinite in a parabolic orbit, the total energy of a particle in a parabolic orbit, according to that equation, must be zero. That is,

$$E_T = 0$$

and this $K + U = 0$ condition must be met at each point in its parabolic orbit, so uniquely in a parabola, the kinetic energy always equals the potential energy:

$$\frac{1}{2}m_2v^2 = \frac{Gm_1m_2}{r}$$

Which is of course consistent with the conclusion that $v = \sqrt{2\mu/r}$. Thus we can always identify an object in a parabolic orbit by its velocity. If its velocity is precisely the necessary escape velocity at its distance, then we can say its orbit is a parabola. This is a very hard condition to exactly satisfy. It represents the special case of unit eccentricity.

In a sense, the parabolic "orbit" should not be called an orbit at all, since the object will never return. It is a trajectory, and it walks a fine line: any less velocity and it will be bound into an elliptical or circular orbit. What if it is more? What if the velocity exceeds that of escape? Then we have the case of the hyperbolic trajectory.

Hyperbolic Trajectories

In orbits where the eccentricity is greater than one, the orbit is also unbound, and is the hyperbola. In this case, if we move a particle infinitely far away, so that its potential energy is zero, it may still have some residual velocity. In the case of the parabola, the total energy is always zero. But in the hyperbola, the lingering kinetic energy at infinity means that the total energy is not zero: it is positive. And the total energy at infinity must (again because of the conservation law) be the total energy everywhere. So for the hyperbola, the total energy is completely a function of that excess velocity: $E_T = \frac{1}{2} mv^2$. It can be shown that this total energy, being always positive, is equal to:

$$E_T = \frac{Gm_1 m_2}{2a}$$

Equating this with the energy conservation expression and solving for velocity, we have:

$$\frac{1}{2} m_2 v^2 - \frac{Gm_1 m_2}{r} = \frac{Gm_1 m_2}{2a}$$

$$v^2 = \mu \left(\frac{2}{r} + \frac{1}{a} \right)$$

This is the velocity in the hyperbolic orbit. If the center of force is the Sun and we desire the units to be in AU and kilometers per second, the equation becomes, for $\kappa = 29.785$,

$$v^2 = \kappa \left(\frac{2}{r} + \frac{1}{a} \right)$$

Compare this with the *vis viva* equation for the bound orbit. The only difference is the plus sign, which in the bound orbit is minus. This is because of the geometry of the hyperbolic orbit, in which the semi-major axis a is at the intersection of the asymptotes and its axis of symmetry, and is negative. The negative value of a can be extremely large. Nevertheless, the equation works to yield hyperbolic velocity dependent, indeed, only upon the value of a and the distance r from the center of force at which we choose to make our inquiry as to velocity.

Summary of Orbital Energy Relationships

The eccentricity of the conic section is a simple indicator of the bound or unbound nature of the orbit. Where $e < 1$ we have a bound orbit, an ellipse or circle. Anything more and the orbit is unbound. Where $e = 1$ the result is a parabola

Summary of Orbital Energy Relationships

and where $e > 1$ the curve is a hyperbola. The difference between the bound orbits of the ellipse and circle, and the unbound parabolic and hyperbolic trajectories near unit eccentricity is a narrow one, yet this is where the eccentricities of many comets lie. A simple graph demonstrates this. Take for example the general equation for any conic section:

$$r = \frac{p}{1 + e \cos \theta}$$

In this equation p is the parameter, known since Newton's day and before as the *semi-latus rectum*, e is the eccentricity and θ is the angle from perihelion counter-clockwise on the arc of the conic. By graphing this equation with values of eccentricity chosen at .9, 1.0, and 1.1, we should see some closely aligned curves of the ellipse, parabola and hyperbola. Here, using *Maple* software is the result:

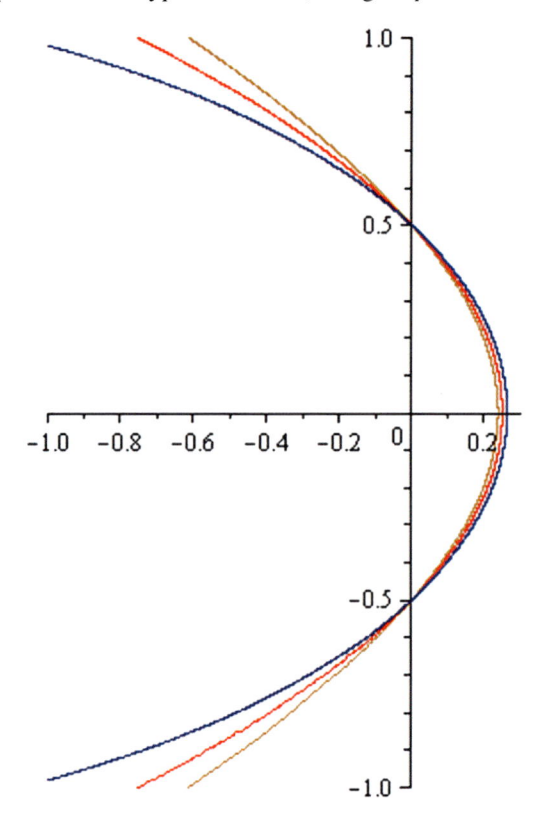

The focus is at the origin and the parameter p has been (arbitrarily) assigned the value of .5. (It is the distance vertically from the origin to that value on the *y* axis.) Looking to the left of the *y* axis, the blue ellipse line is the inner arc, the hyperbola is the gold outer arc, and the parabola is the red arc sandwiched in between. Note that on the right of the *y* axis, the order is reversed, with the elliptical arc being the outer curve, and the hyperbolic arc on the inside.

What is remarkable about the graph is that the differences in the curves near the origin are so slight. If the units were astronomical units, with the Sun at the origin, one can appreciate how exceedingly difficult it would be to tell, simply from the *shape* of the orbit, whether a comet was a periodic comet, due to return after it went to aphelion out in space, or a one-time visitor. Many comets have eccentricities greater than .9 yet less than 1, and none become dramatically bright until they near the Sun. If such a comet were spotted inbound within say, half an AU from the Sun, it would be a challenge to ascertain the nature of its orbit without repeated observations establishing its velocity.

If one can determine the velocity of the comet, then one will have a far a better grip on the nature of the orbit. If the comet has "parabolic" velocity, its path will be a parabola. Less velocity than this, it is an ellipse; more, a hyperbola. Recall this equation for finding the circular velocity in kilometers per second of an object at radius distance *r* astronomical units from the Sun:

$$v = \frac{\kappa}{\sqrt{r}}$$

where $\kappa = 29.785$. The parabolic velocity would be $\sqrt{2}$ greater than this, or

$$v_e = \kappa \sqrt{\frac{2}{r}}$$

Applying this equation, we know that the escape velocity of an object at the distance our Earth is ($r = 1$) from the Sun from its bound orbit would require something to give it a kick to about 42 km/s, and for Mars ($r = 1.524$), to a little over 32 km/s. Similarly, *any* object at 1 AU from the Sun, such as a comet, which is travelling at 42 km/s or greater will be in an unbound orbit. Its orbit will be parabolic if it is moving at exactly escape velocity and hyperbolic if the velocity is in excess of the escape velocity. Naturally, when such any object is nearer or farther from the Sun (or other center of force) its escape velocity from orbit would become greater or less because the gravitational pull of the main attractor is greater or less according to the inverse square law.

It is helpful to see how these energy relationships look when compared against actual solar system bodies. The chart below compares the perihelion velocities of a number of high-eccentricity comets and also the planets in the solar system, using this equation, which we used in the last chapter, to compute the orbital velocities at perihelion, where units are AU and kilometers per second:

Summary of Orbital Energy Relationships

$$v_P = \kappa \left(\frac{1+e}{q}\right)^{1/2}$$

where $\kappa = 29.785$. The graph is divided by the red diagonal, $e = 1$ line which represents exactly parabolic velocity. Objects with eccentricity of 1 would (if the graph resolution were greater) appear precisely on that line. Their total orbital energy in parabolic orbit is zero (K and $- U$ energies being in perfect balance). Objects to the right of the slanting line are in hyperbolic orbits, having eccentricities greater than 1 and positive orbital energies. Objects to the left of the slanted parabola line have eccentricities less than 1, negative orbital energies, and are in bound elliptical (or circular) orbits. These include the planets and asteroids in the solar system, stably bound to revolve around the Sun for eons, represented here by the solid blue diamonds to the left of the diagonal: Mercury, Venus, Earth, Mars, Ceres, Jupiter, Saturn, Uranus and Neptune are represented.

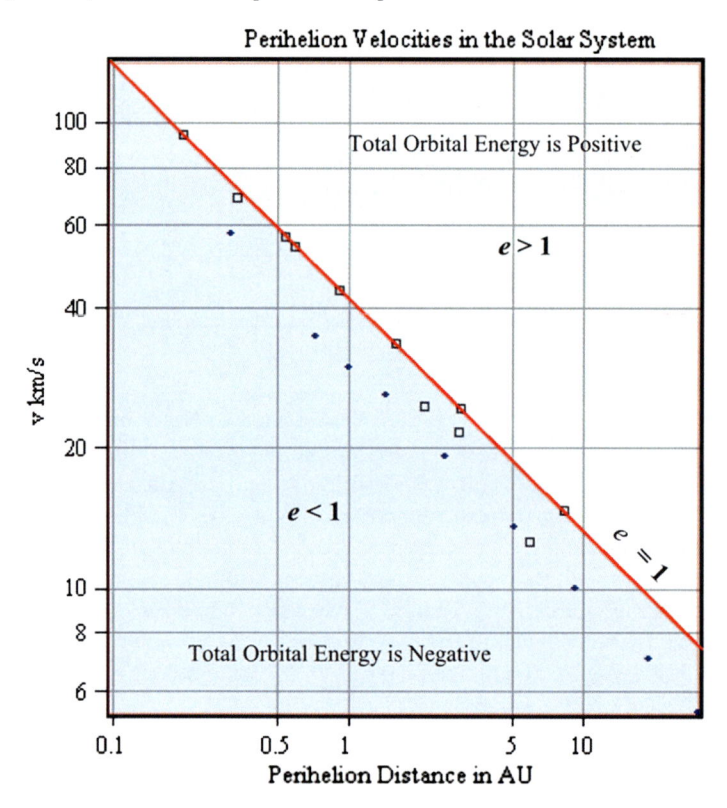

Perihelion Velocities in the Solar System

Why are they in a straight line parallel to the parabolic line? If one multiplied their circular orbital velocities by $\sqrt{2}$, they would be on the line. The small open squares are a random selection of high to very high eccentricity comets listed in the accompanying table.

Comets appearing in the graph[a]				
Comet	Designation	e	q [AU]	v_q [km/s]
Gerradd	C/2009 P1	1.000966	1.55	33.8
Boattini	C/2008 S3	1.000671	8.02	14.9
West	C/1975 VI-A	0.999971	0.20	95.0
Cardinal	C/2010 B1	0.999004	2.94	24.6
Bennett	C/1969 Y1	0.996193	0.54	57.4
Hale-Bopp	C/1995 O1	0.995082	0.91	44.2
Halley	1P	0.967143	0.59	54.6
Encke	2P	0.848332	0.34	69.9
N1	P/2011 N1	0.545582	2.86	21.9
Holmes	17P	0.432878	2.05	24.9
Schwassmann-Wachmann	29P	0.044074	5.73	12.7

[a]The data for these comets was gleaned from http://ssd.jpl.nasa.gov/sbdb.cgi

Determining the Velocity of a Near Earth Asteroid Which Passed by Earth

There is an ongoing effort to detect and track asteroids and comets that pass close to Earth. Depending on its size, an impact from such a Near Earth Object, or NEO, could pose a danger to Earth. NEOs are defined as objects (asteroids and comets) whose perihelion distance is less than 1.3 AU.[1] Asteroids in this group are called Near Earth Asteroids (NEAs). Many NEA's are very small and miss by a long shot, but occasionally one passes uncomfortably close to Earth.

An example of a close pass by a tiny NEO is the object designated 2010 TD54. It is small, with an estimated size of school bus, only 7 m, but had a "miss distance" on October 12, 2010 of a tenth of the distance between the Earth and the Moon. Since the mean lunar distance (LD) is 384,400 km, 2010 TD54 passed about

---

[1] The effort to detect NEOs is part of NASA's Near Earth Object Program, whose website, http://neo.jpl.nasa.gov/neo/groups.html contains a database of NEOs. Some Near Earth Asteroids (NEAs) (the Atens and Apollos) have "Earth-crossing" orbits. Asteroids that are large and bright and pass exceptionally close to Earth are called "Potentially Hazardous Asteroids." There are close to 1,200 of these discovered so far. The site defines these PHAs: "Specifically, all asteroids with an Earth **M**inimum **O**rbit **I**ntersection **D**istance (MOID) of 0.05 AU or less and an absolute magnitude (H) of 22.0 or less are considered PHAs." The absolute magnitude, or intrinsic brightness, of the PHA depends on its reflectivity (or *albedo*). With certain assumptions regarding its albedo, astronomers conclude that an object less than about 150 m in diameter would not be a PHA. A record of recent encounters with NEOs can also be found on the popular site, http://spaceweather.com/. That site states, "Potentially Hazardous Asteroids (PHAs) are space rocks larger than approximately 100 m that can come closer to Earth than 0.05 AU. None of the known PHAs is on a collision course with our planet, although astronomers are finding new ones all the time."

Summary of Orbital Energy Relationships

38,440 km from the center of the Earth.[2] This is in the neighborhood of our geosynchronous satellites! Below is NASA's rendering of this close pass:

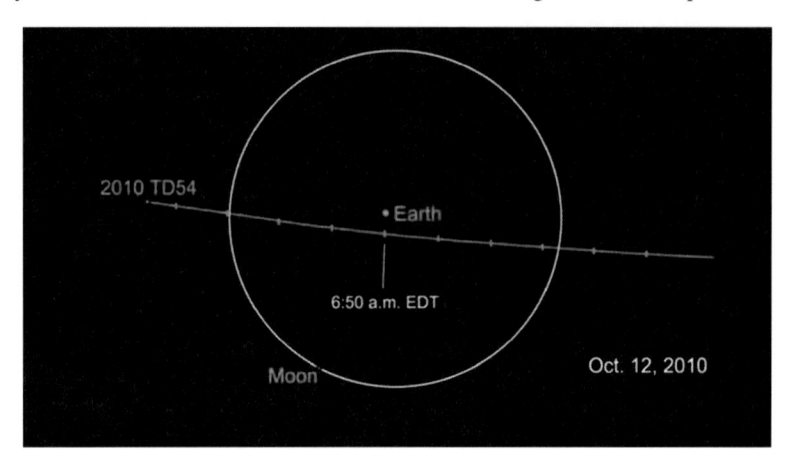

One can imagine that the small asteroid would appear be moving very quickly across our sky. The problem below estimates what the velocity of 2010 TD54 relative to the Sun was on the date of its close encounter with Earth.

Problem Given the information below, find the velocity of 2010 TD54 relative to the Sun as it passed between the Earth and Moon October 12, 2010. Compare that with its aphelion and perihelion velocities

Given

$v = \kappa \sqrt{\left(\frac{2}{r} - \frac{1}{a}\right)}$	The velocity equation for a heliocentric orbit
κ	Modified Gaussian constant ("kappa") of 29.785 that will yield a result in kilometers per second when other units are in astronomical units
r	The radius vector in AU between the Sun and the object in orbit. Since we are determining velocity at the time the asteroid crosses Earth's orbit, $r = 1$
a	Semi-major axis of 2010 TD54: 1.7853 AU
Q	Perihelion distance of 2010 TD54: 2.89 AU

Assumptions We ignore gravitational forces other than the Sun acting on the asteroid, and disregard the slight inclination (about 5°) of the plane of the asteroid's orbit relative to the plane of the Earth's revolution around the Sun (the *ecliptic plane*)

Method Solution of this problem entails straightforward substitution of values into the energy equation. Remember that units here are AU. The issue here will be

---
[2] The NASA news release for the near-Earth passage of 2010 TD54 stated that at closest approach the asteroid would pass over Singapore. See http://www.jpl.nasa.gov/news/news.cfm?release=2010-332.

finding the required value of the radius vectors r at the places in the orbit where it is needed for us to calculate velocity. For the passage close to Earth, we know that its distance from the Sun will be about the same as the Earth's, so for that location in its orbit $r = 1$. The respective values of r at perihelion and aphelion will just be those distances. Since aphelion distance Q is given as 2.89 AU, and the semi-major axis a is 1.7853 AU, the perihelion distance must be the difference between the major axis ($2a = 3.57$ AU) and the aphelion distance, or .68 AU. We will distinguish the magnitudes of the these different *radii vectores*, as the old writers referred to them, by the subscripts E, A and P to stand for the Earth, aphelion and perihelion distances, respectively.

Calculations First solve the velocity equation to find the velocity of the asteroid at the distance of the Earth's orbit of one astronomical unit:

$$v_E = \kappa \sqrt{\left(\frac{2}{r_E} - \frac{1}{a}\right)}$$

$$v = 29.785 \sqrt{\left(2 - \frac{1}{1.7853}\right)}$$

$$v = 35.7 \text{ km/s}$$

To find the perihelion and aphelion velocities, we substitute the magnitudes of the radius vectors for those distances. Beginning with perihelion velocity,

$$v_P = \kappa \sqrt{\left(\frac{2}{r_P} - \frac{1}{a}\right)}$$

$$v_P = 29.785 \sqrt{\left(\frac{2}{.68} - \frac{1}{1.7853}\right)}$$

$$v_P = 46 \text{ km/s}$$

Doing the same process for aphelion velocity yields this result:

$$v_A = 29.785 \sqrt{\left(\frac{2}{2.89} - \frac{1}{1.7853}\right)}$$

$$v_A = 10.8 \text{ km/s}$$

Summary of Orbital Energy Relationships

Observations

1. The accompanying orbit diagram may help to visualize the relationship of the two orbits. Motions of the Earth and the asteroid are counter-clockwise, and the close approach on October 12, 2010 was at the intersection of the orbits on the right.

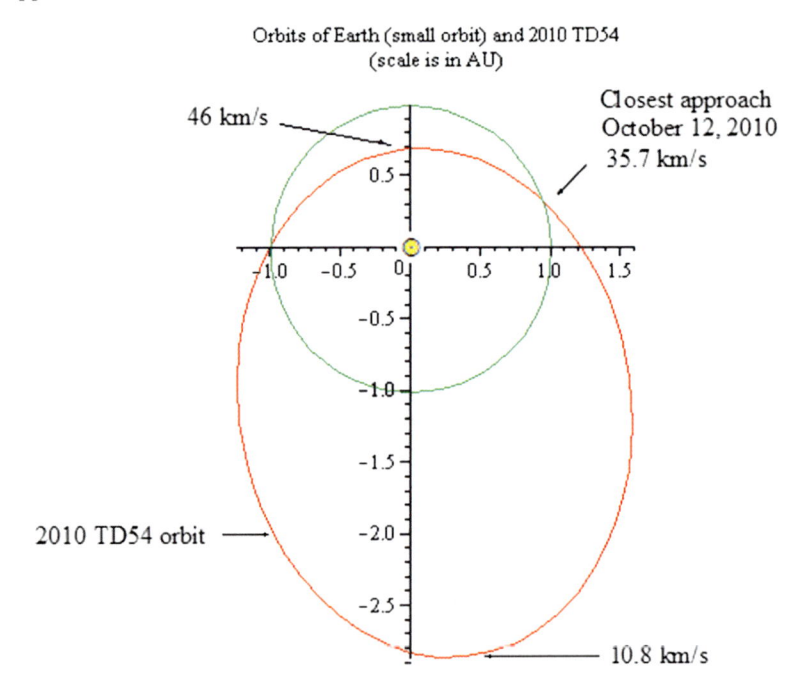

Orbits of Earth (small orbit) and 2010 TD54
(scale is in AU)

46 km/s

Closest approach
October 12, 2010
35.7 km/s

2010 TD54 orbit

10.8 km/s

2. Again we see the characteristic velocity signature of the ellipse, consistent with Kepler's Second Law, where the asteroid has speeded up as it rounds the Sun, more than quadruple the velocity of its far turn. At aphelion it is past the orbit of Mars and well into the asteroid belt where it originated, and where it was likely thrown farther into the inner solar system by the influence of Jupiter at an earlier time.

Newton, Halley and the Great Comet of 1680

The Great Comet of 1680 was one of the great comets of its century. It appears to have been the first discovered using a telescope on November 4th, 1680 by Mr. Gottfried Kirch of Saxony. Its timing coincided with Newton's work on the laws of motion, and his Book III of the *Principia* is filled with many pages of recorded observations, data, analyses and drawings. Newton, "partly by arithmetical operations, and partly by scale and compass" had first assumed that the Comet of 1680 was on a parabolic track. But Halley was able to show from historical records and the similarity of its orbital elements that it was likely the same object that had been seen before, with repeat visits of every 575 years:

Moreover, Dr. *Halley*, observing that a remarkable comet had appeared four times at equal intervals of 575 years (that is, in the month of *September* after *Julius Caesar* was killed; *An. Chr.* 531, in the consulate of *Lampadius* and *Orestes*; *An. Chr.* 1106, in the month of *February*; and at the end of the year 1680; and that with a long and remarkable tail, except when it was seen at *Caesar's* death, at which time, by reason of the inconvenient situation of the earth, the tail was not so conspicuous), set himself to find out an elliptic orbit whose greater axis should be 1382957 parts, the mean distance of the earth from the sun containing 10000 such; in this orbit a comet might revolve in 575 years ... [and with] the equal time of perihelion *Dec.* 7^d. 23^h. 9^m. ... and its conjugate axis 18481.2, he computed the motions of the comet in this elliptic orbit.[3]

With this information, it was apparent that the Comet of 1680 was not parabolic, but was in a high-eccentricity elliptical orbit. The key difference between Newton's first approximation of a parabolic orbit and the ultimate resolution in favor of an elliptical orbit was the existence of a long historical record that proved it.[4] It was a dramatic early demonstration of how difficult it can be to distinguish among any of three types of orbits (elliptical, parabolic and hyperbolic) in that fine zone just slightly this side or the other of eccentricity one.[5]

Problem With the information given below, taken from Newton's account of the Great Comet of 1680, confirm that Halley's use of a 575 year period derived from the historical record does require a major axis of the size stated by Newton; then determine the semi-major and semi-minor axes of the orbit and find the eccentricity, the parameter of the ellipse (i.e., the semi-latus rectum), and the perihelion and aphelion distances of the comet's orbit. Next, calculate its velocity at the distance of the Earth's orbit and at perihelion. Graph the inner orbit of the comet on paper (or computer) at any convenient scale, with the Sun at the origin.

Given

1382957	The "greater axis" of the ellipse described by Newton, "the mean distance of the earth from the sun containing 10,000 such ..."
18481.2	The "conjugate axis" (or minor) of the ellipse, according to Newton
575	The presumed period of the comet's orbit in years

---

[3] *Principia*, Book III, Prop. XLI, p. 515.

[4] This same approach led Halley famously to conclude that the comet of 1682 was also periodic, with the predicted return time of 75 years, which prediction was confirmed after his death; as a tribute to that great man, that comet became known as Halley's Comet.

[5] Newton stated in one corollary that comet orbits "will be so near to parabolas, that parabolas may be used for them without sensible error." *Principia*, Book III, Prop. XL, Cor. 2. Clearly he was not referring to all comets, since in the preceding corollary he had commented on comets with orbits greater than Saturn, and the application of Kepler's Third law to them. Yet, in Corollary 3 of that Proposition, he makes uses the $\sqrt{2}$ times circular velocity method to estimate the velocity of comets at a given distance:

> And, therefore ... the velocity of every comet will always be to the velocity of any planet, supposed to be revolved at the same distance in a circle about the sun, nearly as the square root of double the distance of the planet from the centre of the sun to the distance of the comet from the sun's centre. Ibid., Cor. 3.

Summary of Orbital Energy Relationships

Assumptions We accept the Newton-Halley data, even though, as we will see, current orbital elements differ. The comet has an inclination to the ecliptic plane of about 60°, but we will consider the geometry of the orbit on a two-dimensional xy plane. Influences of other bodies will be ignored, as they were when Newton and Halley investigated this comet.

Method The "greater axis" and "conjugate axis" are the respective major and minor axes of the ellipse. The semi-major axis a and semi-minor axis b are half that. Since an astronomical unit is the mean distance between the Earth and Sun, Newton's phrase, "the mean distance of the earth from the sun containing 10,000 such ..." means that given distances are in ten-thousandths of an astronomical unit. So to convert to AU, the data must be divided by 10,000. With the units in hand being AU and years, we can easily check its period with Kepler's Third Law. The semi-minor axis and parameter of the ellipse can be calculated by the equations relating those to a and b we introduced earlier: $b = a\sqrt{(1-e^2)}$ and $p = a(1-e^2)$, still using AU as our units. Eccentricity can be determined by manipulating the first equation to solve for e. Perihelion distance q can be found by remembering the simple relation: $q = a(1-e)$. Aphelion distance can be found by either $Q = a(1+e)$ or the major axis minus perihelion distance, $Q = 2a - q$. With all that behind us, we can then work through the velocity equations, with this equation for perihelion distance, $v_P = \kappa[(1+e)/q]^{1/2}$ used above; and the *vis viva* equation for velocity at 1 AU. Use the basic conic equation to graph the ellipse with the derived values: $r = p/(1 + e \cos \theta)$.

Calculations Halley chose his major axis to just fit the 575 year period of the Comet of 1680. That period, according to Kepler's Third Law, required one and only one major axis. Halley arrived at the given major axis of 138.2957 AU, which is the same as "1382957 parts, the mean distance of the earth from the sun containing 10000 such ..." The semi-major axis is half that or 69.14785 AU. To confirm this choice of axis, we solve Kepler's Third Law for period where units are years and AU:

$$P = a^{3/2}$$

$$P = (69.14785)^{3/2}$$

$$P = 575 \text{ yr}$$

As to the semi-minor axis b of the ellipse, it is given by Newton that the "conjugate axis" (which is the minor axis) is 18481.2, where we take Newton to be using the same units (of 1/10,000 AU). The minor axis would therefore be 1.84812 AU, and the semi-minor axis half this:

$$2b = 18.4812$$

$$b = 0.92406 \text{ AU}$$

We are starting to get a sense of the shape of this orbit, whose long-to-short axis dimensions appear to be about 69:1.

For the calculation of eccentricity, we may make use of this equation we encountered earlier in the discussion of the geometry of the ellipse:

$$b = a\sqrt{1 - e^2}$$

Solving for e,

$$e = \sqrt{1 - \frac{b^2}{a^2}}$$

Inserting the values for b and a from above, we have,

$$e = \sqrt{1 - \frac{(0.92406)^2}{(69.14785)^2}}$$

$$e = 0.9999107$$

The eccentricity is very close to one, which suggests a very nearly (but not quite!) parabolic orbit. We can get an even better idea of the scale of the ellipse by calculating the parameter, which is the vertical distance (at right angles from the long axis of the ellipse) up from the focus, which in our heliocentric problem is the Sun. To find the parameter of the ellipse, we solve the equation,

$$p = a(1 - e^2)$$

$$p = 69.14785(1 - .9999107^2)$$

$$p = 0.01234871 \text{ AU}$$

Next we are required to find the perihelion and aphelion distances, q and Q, respectively. Using these familiar equations,

$$q = a(1 - e) \qquad Q = a(1 + e)$$

$$q = 69.14785(1 - .9999107) \qquad Q = 69.14785(1 + .9999107)$$

$$q = 0.006175 \text{ AU} \qquad Q = 138.2895 \text{ AU}$$

Summary of Orbital Energy Relationships

This is an extremely close perihelion! Its close shave with the Sun explains why the comet was so brilliant. (Compare its perihelion distance with those of the comets in the table above.)

At last we are able to calculate its velocities at perihelion and at Earth distance from the Sun, or 1 AU. Using the perihelion velocity equation developed above, we have, for the Comet of 1680 at its closest approach to the Sun:

$$v_P = \kappa \left(\frac{1+e}{q}\right)^{1/2}$$

$$v_P = 29.785 \left(\frac{1+.9999107}{0.006175}\right)^{1/2}$$

$$v_P = 536 \text{ km/sec}$$

This is an astonishing velocity! Compare this with the more sedate comets in the table. Let's see what its velocity was at 1 AU Earth-distance:

$$v_{1 \text{ AU}} = \kappa \sqrt{\left(\frac{2}{r_E} - \frac{1}{a}\right)}$$

$$v_{1 \text{ AU}} = 29.785 \sqrt{\left(2 - \frac{1}{69.14785}\right)}$$

$$v_{1 \text{ AU}} = 41.97 \text{ km/s}$$

This is faster than the perihelion velocity of many comets! Let us see what the graph of this remarkable comet looks like. With *Maple* software, the display is as shown in the figure, where units are AU:

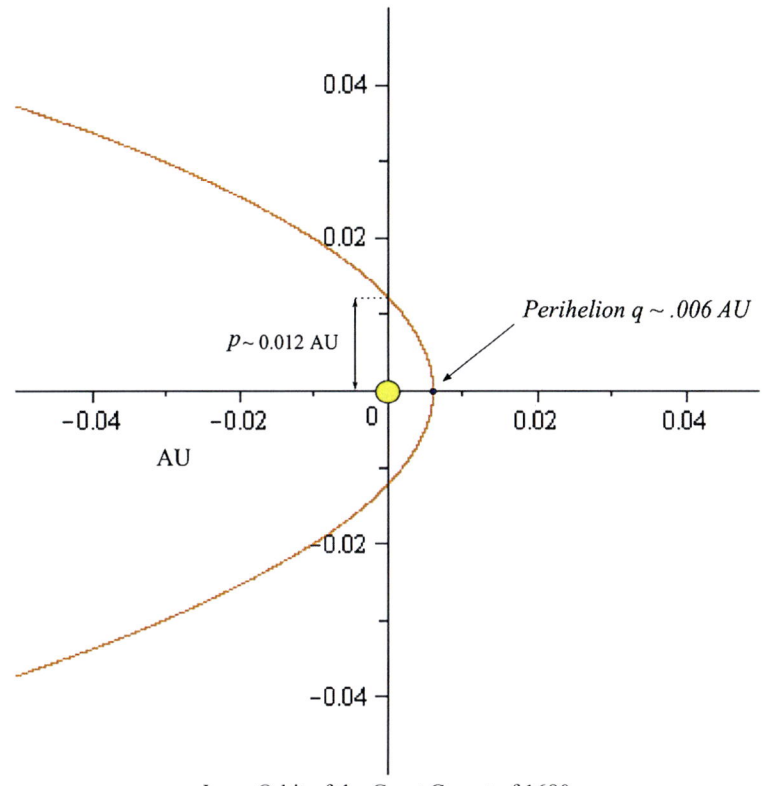

Inner Orbit of the Great Comet of 1680

Observations
1. How close to the Sun is .006 AU? The perihelion distance of this comet was less than a million kilometers from the center of the Sun, whose radius is about 697,000 km. It was thus about 227,000 km above the solar surface. The *Catalog of Cometary Orbits*[6] lists this comet as among the closest to graze the Sun.
2. More modern analyses have refined these numbers. The NASA/JPL website[7] shows the applicable elements to be: $e = .999986$, $a = 444.4285714$ and $q = .006222444.43$. The inner-orbit plot of this comet looks virtually identical, and there are only slight changes to the velocities derived above; yet with these modifications, the major axis of the comet is much larger, with a Q of over 888 AU and the comet's period is almost 9,400 years!

Exercises: The Wonder of Brilliant Comets Brilliant comets have always been the subject of awe and fascination in human history, but "great" comets will forever be the source of intense, world-wide wonder and excitement. Take, for example, the

---

[6] Marsden and Williams [1].
[7] See http://ssd.jpl.nasa.gov/sbdb.cgi#top.

description of the initial reactions to the Great Comet of September, 1882. Here are excerpts from Agnes Clerke's *A Popular History of Astronomy during the 18th Century*[8] describing the interest in that comet:

> The discovery of a great comet at Rio Janeiro, September 11, 1882, became known in Europe through a telegram from M. Cruls, director of the observatory at that place ...
>
> On the forenoon of Sunday, September 17, [Dr. Common, from another location] saw a great comet close to, and rapidly approaching the Sun. It was, in fact, then within a few hours of perihelion...
>
> The comet, of which the silvery radiance contrasted strikingly with the reddish yellow glare of the sun's margin it drew near to, was followed "continuously right into the boiling of the limb" – a circumstance without precedent in cometary history...
>
> On the following morning, the object of this unique observation showed (in Sir David Gill's words) "an astonishing brilliancy as it rose behind the mountains on the east of Table Bay, and seemed in no way diminished in brightness and the sun rose a few minutes afterward. It was only necessary to shade the eye from direct sunlight with a hand at arms length, to see the comet, with its brilliant white nucleus and dense white, sharply bordered tail of quite half a degree and length. All over the world, wherever the sky was clear during that day, September 18, it was obvious to ordinary vision. Since 1843 nothing had been seen like it. From Spain, Italy, Algeria, Southern France, dispatches came in announcing the extraordinary appearance. At Córdoba, in South America, the "blazing star near the sun" was the one topic of discourse. Moreover – and this is altogether extraordinary – the records of its daylight visibility to the naked eye extend over three days.

This global fascination with special comets continues. The marvelous appearance of comet Hale Bopp in the spring of 1997 stirred excitement all around the world. Even lesser comets that appear with a spectacular and surprising splash into view for a brief period can be memorable for a lifetime. A wonderful thing about comets is this: one never knows but that tomorrow, next month, or next year another "great" comet, having been absent for perhaps thousands of years, may enter our tiny little solar neighborhood put on a display of surpassing beauty.

Problems

1. If an object were in circular orbit around the Sun at the same perihelion distance as the Great Comet of 1680, what would its velocity be? Use the current JPL data for the comet in doing your calculation. Compare that velocity to the perihelion velocity of the comet and the escape velocity at that distance. Interpret your results.
2. What is the kinetic energy per unit comet mass of the Comet of 1680 at perihelion? Compare that with the kinetic energy per unit mass of a body in the circular orbit described in the previous problem. What is their ratio?
3. Comet Elenin C/2010 X1 has the following relevant elements: $e = 1.000064$, $a = -7532.317$ AU, $q = .482431$ AU. On August 1, 2011 it was 1.038218 AU from the Sun. What sort of orbit is this? Find the parameter of the orbit and its velocity on that date.

---

[8] Clerke [2].

4. Plot the inner orbit (within 2 or 3 AU of the Sun) of Comet Elenin C/2010 X1 given the information above. Use AU as your units.
5. Considering the plane of the Earth's orbit as the xy plane with the Sun at the origin, the heliocentric xyz velocity vectors, in kilometers per second, of Comet Elenin on August 2, 2011 were as follows: v[x] = 41.19707592863396; v[y] = −6.868007304284843; v[z] = .6153558567863349. Use the Pythagorean Theorem to find the velocity of the comet on that date.
6. Comet 1/P Halley has an eccentricity of 0.967143 and a semi-major axis of 17.8341 AU. Find its perihelion distance and the approximate velocities of Comet Halley as it reaches the orbital distances of Jupiter (5.2 AU), Mars (1.52 AU), Earth (1 AU), Venus (.723 AU), Mercury (.387 AU), and at perihelion.
7. What is the total orbital energy of Comet Halley, assuming its mass is approximately 3×10^{14} kg?
8. From the *vis viva* equation, derive the equations for apsidal velocities when just the semi-major axis and eccentricity are known.
9. Comet 17P Holmes has a perihelion distance of 2.053365 AU, a semi-major axis of 3.62067 AU, and a semi-minor axis of 3.263863 AU. Find its eccentricity, orbital parameter, apsidal distances, and velocities. Graph the orbit of the comet at any convenient scale you choose.
10. Assume that Comet Halley and Comet Holmes have the same mass. Describe a simple way of comparing their total orbital energies. Which has the greater total orbital energy, and by how much?

References

1. Marsden BG, Williams GV (2008) Catalogue of cometary orbits, 17th edn. Central Bureau for Astronomical Telegrams & Minor Planet Center, Smithsonian Astrophysical Observatory, Cambridge, MA
2. Clerke AM (2003) A popular history of astronomy during the 18th century. Sattre Press, Decorah, pp 358–359, first printed in 1902

Chapter 14
Introduction to Spaceflight

Toward the end of his classic book, *Interplanetary Flight*,[1] Arthur C. Clarke in 1950 was looking ahead to the end of the twentieth century and the prospects for the future of space travel. He noted that to the ordinary man "this planet is still the whole of the universe: he knows that other worlds exist, but the knowledge does not affect his life and therefore has little real meaning to him."[2] But this, Clarke predicted, would soon be a thing of the past:

> All this will be changed before the 20th century draws to an end. Into a few decades may be compressed more profound alterations in our world picture than occurred during the whole of the Renaissance and the age of discovery that followed. To our grandchildren the Moon may become what the Americas were four hundred years ago – a world of unknown danger, promise and opportunity. No longer will Mars and Venus be merely the names of wandering lights seldom glimpsed by the dwellers in the cities. They will be more familiar than ever they were to those eastern watchers who first marked their movements, for they will be the new frontiers of the human mind.[3]

It is indeed amazing how far things have progressed in the human adventure of space travel in the more than half-century since he wrote the book. His prediction was accurate. We have gone to the Moon; launched robotic explorers to circle planets, moons and asteroids; sent vehicles to explore or impact Venus, Mars, Titan, and several comets; and have enjoyed stunning close-up images of Mars' ancient waterways, Saturn's rings, Io's volcanoes, and the scratched surface of Uranus' moon Miranda, to name just a few examples. We now routinely see fascinating and mysterious images transmitted to us from other worlds that people only a generation or two ago could not have imagined.

Clarke intended his book as a "survey of the possibilities and problems of interplanetary flight, as far as they can be foreseen at the present day."[4] He noted

---

[1] Arthur C. Clarke [1].

[2] Ibid., 145.

[3] Ibid.

[4] Ibid, xiii.

that many of the fundamental techniques already existed to analyze spaceflight: "It is, for example, possible to calculate by quite simple methods the velocities and durations required for interplanetary journeys, irrespective of the physical means that may be used to accomplish them."[5] He was of course correct there too, and we will do just that in this chapter. There are almost limitless fascinating problems in the orbital mechanics of space travel. Most problems in the area, however, at least begin with the basic techniques pioneered by Kepler, Galileo, and especially Newton. These concepts have been developed to a fine art by their successors, and have many complex nuances; we consider below only a few of the most fundamental ideas in interplanetary travel.

Using a Hohmann Transfer to Achieve a Geostationary Orbit

Imagine an artificial satellite that has been launched into an equatorial, circular orbit 500 km above the surface of the Earth, and its velocity is 7.61 km/s. Assume that our goal is to transfer the satellite to another circular orbit far higher – a geosynchronous orbit approximately 42,164 km above the center of the Earth, where the orbital velocity is about 3.1 km/s. The procedure will be to use an intermediate elliptical "transfer" orbit that connects the two circular orbits, a method developed by German engineer and rocket enthusiast Walter Hohmann in 1925. The satellite has a rocket propulsion engine aboard that can be fired from Earth either to speed the satellite up or slow it down.

Problem What must the injection velocity of the satellite be from its low, 500 km above the equator Earth orbit to place it into a elliptical transfer orbit whose apogee is 42,164 km? Given the velocities noted above, what must the changes in satellite velocity then be to convert the satellite's elliptical transfer orbit into the desired circular geosynchronous orbit?

Given

$e = (Q - q)/(Q + q)$	Equation for eccentricity of an elliptical orbit, derived from apogee Q and perigee q distances
$v_P = \sqrt{\frac{\mu}{q}(1 + e)}$	Equation for perigee velocity in a geocentric elliptical orbit
$v_A = \sqrt{\frac{\mu}{Q}(1 - e)}$	Equation for apogee velocity in a geocentric elliptical orbit
μ	Shorthand symbol for GM_E, the product of the gravitational constant G and mass of the Earth: 3.99×10^{14}
R_E	Equatorial radius of the Earth: 6.38×10^6 m

---

[5] Ibid.

Using a Hohmann Transfer to Achieve a Geostationary Orbit

Assumptions It is assumed that the Earth is a spherical body of uniform density, so the paths of the satellite follow an idealized circle or ellipse, without distortions in its velocity caused by gravitational anomalies due to uneven mass distribution. The points where the satellite gets its boosts will be at exactly the desired perigee and apogee of the transfer orbit, and tangential to the radius vector. All the orbits are "coplanar" (lying along the same plane), above the Earth's equator in the direction of the Earth's rotation.

Method The perigee and apogee of the elliptical "transfer" orbit are known: the perigee is the radius (from Earth's center) of the 500 km orbit, and its apogee is the radius of the geostationary orbit, approximately 42,164 km. Since the perigee and apogee are given, the first task is to determine the eccentricity of the transfer orbit (but first making all distances refer to the center of the Earth). From the eccentricity of the planned transfer orbit, we can use the velocity equations to calculate the perigee and apogee velocities needed to produce the transfer orbit. Then it is necessary to find how to get the satellite into and out of the transfer orbit. Since the velocities of the circular "before" and "after" orbits are known, we can calculate the velocity changes needed, from perigee first to move the satellite from the 500 km circular orbit to the transfer orbit; and then at apogee to move the satellite from the transfer orbit to the geosynchronous orbit.

Calculations The satellite height from the center of the Earth is 500,000 m + R_E:

Perigee: $\qquad$ Apogee:

$$(5 \times 10^5) + (6.38 \times 10^6) = 6.88 \times 10^6 \text{m} \qquad 4.2164 \times 10^7 \text{m}$$

To find eccentricity,

$$e = \frac{Q - q}{Q + q}$$

$$e = \frac{4.2164 \times 10^7 - 6.88 \times 10^6}{4.2164 \times 10^7 + 6.88 \times 10^6}$$

$$e = .7194$$

To find velocities at perigee and apogee we solve these equations:

Perigee: $\qquad\qquad$ Apogee:

$$v_P = \sqrt{\frac{\mu}{q}(1 + e)} \qquad\qquad v_A = \sqrt{\frac{\mu}{Q}(1 - e)}$$

$$v_P = \sqrt{\frac{(3.99 \times 10^{14})}{(6.88 \times 10^6)}(1.7194)} \qquad v_A = \sqrt{\frac{(3.99 \times 10^{14})}{(4.22 \times 10^7)}(1 - .7194)}$$

$$v_P = 9.99 \text{km/s} \qquad\qquad v_A = 1.63 \text{km/s}$$

These are velocities that the satellite needs to obtain at perigee and apogee in the transfer orbit. How do they compare with the given circular orbital velocities v_c at 500 km and at 42,164 km?

$v_c = 7.61 \text{km/s}$ for the 500km orbit $v_c = 3.07 \text{km/s}$ for the 42,164km orbit

What must the change in velocities be, from 500 km circular orbit to elliptical transfer orbit, then from transfer orbit to 42,164 km geostationary orbit?

$$\Delta v_P = v_P - v_{C500} \qquad \Delta v_A = v_{C42,200} - v_A$$

Needed $v_P = $ 9.99 km/s $v_{C42,200} = $ 3.08 km/s

From $-v_{C500} = $ 7.61km/s $-v_A = $ 1.63km/s

Change $\Delta v_P = $ +2.38km/s $\Delta v_A = $ +1.45km/s

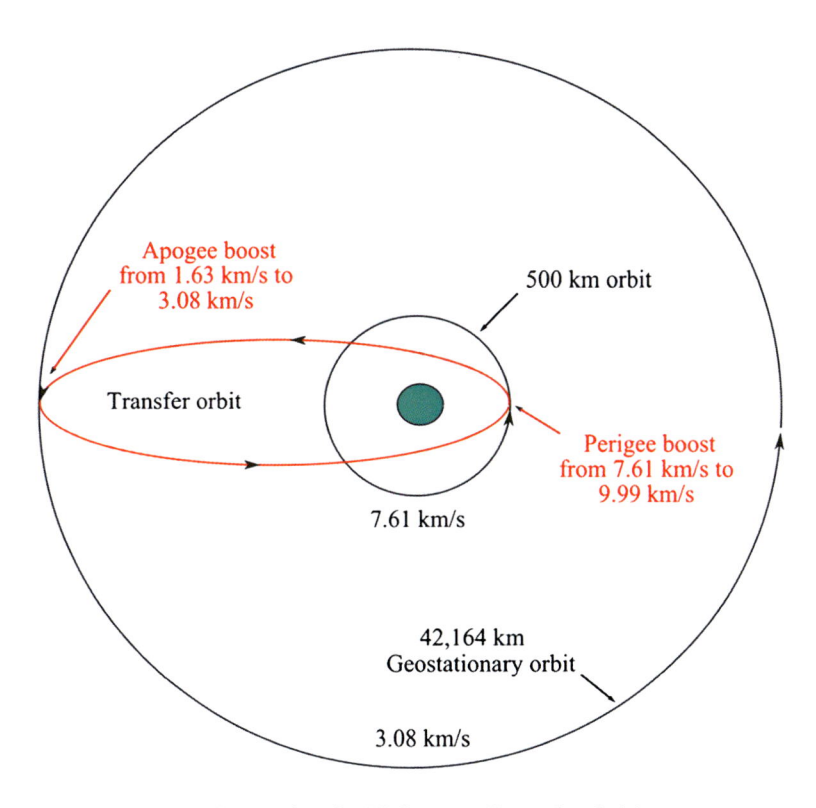

Example of a Hohmann Transfer Orbit

Designing an Orbit for a Lunar Mission   293

Hence the satellite velocity must be boosted by 2.38 km/s at a designated perigee injection point to get it into the elliptical transfer orbit, with a rocket burn timed to achieve just that velocity. Then at the apogee of the transfer orbit, the satellite must be given an apogee "kick" of 1.45 km/s to place it into the circular geostationary orbit.

Observation This is only one example of such an orbital transfers, which can be shown to involve the least amount of energy.

Designing an Orbit for a Lunar Mission

Suppose you are to help plan a robotic mission to the Moon. For this trip, you will want to simply have the spacecraft go around the Moon and return, to check feasibility of a later robotic landing. You decide to do it in two steps. Inspired by the above example, you advise launching the vehicle into a circular orbit 500 km above the surface of the Earth (thus, 6.88×10^6 m from the center of the Earth). Its velocity at that altitude will be 7.61 km/s. You goal is to plan for a spacecraft orbit that just rounds the Moon and returns into circular Earth orbit, before initiating re-entry.

Problem Find the (1) eccentricity of the spacecraft orbit; (2) its perigee and apogee velocities of the orbit; (3) the injection velocity at perigee necessary to attain this orbit; and (4) the approximate travel time to the Moon.

Given

$e = (Q - q)/(Q + q)$	Equation for eccentricity of an elliptical orbit, derived from apogee Q and perigee q distances
$v_P = \sqrt{\frac{\mu}{q}(1+e)}$	Equation for perigee velocity in an elliptical orbit
$v_A = \sqrt{\frac{\mu}{Q}(1-e)}$	Equation for apogee velocity in an elliptical orbit
μ	Shorthand symbol for GM_E, the product of the gravitational constant G and mass of the Earth: 3.99×10^{14}
Q	Apogee of the spacecraft orbit, which equals the semi-major axis distance from Earth to Moon: 384,400 km
q	Perigee of the spacecraft orbit: 500 km from the center of the Earth, or 6.888×10^6 m
a	Semi-major axis of the spacecraft orbit, determined from the equation $a = (q + Q)/2$
$P = \frac{2\pi}{\sqrt{\mu}} a^{3/2}$	Period of the spacecraft in its orbit

Assumptions All the assumptions of the prior problem. We'll also assume that the mean distance from the Earth to the Moon is the distance the Moon will in fact be from the Earth at the time of arrival at the Moon. We will also ignore for the sake of this problem ignore the Moon's own gravitational pull, its "terminal attraction" on the spacecraft as it nears the end of its voyage, as it approaches the Moon. We also assume no in-flight changes in velocity. Finally, we'll assume that the spacecraft's mass is negligible relative to the other bodies.

Method The perigee and apogee of the elliptical orbit are respectively given as 6.88×10^6 and 3.844×10^6 m. We can therefore calculate the eccentricity of the orbit in the same manner as was done in the previous problem. From this e, the velocity equations may again be used to calculate the perigee and apogee velocities needed to produce the spacecraft trans-lunar orbit. The task is to find how to get the satellite into and out of that orbit. Using the same method as shown before, it will be necessary to calculate the velocity change needed at perigee to move the satellite from the 500 km circular orbit to its orbit to the Moon. For the estimated length of the voyage to the Moon, one approach is to solve Kepler's Third Law for P then divide by 2, since we want the one-way travel time.

Calculations

Perigee: Apogee:

6.88×10^6 m 3.844×10^8 m

To find eccentricity,

$$e = \frac{Q - q}{Q + q}$$

$$e = \frac{3.844 \times 10^8 - 6.88 \times 10^6}{3.844 \times 10^8 + 6.88 \times 10^6}$$

$$e = \frac{3.775 \times 10^8}{3.913 \times 10^8}$$

$$e = .965$$

To find velocities at perigee and apogee in this highly elongated transfer orbit we need to solve these equations:

Perigee: Apogee:

$v_P = \sqrt{\frac{\mu}{q}(1 + e)}$ $v_A = \sqrt{\frac{\mu}{Q}(1 - e)}$

$v_P = \sqrt{\frac{(3.99 \times 10^{14})}{(6.88 \times 10^6)}(1.965)}$ $v_A = \sqrt{\frac{(3.99 \times 10^{14})}{(3.844 \times 10^8)}(1 - .965)}$

$v_P = 10.68$ km/s $v_A = .191$ km/s

Designing an Orbit for a Lunar Mission

The first is the perigee velocity the spacecraft needs to obtain to achieve the lunar flyby. How does it compare with the circular orbit velocity v_c at 500 km?

$v_c = 7.61$ km/s for 500 km orbit vs. $v_P = 10.68$ km/s for lunar orbit

The change in velocity must therefore be this difference:

$$\Delta v_P = v_P - v_{c_{500}}$$

$$\text{Needed} \quad v_P = 10.68 \text{km/s}$$

$$\text{From} \quad -v_{c_{500}} = 7.61 \text{km/s}$$

$$\text{Change} \quad \Delta v_P = +3.07 \text{km/s}$$

Hence we must boost the space vehicle's velocity at perigee by 3.07 km/s at an injection point opposite of the intended apogee to get it into the elliptical transfer orbit to the moon.

The final task is to find the approximate time of travel to the Moon, given the facts and assumptions of this problem. The period of the orbit can be determined from its semi-major axis applying Kepler's Third Law. The semi-major axis for the spacecraft orbit will be the average of the perigee and apogee distances.

$$a = \frac{q + Q}{2}$$

The apogee Q is the mean lunar distance from the center of the Earth, of 384,400 km; the perigee q is the 500 km circular parking orbit's distance from the center of the Earth. The semi-major axis of the trans-lunar orbit is:

$$a = \frac{6.888 \times 10^6 + 3.884 \times 10^6}{2}$$

$$a = 1.95644 \times 10^8 \text{ m}$$

or about 195,644.4 km. From this we can determine the period of the spacecraft orbit using Kepler's Third Law:

$$P = \frac{2\pi}{\sqrt{\mu}} a^{3/2}$$

$$P = \frac{2(3.14..)}{\sqrt{3.99 \times 10^{14}}} (1.95644 \times 10^8)^{3/2}$$

$$P = 8.607820876 \times 10^5 \text{ seconds}$$

$$P \approx 9.96 \text{ days}$$

The one-way travel time will be half this, or,

$$T_{One\ way} \approx 4.98 \text{ days}$$

$$T_{One\ way} \approx 119.55 \text{ hours}$$

Observations

1. Here is the outward path of the spacecraft as graphed by *Maple* software using the parameters given:

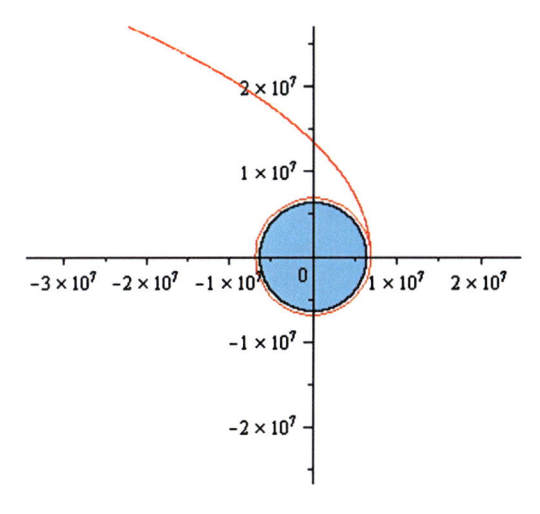

2. On first examining the problem of travel time one might be tempted to imagine computing the travel time to the Moon as one of determining straight line distance over velocity, somehow taking into account the gradual, inverse square reduction of the Earth's gravitational field as the rocket headed for the Moon. While approximations can be obtained this way (using calculus), the important insight here is that any motion in a gravitational field short of escape velocity traces a closed, bound, *orbit*, and travel time is computed by Kepler's Third Law in terms of a portion of the overall orbital period.

3. The calculation of travel time can be approximated in a back-of-the-envelope fashion by assuming that the major axis of the spacecraft's orbit is about equal to the semi-major axis of the moon's orbit. Begin with Kepler's Third or Harmonic Law, a^3: a'^3::P^2:P'^2, where a and a' are the respective semi-major axes of the Moon and spacecraft, and P and P' are their respective periods. The relation can be written this way:

$$\frac{a^3}{a'^3} = \frac{P^2}{P'^2}$$

Now isolate P' (the period of the spacecraft orbit) so that

$$P' = \sqrt{\frac{P^2 a'^3}{a^3}}$$

Because the major axis of the spacecraft's orbit is about equal to the semi-major axis of the moon's orbit, the value of a' is half of a, and the equation becomes,

$$P' = \sqrt{\frac{P^2}{8}}$$

$$P' = \frac{27.32}{\sqrt{8}}$$

$$P' \simeq 9.66 \text{ days}$$

Time of one-way travel is therefore half of this, or about 4.83 days or 116 h. Because the approximation ignores the actual perigee of several thousand kilometers from Earth's center on the opposite side of the Earth, making the major axis of the orbit shorter, the travel time is logically a little less.

4. Note that the apogee velocity of .191 km/s is slower than the Moon's circular velocity of 1.02 km/s. If we wanted to put the rocket into a circular moon-like orbit around the Earth at the Moon's distance would, under the idealized assumptions of this problem, take a slight apogee kick of about .83 km/s (830 m/s) in the direction of motion tangent to the apogee point.

5. The perigee velocity is close to, but does not attain, escape velocity. How do we know this? Since escape velocity is $\sqrt{2}$ faster than circular velocity, escape velocity at 500 km is 7.61$\sqrt{2}$ km/s or 10.76 km/s. Perigee velocity is 10.68 km/s or about 99 % of escape velocity.

6. The Moon's mass is .012 the mass of the earth, and has a radius of 1,740 km. From this information one can calculate the velocity needed to put the spacecraft in a circular orbit around the moon at altitude r from the lunar center. The equation is,

$$v_c = \sqrt{\frac{\mu_{moon}}{r}}$$

where $\mu_{moon} = .012 \, \mu$. At the 110 km orbital altitude of the Apollo Command and Service Module (when added to the lunar radius of 1,737.5 km), the circular orbital velocity is about 1.6 km/s. Since the apogee velocity of the transfer orbit is .191 km/s, a rocket burn is needed to boost the craft's speed to circular orbital velocity. The burn must last just long enough to change the module's apogee velocity by that difference, or, under our assumptions, about 1.4 km/s. To return to Earth we must reverse the process, slowing the spacecraft down to get it back to the apogee velocity of the transfer orbit.

7. The accompanying sketch shows the relative orbits of involved in the mission, not to scale:

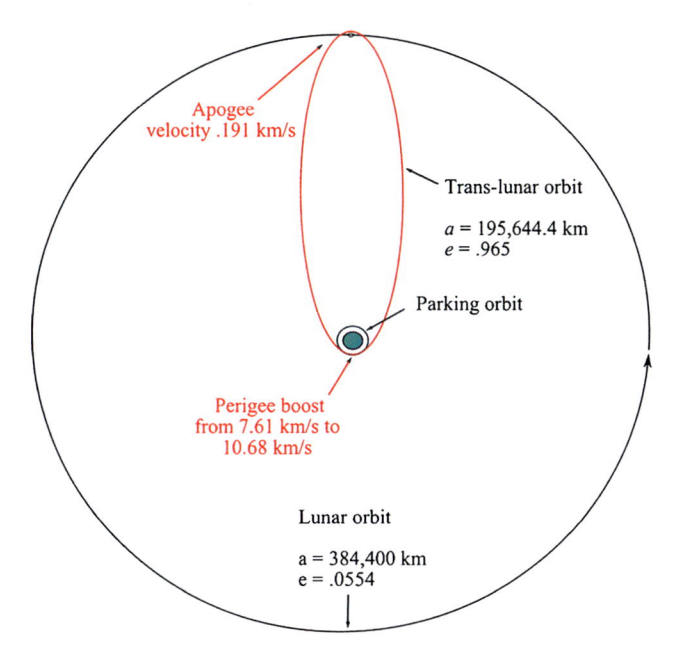

Apogee velocity .191 km/s

Trans-lunar orbit
$a = 195{,}644.4$ km
$e = .965$

Parking orbit

Perigee boost from 7.61 km/s to 10.68 km/s

Lunar orbit
$a = 384{,}400$ km
$e = .0554$

An Idealized Moon Mission

8. What if, in considering a manned mission, we are impatient to get to the Moon and the idea of sitting in a cramped rocket ship for many days on each leg of the trip is unappealing? We could of course go to the Moon faster by increasing the kick at perigee of the transfer orbit. This would shorten the amount of provisions needed to keep a crew sustained. But by going faster, the spacecraft will likely approach or even attain the escape velocity of the Earth's gravitational field – remember under the assumptions of the problem the rocket is already at 99 % of escape velocity as is. This would necessitate carrying more fuel, hence added weight, to slow the rocket down as it neared the Moon. More importantly, however, if there were a failure of the rocket engines en route, the spacecraft would go off into space and not be capable of returning. With the spacecraft velocity no more than the transfer orbital velocity, however, a system failure would mean that the craft would round the moon and do a "free return" to the Earth naturally.

Planning a Mission to Mars

The next task is to determine the travel time for a voyage to Mars, whose mean distance from the Sun (its semi-major axis distance) is about one and a half times that of the Earth. Significantly more than is required for a trip to the Moon, the spacecraft now must gain sufficient energy to at least begin the climb out of the Sun's gravitational well. It can get a good start, however, by launching in the direction of the Earth's rotation around the Sun.

Problem Given the information below, create a concept-level plan for a transfer orbit from Earth to Mars and approximate the travel time from Earth to the red planet along this orbit.

Given

$P = \frac{2\pi}{k} a^{3/2}$	Kepler's Third Law for determining the period P
k	Gaussian constant: .01720209895
P	Period of the spacecraft in orbit to Mars
a	Semi-major axis of spacecraft orbit
a_{Mars}	Mars' mean distance from the Sun in AU: 1.52371034[6]

Assumptions We will assume that the orbits of Earth and Mars are coplanar, so we don't need to make any adjustments out of the exact plane of the ecliptic. We'll take the mean distance from the Sun to Mars as Mars' distance from the Sun at the time of the rocket's arrival at Mars. Finally, we'll assume that the spacecraft's and the planets' masses are all negligible relative to the Sun and to each other.

Method It will require the least energy to launch the spacecraft along a Hohmann transfer orbit to get to Mars. The apsides of the elliptical transfer orbit can be calculated by comparison of the two orbits. From this we can calculate the semi-major axis of the transfer orbit and use Kepler's Third Law to find the period of the transfer orbit. Time of travel will be half the period. Units will be astronomical units, with the mean distance from Earth to Sun and the Sun's mass each taken as unit distance and mass, respectively.

Calculations Perihelion of the orbit will be earth's orbit, at 1 AU and aphelion will be at Mars' orbit at 1.52371034 AU from the Sun. The major axis of the transfer orbit, $2a$ will be sum of these, so the semi-major axis will be half that: $2a = 1 + 1.52371034 = 2.52371034$, so $a = 1.26185517$ AU. To find the period:

$$P = \frac{2\pi}{k} a^{3/2}$$

---

[6] From the NASA/JPL website, http://ssd.jpl.nasa.gov/txt/p_elem_t1.txt.

$$P = \frac{2(3.14..)}{.01720209895}(1.26185517)^{3/2}$$

$$P \simeq 518 \text{ days}$$

This is the period of the transfer orbit to Mars and back (to the position the Earth was in at the time of launch). One-way travel time is half this, or 259 days.

This is over 8½ months in space for the trip to Mars.

Observations

1. Since this is now an interplanetary voyage, we must use the Sun's mass as the gravitational focus of the ellipse, rather than the Earth's. Hence we did not employ the geocentric inertial constant, μ. Using unit solar mass with the Gaussian constant and AU as our units greatly simplify the calculations.
2. Critical here is timing of the launch from Earth: it is important to be sure that Mars will be exactly at the apogee of the transfer orbit when the spacecraft arrives 8 months later!
3. Likewise, in order for the Earth to be in the right place at the end of the return voyage, the spacecraft will have to depart Mars at a time that will assure that the Earth will be in the right spot 258 days later. This means it will have to stay on Mars until the time is right for return.

Calculating the Velocity Needed for the Trip to Mars

Now that we have a rough idea of the orbit to Mars, we can work out how what the initial velocity conditions must be for the trip.

Problem The given the facts and assumptions of the previous problem, approximate: the (1) eccentricity of the transfer orbit, (2) its perihelion and aphelion velocities, (3) the injection velocity at spacecraft perihelion (from Earth's orbit) necessary to attain this orbit, and (4) the velocity needed upon arrival at Mars to match its orbital velocity. Find velocities in Au/day and km/s.

Given

$e = (Q - q)/(Q + q)$	Eccentricity of an elliptical orbit, derived from aphelion Q and perihelion q distances
$v_P = k\sqrt{\frac{1+e}{q}}$	Perihelion velocity for a heliocentric orbit
$v_A = k\sqrt{\frac{1-e}{Q}}$	Aphelion velocity for a heliocentric orbit
q	Perihelion of the spacecraft orbit, from the previous problem: 1 AU
Q	Aphelion of the spacecraft orbit, from the previous problem, 1.52371034 AU

(continued)

k	Gaussian constant: .01720209895
a	Semi-major axis of the spacecraft orbit, 1.26 AU
$v_c = \frac{k}{\sqrt{a}}$	Heliocentric circular orbital velocity equation

Assumptions Same assumptions as above. In comparing orbital velocities, we will also make the simplifying assumption that the orbits of Earth and Mars are circular. More accurate calculations would account for the eccentricity of the orbits.

Method Determine eccentricity as before from the given equation, and then find the respective perihelion and aphelion velocities. Compare these transfer orbit velocities with the mean orbital velocities of Earth and Mars, which may be easily approximated from the circular velocity equation, and thus find the velocity changes needed for injection and arrival at Mars. We'll use Gaussian constant for simplicity then convert to m/s.

Calculation To find eccentricity, we again follow this procedure,

$$e = \frac{Q - q}{Q + q}$$

$$e = \frac{1.52371034 - 1}{1.52371034 + 1}$$

$$e = .2075$$

To find velocities at perihelion and aphelion in this transfer orbit we need to solve these equations:

Perihelion:

$$v_P = k\sqrt{\frac{1+e}{q}}$$

$$v_P = .01720209895\sqrt{\frac{1+.2075}{1}}$$

$$v_P = .0189$$

Aphelion:

$$v_A = k\sqrt{\frac{1-e}{Q}}$$

$$v_A = .01720209895\sqrt{\frac{1-.2075}{1.52371034}}$$

$$v_A = .0124$$

Multiplying each by the conversion factor[7] of 1731.5 for km/s gives,

$$v_P = 32.73 \text{ km/s} \quad v_A = 21.48 \text{ km/s}$$

---

[7] See Chap. 12's discussion on the Gaussian constant for the derivation of this number.

These are the perihelion and aphelion velocities the spacecraft needs to achieve the transfer orbit. How do they compare with each circular orbit velocity v_c of Earth and Mars?

Earth's mean orbital velocity

$$v_E = \frac{k}{\sqrt{a_{Earth}}}$$

$$v_E = \frac{.01720209895}{\sqrt{1}}$$

$$v_E = .01720209895$$

Mars' mean orbital velocity

$$v_M = \frac{k}{\sqrt{a_{Mars}}}$$

$$v_M = \frac{.01720209895}{\sqrt{1.52371034}}$$

$$v_M = .01393574639$$

These are in units of Au/day. Multiplying each by the conversion factor of 1731.5 for km/s gives,

$$v_E \simeq 29.8 \text{ km/s} \quad v_M \simeq 24.1 \text{ km/s}$$

The approximate change in velocity from Earth's orbit to the transfer orbit (at the spacecraft perihelion) must therefore be,

$$\Delta v_P = v_P - v_E$$

$$\Delta v_P = 32.7 - 29.8$$

$$\Delta v_P = +2.9 \text{ km/s for perihelion boost}$$

The approximate change in velocity from the transfer orbit to Mars' orbit (at the spacecraft apogee) must therefore be

$$\Delta v_A = v_M - v_A$$

$$\Delta v_A = 24.1 - 21.5$$

$$\Delta v_A = +2.6 \text{ km/s for aphelion boost}$$

Hence we must increase the space vehicle's velocity at these two points to get it into the elliptical transfer orbit to Mars, and then to make it stay with Mars for a landing.

Observation The graph below shows the relationship among the orbits of Earth and Mars and the transfer orbit (in red) (distances are in AU from the Sun in the center). For simplicity we have assumed the orbit of Mars is circular, when in fact it has an eccentricity of about .093, which makes it important to incorporate Martian perihelion time and location in planning the mission.

Transfer Orbit to Mars

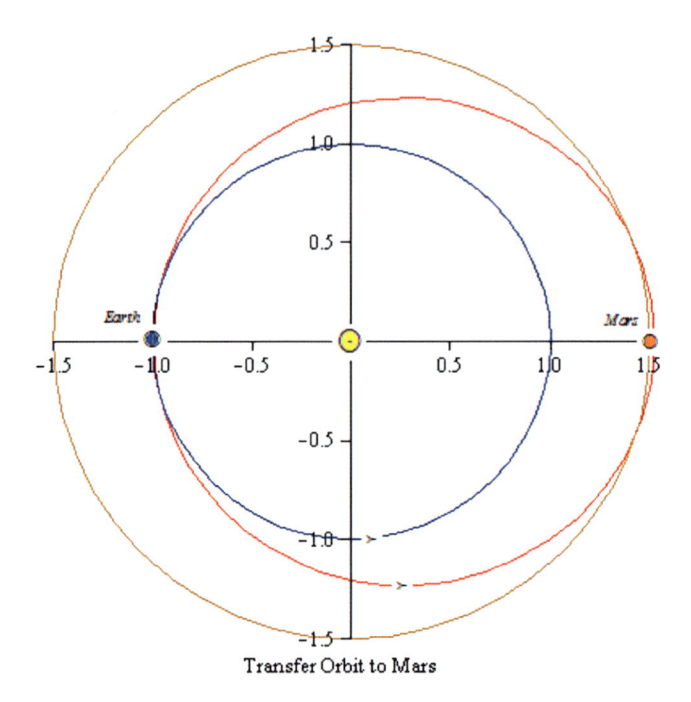

Using the Energy Equations to Define the Orbit to Mars

Now let us compare the above results with an alternative approach, where we use the energy concepts developed in the previous chapters and tie them even more tightly into the geometry of the orbit. We will begin by summarizing some of the key concepts and equations.

Let us regard the primary's mass as M, and since the secondary mass is insignificant, use μ as the product of GM throughout. Total orbital energy in an elliptical orbit therefore equals

$$E_T = -\frac{\mu}{2a}$$

Knowing the orbital energy, the semi-major axis of the elliptical orbit can then be found by,

$$a = -\frac{\mu}{2E_T}$$

We can find that total energy by assessing its value anywhere in the orbit, provided there have been no changes during flight. During the Apollo program, lunar launches occurred from a low Earth parking orbit. From there the craft was blasted into a lunar orbit, with an *injection velocity*, vo, that set the stage for the remainder of the flight unless and until mid-course corrections were necessitated. The altitude of the parking orbit, ro (the distance from the primary mass) determines the starting potential energy – U and the injection velocity determines the starting kinetic energy K. For the trip to Mars for our problem, this jumping-off orbit is the Earth's own orbit around the Sun. These initial distance and velocity conditions define the whole orbit. This is so because, apart from external influences and internal course or speed adjustments, the total orbital energy is constant throughout. Of course, if vo is √2 times the circular velocity at the height of the parking orbit, then the orbit is unbound and parabolic and the craft escapes. If vo were still higher, the craft's path with its excess velocity would follow a hyperbolic trajectory.

Summary of Some Key Energy-Derived Equations

The *total energy ET* of the orbit will be given by the initial conditions; it is dependent upon the injection velocity vo, and the distance from the primary ro of the parking orbit from which injection into the trans-lunar or trans-Martian trip begins:

$$E_T = \frac{v_o^2}{2} - \frac{\mu}{r_o}$$

The *parameter* of the ellipse can be derived from μ and the angular momentum, L:

$$p = \frac{L^2}{\mu}$$

And, for the situation where the orbital velocity is tangent to the radius vector (at the apsides in an elliptical orbit), the *angular momentum*, L is found by,

$$L = r_o v_o$$

This provides alternative derivation of the parameter of the ellipse,

$$p = \frac{r_o^2 v_o^2}{\mu}$$

Once we have the parameter, the *eccentricity* of the ellipse can be teased out by this equation, which is a variation on the equation for the parameter we saw in Chap. 4:

$$e = \sqrt{1 - \frac{p}{a}}$$

Using the Energy Equations to Define the Orbit to Mars

Thus the key elements of the orbit are presented using energy concepts. Finally, there is the *vis viva* equation that reveals of the *velocity* of the spacecraft at any point in its orbit:

$$v = \sqrt{\mu\left(\frac{2}{r} - \frac{1}{a}\right)}$$

Now that we have a rough idea of the orbit to Mars, we can work out how what the initial velocity conditions must be for the trip.

Problem Use the energy-derived equations reviewed above to confirm the semi-major axis and find the eccentricity and parameter of the transfer orbit to Mars.

Given

$E_T = \dfrac{v_o^2}{2} - \dfrac{\mu_{sun}}{r_o}$	Equation for the total energy of the heliocentric orbit
$a = -\dfrac{\mu_{sun}}{2E_T}$	Equation for the semi-major axis of the heliocentric orbit
$p = \dfrac{(r_o v_o)^2}{\mu_{sun}}$	Equation for the parameter of the heliocentric orbit
$e = \sqrt{1 - \dfrac{p}{a}}$	Equation for the eccentricity of the orbit
r_o	Distance from Sun of the initial orbit. Here it is the Earth's mean distance: one AU or 149,597,900 km
v_0	Injection velocity from Earth orbit, from the previous problem: 32.73 km/s
μ_{sun}	The gravitational parameter for the Sun, 1.3275×10^{20}. It is GM_{Sun}, the product of the gravitational constant G (6.674×10^{-11}) and the mass of the Sun (1.981×10^{30} kg)

Assumptions We will incorporate the assumptions of the previous problems here.

Method Begin with the initial conditions of distance and velocity to define the total orbital energy. From there, confirm the semi-major axis, the parameter and eccentricity of the orbit. For this it is necessary to use SI units.

Calculations The total potential and kinetic energy E_T is, after inserting the given variables,

$$E_T = \frac{v_o^2}{2} - \frac{\mu_{sun}}{r_o}$$

$$E_T = \frac{3273^2}{2} - \frac{1.3275 \times 10^{20}}{1.49597900 \times 10^{11}}$$

$$E_T = -3.52751 \times 10^8 \text{ Joules}$$

From this we can confirm the semi-major axis a:

$$a = -\frac{\mu_{sun}}{2E_T}$$

$$a = -\frac{1.3275 \times 10^{20}}{2(-3.52751 \times 10^8)}$$

$$a = 1.88771 \times 10^{11} \text{ km}$$

This is about 1.26 AU, which appears correct. The parameter of the Mars transfer orbit is,

$$p = \frac{r_o^2 v_o^2}{\mu_{sun}}$$

$$p = \frac{(1.495979 \times 10^{11})^2 (3273)^2}{1.3275 \times 10^{20}}$$

$$p = 1.8026 \times 10^{11} \text{ km}$$

The parameter is thus about 1.21 AU. Finally, we can derive the eccentricity from,

$$e = \sqrt{1 - \frac{p}{a}}$$

$$e = \sqrt{1 - \frac{1.8026 \times 10^{11}}{1.88771 \times 10^{11}}}$$

$$e = .2050$$

Which value is reasonably close to the value of .2075 we had approximated earlier.

Observation We have thus derived all the key elements of the orbit to Mars using just two initial conditions: the injection velocity and the distance of the launch orbit from the primary mass, which in this case was the Sun.

Notes on the Apollo 11 Moon Mission

NASA's Apollo 11 Moon mission in July, 1969 put the first man on the Moon. The vehicle that got it there was the colossal three stage Saturn rocket. An abbreviation of NASA's concise report of the mission follows.[8] It lifted off from the Kennedy

---

[8] This summary was drawn from NASA's excellent resource, http://history.nasa.gov/SP-4029/Apollo_11a_Summary.htm.

Space Center Florida and ascended East, in the direction of the Earth's rotation, and arrived in an almost perfectly circular parking orbit 100.4 nautical miles[9] (185.94 km) by 98.9 nautical miles (183.16 km).[10] Its orbital period, according to NASA, was 88.18 min at 25,567.8 ft/s (all of NASA's velocity units in those days were in feet per second and nautical miles). This translates to 7.79 km/s. After one and a half orbits its third stage, the S-IVB rocket, from a height of 180.581 nautical miles (334.436 km) fired for almost 6 min to put the spacecraft into a trans-lunar injection (TLI) orbit at a spaced-fixed velocity of 35,545.6 ft/s, or 10.8343 km/s.[11] This was the velocity which, but for a minor mid-course correction, kicked it all the way to the Moon on a ballistic path. After about 73 h of coasting flight, the craft was captured into a lunar orbit of about 314.3 km by 111.1 km above the lunar surface, having been slowed by an almost 6 min engine burn. It was a dramatic and almost flawlessly executed trip. After adjusting the lunar orbit to circularize it, the lunar module descended to the surface of the Moon, and landed on Mare Tranquilitatis, the Sea of Tranquility, on July 20, 1969. After the astonishing and industrious 21½ hours of activities on the Moon, the lunar module (LM) lifted off the next day into an elliptical orbit, which again was circularized and became the orbital platform for the return to Earth. The trans-Earth injection velocity was 8,589 ft/s, or 2.618 km/s, and reentered the Earth's atmosphere (at 400,000 ft or about 122 km) at a velocity of 36,194.4 ft/s. or 11.032 km/s. The trans-Earth coast lasted 59 h 36 min and 52 s, a much shorter return trip.

A question that will strike the student almost immediately is how is it that the Apollo 11 crew reached the Moon in just 73 h when the idealized lunar mission discussed above seemed to mandate a much longer flight time, almost 120 h? What sort of orbit could, by the application of Kepler's Third Law, have produced a total period of about 146 h or just 6 days? The answer is that the Apollo 11 traveled faster than the spacecraft following the orbit in our idealized lunar mission. Apollo 11's injection velocity was greater, more than 10.8 km/s vs. 10.68 for the hypothetical case. Its initial parking orbit for TLI was also lower (about 334 km vs. the 500 km in the hypothetical). The greater initial velocity means greater initial kinetic energy, which meant a larger semi-major axis. The result was a longer, slightly more eccentric orbit that we approximated above. NASA's reported eccentricity was .97696. The apogee of the Apollo 11 orbit did not just touch the Moon, as in our hypothetical; it went well beyond it. But by the time it reached the Moon, it had slowed down to a fraction of its initial velocity, of course, to somewhere in the neighborhood of 800 m/s, and was able to be easily captured by the lunar gravity,

---

[9] One nautical mile, the unit used by NASA, is 1.852 km. See http://en.wikipedia.org/wiki/Nautical_mile. Using this measure, NASA used a value for the Earth's radius of 6,378.159656 km.

[10] See http://history.nasa.gov/SP-4029/Apollo_18-21_Earth_Orbit_Data.htm for the parameters of the Earth orbits for the Apollo spacecraft.

[11] See http://history.nasa.gov/SP-4029/Apollo_18-24_Translunar_Injection.htmv for a comprehensive summary of the launch and injection parameters of all the spacecraft in the Apollo program.

and with propulsion assists, to navigate into an orbit that ultimately resulted in the circularized lunar parking orbit mentioned above.[12]

Let us see if we can model the Apollo 11 orbit. It will not be NASA-precise, because we will deliberately ignore refinements that are necessary to convey high accuracy. (We will adopt, for example, all the assumptions noted in the previous problems of this chapter.) We are interested in exploring just the basic parameters of the orbit. That said, we will use NASA's own data on the radius of the Earth, the parking orbit, velocities and other elements to the level of precision reported by NASA.[13] Taking just the injection velocity of 10.8343 km/s, and the semi-major axis of the parking orbit, of 180.581 km or 6,712.595 km from the center of the Earth, we have the initial conditions that allow us to compute the parameters of the trans-lunar orbit.

We find total energy of the orbit using the equation,

$$E_T = \frac{v_o^2}{2} - \frac{\mu_{earth}}{r_o}$$

$$E_T = \frac{(10834.298)^2}{2} - \frac{3.9857 \times 10^{14}}{6.712595 \times 10^6}$$

$$E_T = -6.8542835 \times 10^5 \text{ Joules}$$

From this we can derive the semi-major axis distance of the trans-lunar orbit of Apollo 11:

$$a = -\frac{\mu_{earth}}{2E_T}$$

$$a = -\frac{3.9857 \times 10^{14}}{2(-6.8542835 \times 10^5)}$$

$$a = 2.907451960 \times 10^8 \text{ m}$$

$$a \simeq 290,745 \text{ km}$$

---

[12] Robert A. Braeunig has calculated the velocities at each point in the mission at http://www.braeunig.us/apollo/apollo11-TLI.htm.

[13] NASA's value for the radius of the Earth (converted from nautical miles) is 6,378.159656 km, which we use for this discussion. The value of μ is slightly more refined here: $GM = 3.9857 \times 10^{14}$, which is a slightly more precise value than the 3.99×10^{14} value used for most of the problems. This is found by the product of the mass of the Earth of 5.97219×10^{24} kg, taken from the NASA site, http://ssd.jpl.nasa.gov/?planet_phys_par, and the CODATA value of the Newtonian Gravitational Constant of 6.67384×10^{-11} m^{3}kg^{-1} s^{-2}. See http://physics.nist.gov/cgi-bin/cuu/Value?bg|search_for=G. The result was then rounded to four decimal places.

Notes on the Apollo 11 Moon Mission

This compares with the 195,644 km we found for the idealized lunar orbit in problem 14.2. The Parameter of the ellipse of the Apollo 11 trans-lunar orbit is given again by this equation:

$$p = \frac{(r_o v_o)^2}{\mu_{earth}}$$

$$p = \frac{(6.712595 \times 10^6)^2 (10834.298)^2}{3.9857 \times 10^{14}}$$

$$p = 1.327021391 \times 10^7 \text{ m}$$

$$p \simeq 13,270 \text{ km}$$

The eccentricity then follows,

$$e = \sqrt{1 - \frac{1.327021391 \times 10^7}{2.907451960 \times 10^8}}$$

$$e = .97691$$

This compares rather favorably to the eccentricity value of .97696 reported by NASA.[14] It will be interesting now to find the approximate velocity of Apollo 11 as it reached the Moon's orbit. When the Apollo 11 crew arrived at lunar orbit the day before the landing, July 19, 1969, the Moon was about 394,193 km from the Earth.[15] The semi-major axis of the lunar insertion orbit was about 1,950 km, which makes the magnitude of the radius vector r for which velocity is to be calculated about 392,243 km. Using what we have for the velocity equation, we can approximate the velocity of the spacecraft when it arrived at its circumlunar station:

$$v = \sqrt{\mu_{earth} \left(\frac{2}{r} - \frac{1}{a}\right)}$$

$$v = \sqrt{3.9857 \times 10^{14} \left(\frac{2}{3.94193 \times 10^8} - \frac{1}{2.907451960 \times 10^8}\right)}$$

$$v \simeq 813 \text{ m/s}$$

---

[14] From http://history.nasa.gov/SP-4029/Apollo_18-24_Translunar_Injection.htm
[15] This result was obtained using *TheSky* software.

This velocity would be somewhat greater if we had factored into the calculations the attractive pull of the Moon on the Apollo 11 spacecraft during the latter part of the orbit.

Exercises: The Magic Spell of Space Travel Arthur C. Clarke's 1950 book *Interplanetary Flight* that we referred to at the beginning of this chapter was a clear and lively exposition of the nuts and bolts of space travel that made it seem truly possible. It began his book by this eloquent exposition:

> The dream of interplanetary travel is as old as the dream of flight: indeed, for many centuries both were inextricably entangled. If one could fly at all, men believed, then presumably it would be possible to go to the Moon, or even to the Sun. So it was thought in the days before Galileo and Newton, when the old medieval ideas of the universe still held sway. The Moon might be fairly distant, it was true; but it could hardly be more remote than the fabulous lands of Hindustan or Cathay.[16]

Space travel has, it seems, always fascinated Man ever since he became aware of its possibilities. Now it is no longer a dream of course and we can share information about the actual *experiences* of space missions on a regular basis. The solar system, its physics and geology are gradually becoming as familiar as the Earth's. The satisfaction of knowing how rockets get to their destinations, how they orbit, and how they (sometimes) return, enables a deeper participation in the present scientific revolution than is possible from being a passive spectator. The problems below, a few of which touch on actual rocket launches and events in our history, should further that understanding.

Problems

1. Referring to the energy-derived equations in this chapter, find the total energy of the idealized orbit to the Moon, and use those equations to confirm the semi-major axis, the parameter, and the eccentricity of the orbit. Use 10.6748 km/s as the initial velocity from 6,878 km. Experiment with very slight adjustments to initial velocity to see how sensitive the parameters are to initial conditions.
2. Calculate the outbound velocity of the Apollo spacecraft at 200,000 km away from Earth in the trans-lunar orbit of the apollo mission.
3. Suppose you are an astronaut in a rocket on Mars' moon Phobos. You want to calculate an orbit that will take you from Phobos to the planet's other moon Deimos. The semi-major axes of Phobos and Deimos are 9,376 and 23,458 km respectively, and you may regard their axes as circular, which they nearly are. Mars' mass is 6.41693×10^{23} kg. Design a Hohmann transfer orbit (including semi-major axis and eccentricity) that will accomplish your task, including necessary periapsis is and apoapsis velocities changes.

---

[16] Arthur C. Clarke [1], p. 1.

Notes on the Apollo 11 Moon Mission   311

4. Find the total energy of the transfer orbit in the previous problem, and confirm the orbital elements of that orbit by using the energy-derived equations.
5. Design an orbit for a Jupiter mission. Jupiter is 5.2 AU from the Sun. You may assume a circular Jovian orbit for the purposes of this problem. Approximate the needed perihelion and aphelion distances, the semi-major axis, eccentricity, velocity changes and one way travel time to the planet. Ignore the (considerable) attractive pull of the planet as the spacecraft nears it.
6. During the Apollo 11 flight for the first lunar landing in July, 1969, the third stage Saturn S-IVB engine was ejected a little more than 4 h into the mission. The S-IVB kept on drifting, passed the lunar surface at about 3,380 km and went into solar orbit. Its aphelion was 82,000,000 nautical miles (151,864,000 km) and its perihelion was 72,520,000 nautical miles (134,307,040 km). Using the Gaussian constant and AU as the units, find the semi-major axis, eccentricity, and parameter of its orbit, and its orbital period in days. One AU is 1.49598×10^8 km.
7. When the Apollo 11 Command Service Module (CSM) with its attached Lunar Module (LM) arrived at the Moon, the service propulsion engine fired for almost 6 min to insert the spacecraft into a lunar orbit of 169.7 by 60.0 nautical miles, or 314.28 by 111.1 km. The Moon's mass is 7.346×10^{22} kg and its radius is 1737.5 km. Find the eccentricity of that orbit, its semi-major axis, parameter, the orbital period in minutes, and periapsis and apoapsis velocities in km/s.
8. Find the total energy of the initial CSM orbit in the previous problem and confirm the elements of that orbit by using the energy-derived equations.
9. The Mars Global Surveyor spacecraft was launched in late 1996 and rendezvoused with Mars in September, 1997. It is one of the most successful planetary missions in NASA history. Upon its arrival, the spacecraft executed a Mars Orbital Insertion (MOI) burn to slow it down and allow itself to be captured by the Martian gravitational field. The plan was for its initial capture orbit to have a periapsis of 300 km and apoapsis of 56,675 km above the Martian surface. Then, because propellant was limited, the capture orbit would be gradually circularized by allowing the Martian atmosphere to slow it down in each close pass by a process known as "aerobraking." This drag process initially was to cause the apoapsis altitude to shrink to 2,000 km and then ultimately to 450 km, with periapsis at 400 km above the Martian surface. Determine the amount of energy loss that atmospheric drag would have to impart upon the spacecraft to reduce its apoapsis from 56,675 km to the nearly circularized orbit of 450 by 400 km. The radius of Mars is 3,389.5 km and its mass is $6.41693\ 10^{23}$ kg.
10. The Mars Global Surveyor ultimately was placed into a mapping orbit above the Martian surface, with a semi-major axis of 3,774.998 km, an eccentricity of .00953 and a nearly polar inclination with the periapsis close to the Martian south pole. From there the on-board Mars Orbiter Camera and Mars Orbiter

Laser Altimeter returned spectacular images of the red planet along with accurate topological data. Find the parameter of the elliptical mapping orbit, the velocities at the apsides, and calculate the sum of the kinetic and potential energies of the spacecraft at the apsides, demonstrating that they are the same.

Reference

1. Clarke AC (1950) Interplanetary flight. Harper & Row, New York

Chapter 15
Getting Oriented: The Sun, the Earth and the Ecliptic Plane

There are many coordinate systems one may use as references for the placement of heavenly objects. A standard reference that we will employ is the *heliocentric ecliptic* coordinate system.

Visualizing the Heliocentric Elements of an Elliptical Orbit

Imagine a reference plane that is like a plate or disk with the Sun at the center – it is helio- (for Sun) centric. The Earth's mean or long-term average orbital plane for a given *epoch* (time) defines the plane of the plate: it is called the *ecliptic* plane, and is invariable. For the most part the planets of the solar system stick pretty closely to this plane. (The extent to which a planet does not is called the *inclination* of its orbit.) We will begin by examining the orbit of the Earth and defining the first two important reference points on the ecliptic plane: the *vernal equinox* and the *longitude of the perihelion*.

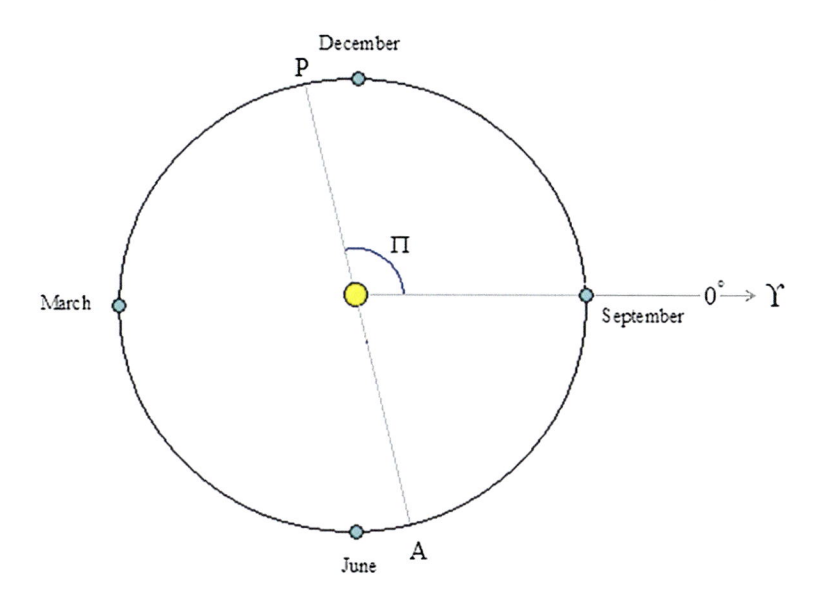

The Earth in its Orbit

The *vernal equinox* is a fixed reference point among the stars, *outside* the orbit, which helps us anchor the orbit and compare it with the location of other orbits. It is sometimes called the *first point of Aries* because the vernal equinox was originally in that constellation. It is traditionally represented by the zodiacal sign of Aries (resembling the Greek letter ϒ) even though due to precession, the vernal equinox is now in the constellation of Pisces. The Sun's apparent position lies upon the vernal equinox on the first day of spring in the northern hemisphere. The line of the equinoxes itself is where the plane of the celestial equator, tilted as it is by about 23.4° from the ecliptic plane, intersects the ecliptic plane.[1] It varies over the years but is commonly about March 22nd; if we could look through the Sun on that date we would have a fix on the vernal equinox on the other side of the orbit. Now think of a line from the Sun projected out to the vernal equinox at midnight 6 months later, on the first day of Earth's northern hemisphere autumn, about September 22nd: this is the location among the stars that is the beginning point in the orbit for longitude in this plane. It is shown on the diagram as zero degrees *heliocentric ecliptic longitude*. If we were to examine a sky map of this region in mid-March, we would see the Sun as it was approaching the vernal equinox. The Sun (from our geocentric view) appears to be moving upward along the ecliptic toward the

---

[1] The celestial equator is the projection of the Earth's equator into the sky. Since the Earth's axis is titled 23.4° from the ecliptic plane, the plane of the celestial equator and the plane of the ecliptic are correspondingly non-parallel by this amount. The intersection of the two planes creates a nodal line which points to the equinoxes.

Visualizing the Heliocentric Elements of an Elliptical Orbit 315

intersection of the ecliptic and the celestial equator, that is, toward the vernal equinox. Below is such a map, which on the arbitrary date picked (March 11, 2011) happens also to show four planets very close to the equinox. Note how they are essentially aligned along the ecliptic plane. You can also confirm that the angle between the ecliptic and the celestial equator is correctly represented as nearly 23½°. Having thus established an external anchor among the stars to fix our celestial calculations, we return to the orbit itself.

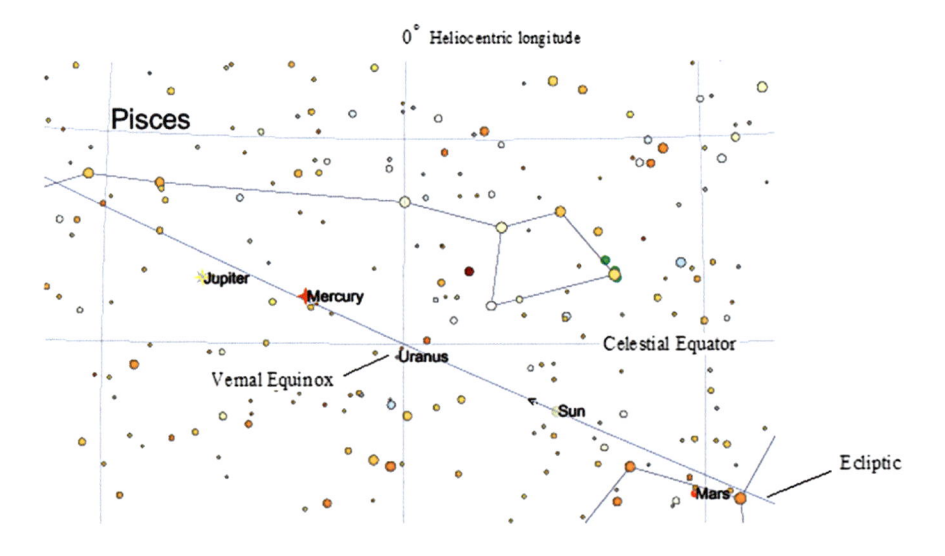

Configuration of planets and Sun on March 11, 2011
Vernal Equinox is at intersection of Celestial Equator and Ecliptic

Perihelion is a reference point within the orbit itself. It is of course the point of closest approach to the Sun, and is marked on the diagram with a P. Aligning the semi-major axis along the Sun-planet line (the diagonal in the first diagram), the focus of the ellipse containing the Sun will be the focus nearest the perihelion point. While the Earth's orbit's fairly slight eccentricity is not apparent in the diagram, the point P is the point closest to the Sun, and this occurs in early January. If we measure counterclockwise (the direction of the Earth's motion in its orbit) from the equinox *Y* we arrive at the perihelion point. The angle between the equinox and perihelion, between *Y* and P is called the *longitude of the perihelion*, and is commonly represented by the symbol Π. For Earth the value of Π is about 103°. This number (as with other elements) does vary minutely due to the perturbations of the other planets, but it does so exceedingly slowly, and for our purposes may be regarded as fixed.

Now we introduce a third and fourth reference in the alignment of an orbit in space. They are called the *inclination* of the orbit and the *line of the nodes*. Where the plane of an orbit tilts with respect to the ecliptic plane, its orbit is said to have *inclination*. For example, the planet Saturn has an orbital inclination of about 2½°. This means its orbital plane is tilted or inclined by that much with respect to the ecliptic plane. The angle of inclination is represented by the symbol i. The two intersecting planes geometrically create a line. This line is called the *line of the nodes*.

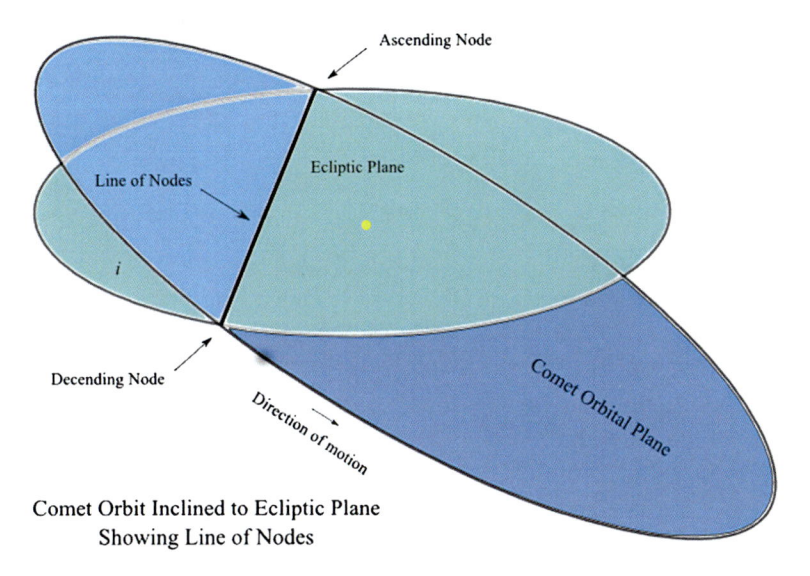

**Comet Orbit Inclined to Ecliptic Plane
Showing Line of Nodes**

For the Earth's orbit, we are using a heliocentric-reference system, so the orbital inclination of the Earth is zero. But as we will see, the orbiting objects of the solar system do typically have inclinations with respect to the ecliptic and therefore nodes. For example, the inclination of comet Schwassmann-Wachmann 3 discussed earlier has an inclination of about 11.4°, and many have far more extreme inclinations. It should be apparent that with respect to the ecliptic plane, part of the comet's orbit lies above the ecliptic plane, and part lies below it. The line of nodes is the dividing line. The point where the comet (or planet or asteroid or orbiting spacecraft) traveling counterclockwise, rises above the ecliptic plane is called the *ascending node*. The ascending node is the most important nodal reference point in the orbit. The place where the orbit goes below the ecliptic is called the *descending node*. The ascending node's angular distance going counter-clockwise from the vernal equinox is called the *longitude of the ascending node* and is represented by the symbol Ω. Thus the orbit is fixed in this way to the external reference among the stars. The perihelion point can in turn be referred to the ascending node (rather than the equinox) and this angle (again running counter-clockwise from the ascending node) is called the *argument of the perihelion*; it is usually (not always) denoted by the symbol ω. Here is a schematic of the orbit and ecliptic as seen from the Sun:

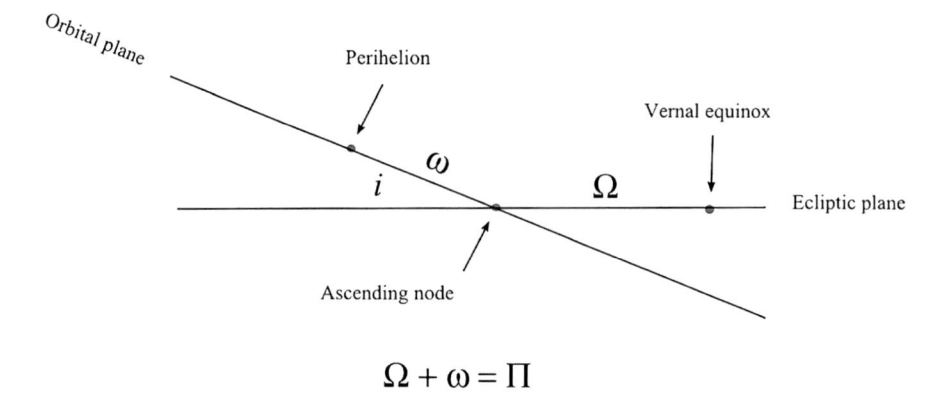

Schematic of Selected Elements of the Orbits of Mars, Jupiter and Saturn 317

[Schematic diagram showing orbital plane intersecting ecliptic plane, with labeled angles: Perihelion, Vernal equinox, inclination i, argument of perihelion ω, longitude of ascending node Ω, and Ascending node]

$$\Omega + \omega = \Pi$$

Schematic Diagram of Ecliptic and Orbit of Planet or Other Body

Let us summarize these anchor points of an orbit:

- The angle between the orbital plane and the ecliptic is the *inclination* of the orbit.
- Longitude angles on the planes are measured in the counter-clockwise direction (going East).
- Any angle measured from the vernal equinox will be called a *longitude*.
- The angle from the equinox to the ascending node is Ω, the *longitude of the ascending node*.
- The angle from the equinox to the perihelion is Π, the *longitude of the perihelion*.
- The angle from the ascending node to the perihelion is ω, the *argument of the perihelion*.

Note that the angle Π is the sum of angles in two different planes. Thus $\Omega + \omega = \Pi$, as shown.

Schematic of Selected Elements of the Orbits of Mars, Jupiter and Saturn

It is sometimes helpful schematically to depict the ecliptic and orbital planes in schematic fashion, angled from each other by the orbital inclination i and with the vernal equinox being the point of their intersection. The longitude of the ascending node Ω and longitude of the perihelion Π can then be plotted and compared. Consider for example the inclination, longitude of the ascending node and longitude of the perihelion for the orbits of Mars, Jupiter and Saturn. Here is a table of those orbital elements:

Description of element	Selected elements of the orbits of Mars and Jupiter (Epoch 2000.0) (longitudes of Ω and Π in degrees east from vernal equinox)		
	Mars	Jupiter	Saturn
i	1.85	1.3	2.5
Ω	49.6	100.5	113.7
Π	336.1	14.7	92.6

With this information we can sketch the schematic[2]:

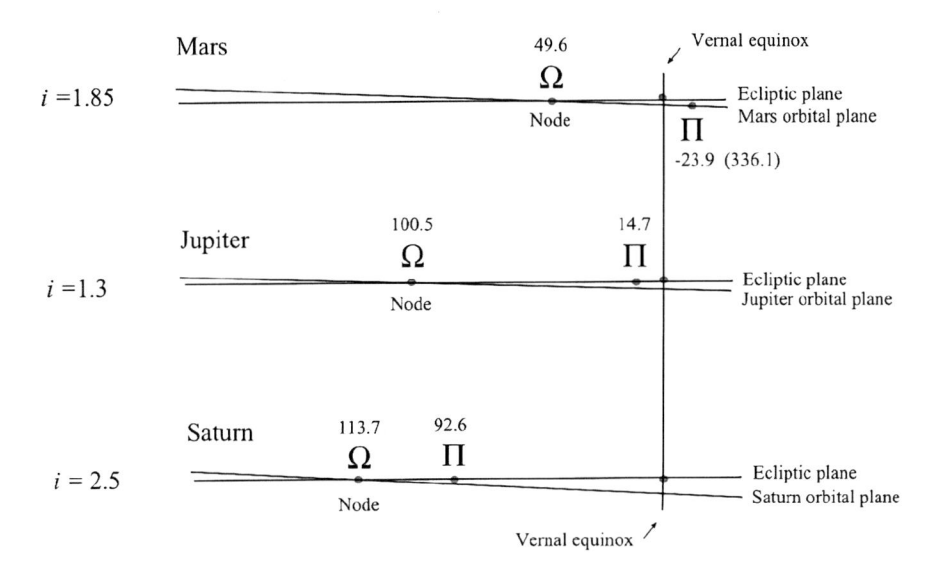

```
                  Mars                    49.6              Vernal equinox
                                           Ω              /
i =1.85  ═══════════════════════════════════════════════•═══════ Ecliptic plane
                                          Node         Π         Mars orbital plane
                                                      -23.9 (336.1)


                  Jupiter                100.5       14.7
                                          Ω           Π
i =1.3   ═══════════════════════════════════════════════•═══════ Ecliptic plane
                                          Node                   Jupiter orbital plane


                  Saturn         113.7    92.6
                                   Ω       Π
i = 2.5  ═══════════════════════════════════════════════•═══════ Ecliptic plane
                                  Node                           Saturn orbital plane

                                        Vernal equinox /
```

Longitudes (in degrees from the vernal equnox) of the ascending node and perihelions for Mars, Jupiter and Saturn (not to scale)

The vernal equinox is the common "anchor" to all three schematics. Distances in degrees (out of 360) are estimated from the vernal equinox, moving counter-clockwise (to the left), first along the ecliptic to the ascending node, then along the orbital plane. Planar lines for the orbits can be constructed to extend to some point to the left, say 270°, then return on the right. Begin by drawing a vertical line for the vernal equinox. Then key the longitudes off that. Drawing schematics like this can assist in visualizing the intersection of planetary orbits with those of comets and asteroids.

---

[2] Of course the schematic will not purport to be to scale or show actual inclinations in degrees, since the quantities are too small to be accurately depicted. The values of the elements are taken and rounded from the NASA/JPL site, http://ssd.jpl.nasa.gov/txt/p_elem_t1.txt, epoch 2000.0.

The Heliocentric Longitude of a Body

The semi-major axis, eccentricity, inclination, longitude of the ascending node and the longitude of the perihelion are called the *elements* of the orbit. The ellipse in the abstract has a scale and shape determined by the semi-major axis and the eccentricity; the task here has been to fix the orientation of the ellipse with reference to the stars, the vernal equinox, and if it has a tilt, measure that tilt off a flat plane known as the ecliptic plane. The heliocentric orbit is thus aligned in space with reference to the vernal equinox, and inclined with respect to the ecliptic, from which another reference point, the ascending node, is given. From these, the perihelion of the ellipse can easily be fixed. But there is one more element to remember, the *mean longitude* of the orbiting planet (or other body). This tells us where it is at a given moment on its continuous journey around the Sun, *as if the planet were moving in a circular orbit*. This virtual orbit is sometimes referred to as the *fictitious circle*, and we will encounter it again later. We know that the Earth is at zero degrees longitude at the moment of the vernal equinox in a given year, but where is it in its orbit on another given date, and how do we find out? If you are thinking it is simply a function of its mean daily motion (i.e., $360°$ divided by its period in days) from a certain reference date to the date of interest, you are correct. But before getting into the actual calculation of heliocentric longitude, and because the mélange of Greek letters in this subject can be confusing, we will briefly revisit the order of the heliocentric elements to help fix them in mind.

Summary of the Elements of the Heliocentric Orbit

Suppose you want to find approximately where a planet is; that is, you desire to find the mean heliocentric longitude of a planet, its degrees-around-the-Sun progression from the vernal equinox as of a certain date as if its orbit were a circle. You follow an imaginary itinerary as if we were on a fast-travelling spaceship. Your pilot is instructed to stick closely to the flight plan while you with your clipboard keep track of the angular distances from point to point. You begin the journey by venturing counter-clockwise (going east) from the first point of Aries, the *vernal equinox* ϒ. That is your launching point. You fly east circularly around the ecliptic plane sticking close and low to that plane until you arrive at a place where the orbital plane of the target planet rises up ahead of you. That place is the ascending node. You label it N at the place where the up-rising orbit intersects the equinoctial plane along which you have been travelling, and you go on up. Now, like switching metro lines, the spaceship transfers to the plane of the planet. On your clipboard you note the distance in degrees from the start of your trip to this intersection of planes, from ϒ to N: this amount of circular travel in degrees is the *longitude of the ascending node*, Ω. That trip-to-the-node concludes the first phase of your itinerary. From N you travel along the up-tilted plane of the planet to the perihelion P of the elliptical orbit, where the orbit is closest to the Sun. You now note the angular distance from N to perihelion P, which you label the *argument of the perihelion* ω.

Continuing your spacecraft's journey from the perihelion P we finally arrive at the *mean planet*, X. You record the angular distance from P to X as the *mean anomaly*, M. This point X is where a line from the Sun would intersect a fictitious circle, on the orbital plane, along which the planet would move according to its mean daily motion. Now you add up all the angular distances. The total angular distance you have travelled on these two jointed planes from ♈ to M is the *mean heliocentric longitude* (or just mean longitude if you know you are working with heliocentric coordinates) of the planet, and is usually symbolized by the letter L. Concluding the trip, you review the symbols.[3] The angular distance from vernal equinox ♈ to the mean planet is the mean longitude: $L = \Omega + \omega + M$:

$$\underbrace{L}_{\substack{\text{Mean longitude of the planet}\\ \text{(degrees east of planet}\\ \text{from vernal equinox)}}} = \underbrace{\Omega}_{\substack{\text{Longitude of the ascending node}\\ \text{(degrees east of node}\\ \text{from vernal equinox)}}}$$

$$+ \underbrace{\omega}_{\substack{\text{Argument of the perihelion}\\ \text{(degrees east of perihelion from node)}}} + \underbrace{M}_{\substack{\text{Mean anomaly of the planet}\\ \text{(degrees east of planet}\\ \text{from perihelion)}}}$$

Remember also that the longitude of the perihelion, Π, is the distance from ♈ to perihelion, or

$$\Pi = \Omega + \omega$$

Therefore,

$$L = \Pi + M.$$

Before celebrating the end of your trip, it is important here to realize that you have arrived only at the mean planet on the orbital plane: you are still not at the true position of the planet as it is located on its ellipse. So you can pause here in space, at X, before actually landing on the planet; that final leg of the voyage to the planet will follow.

---

[3] Often symbols particular to the source may sometimes be used instead of the above ones. For example the Jet Propulsion Laboratory uses the words "node" or OM (for omega) to render Ω, the longitude of the ascending node; "peri" or W or "argument of perifocus" for ω, the argument of the perihelion; and MA for mean anomaly M. Other elements commonly used, are P, the period; q, the perihelion distance; t_p the time of perihelion passage which is often given with a Julian day and calendar date; and n, the mean daily motion of the body. Sometimes the source will also provide the aphelion distance Q.

Method for Determining Earth's Heliocentric Longitude

Determining the Earth's heliocentric longitude is straightforward. We must, however, use a very particular convention in accounting for time. That is the *Julian Day* system, which is universally used by astronomers. It is a continuous count of days (and decimal parts of a day) from – 4,712. The Julian Day begins at noon, Greenwich Mean Time. A Julian century is 36,525 days. The date of the beginning of the current century (the *epoch J2000.0*) is JD2451 545.0. This corresponds to January 1.5, 2000. The trick is to find the time forward (or backward) from that point to the date you are interested in. There are various programs commonly available for calculating Julian days. If one desires to get right to the result and move on to the other calculations, an easy on-line source is the United States Naval Observatory site, http://aa.usno.navy.mil/data/docs/JulianDate.php.

There are several ways of calculating a planet's heliocentric longitude to any degree of accuracy desired. The one we have used relies upon the NASA/JPL Solar System Dynamics website.[4] Other sources too are available.[5] It provides approximate Keplerian elements for long intervals, and its HORIZONS system allows for the very precise computation of elements and ephemerides (positions) for thousands of objects. Since our present focus is the determination of heliocentric longitude, the accompanying table, put together from the site's *Keplerian Elements for Approximate Positions of the Major Planets*,[6] gives just the mean heliocentric longitude L for the planets and Pluto.

Coefficients for the approximate mean longitude L of the planets and Pluto[a] (mean ecliptic and equinox of J2000, valid for the time-interval 1800–2050 A.D.)

Planet	a_0	a_1
Mercury	252.2503235	149,472.67411175
Venus	181.97909950	58,517.81538729
Earth (EM barycenter)	100.46457166	35,999.37244981
Mars	355.4465680	19,140.30268499
Jupiter	34.39644051	3,034.74612775
Saturn	49.95424423	1,222.49362201
Uranus	313.23810451	428.48202785
Neptune	304.8799703	218.45945325
Pluto	238.92903833	145.20780515

[a]From http://ssd.jpl.nasa.gov/txt/p_elem_t1.txt

The coefficients a_0 and a_0 are the coefficients of the polynomial,

$$L = a_0 + a_1 T$$

---

[4] Found at: http://ssd.jpl.nasa.gov/?planet_pos.

[5] One could also use the approach set forth in Jean Meeus' book [1]. It has set of tables of polynomial coefficients for calculating the elements of any planetary orbit. See also, Jean Meeus [2].

[6] http://ssd.jpl.nasa.gov/?planet_pos.

Where L is the mean heliocentric longitude and T is the time measured in Julian centuries measured from epoch J2000.0, determined by this equation:

$$T = \frac{JD - 2451545.0}{36525}$$

The quantity is negative before J2000.0. The mean longitude for the Earth, for example, can be calculated by plugging in the appropriate coefficients to the polynomial equation: For Earth (or more precisely, the Earth–Moon barycenter), for example, the coefficients are, for the mean equinox of J2000.0:

$$L = a_0 + a_1 T$$

$$L = 100.46457166 + 35999.37244981 T$$

All that is then needed to find a quite good position of the Earth from is to put in the time in centuries T calculated as noted above.

Finding the Earth's Heliocentric Longitude at Close Encounter with Asteroid 2010 TD54

Problem Determine the number of Julian days from epoch J2000.0 until noon on October 12, 2010, about when asteroid 2010 TD54 made its close approach to the Earth. From this and the information below, calculate the Earth's heliocentric longitude for that moment.

Given

JD 2455482	Date of the closest approach of the asteroid
T = (JD − 2451545.0)/36525	Time in centuries from epoch J2000.0 to JD
L = 100.46457166 + 35999.37244981T	Mean heliocentric longitude of Earth at T

Assumptions We assume the accuracy of the coefficients for our purposes, and the information regarding the asteroid. We are not seeking high level of computational accuracy for ephemeris purposes. We are determining *mean* heliocentric longitude, as if the Earth's orbit were a circle, and therefore there will be slight differences between this longitude and the Earth's actual position in its orbit, since the Earth's orbit is elliptical. Calculating the *true anomaly*, the planet's actual position on the ellipse, will be covered later. For the Earth, with its very slight eccentricity, the mean longitude will give us a pretty fair approximation of where it is at a given time.

Method We will calculate the time elapsed since J2000.0 then insert that value T into the equation for determining heliocentric longitude. Then substitute the value into the longitude equation. If the result is more than 360°, find the remaining fraction of the orbit in degrees.

Calculations First find the elapsed time from epoch J2000.0, by inserting the date of the asteroid encounter into the time equation:

$$T = \frac{JD - 2451545.0}{36525}$$

$$T = \frac{2455482 - 2451545.0}{36525}$$

$$T = 0.10778918$$

That is about a tenth of a century, which looks right. Now we substitute this value for time in the longitude equation:

$$L = 100.46457166 + 35999.37244981T$$

$$L = 100.46457166 + 35999.37244981(0.10778918)$$

$$L \simeq 3980.81$$

This is more than 360° so it is necessary to divide by 360 and truncate the integers to find the remainder portion of the orbit:

$$L = \frac{3980.81}{360} = 11.05779836 = .05779836 \text{ of a circle}$$

$$L = 360(.05779836)$$

$$L \simeq 20.81°$$

Observations

1. Does this result square with our intuition? We would expect the mean longitude on October 12th to be about 20 days after equinox (~9/22), which is point ϒ. Earth's mean daily motion eastward in its orbit is 360/365.25 = .9856°. The result is very close to 20° of movement. The answer appears to be the intuitively correct mean longitude on the fictitious circle with the Sun at its center.
2. Locating heliocentric longitude can be very practical. If we can determine the heliocentric longitude of an asteroid or comet for a given range of dates, and we can also can also ascertain where the Earth is in its orbit, we can find out if the asteroid or comet will pass close to the Earth, and when. From the mean longitude we can then calculate where it is on its elliptical path (the true anomaly), but this is a step later.

Calculating the Mean Anomaly and Heliocentric Longitude of the Near Earth Asteroid 2010 TD54

We will take the published orbital elements of the asteroid 2020 TD54 as a good case study in extracting orbital information from the elements of the orbit. We will focus on the elements that are essential to the calculation of the true anomaly, which we encounter in the next chapter – the determination of a body's actual position in the orbit. In working with these angles, the direction of measurement of the longitudes is always counter-clockwise. And keep in mind the important distinctions as to the starting points. The distance from the vernal equinox to the ascending node, Ω, is measured counter-clockwise from the vernal equinox. The distance ω from there to perihelion is measured counter-clockwise from the node. The sum of those distances, Π, is a number that thus originates at the vernal equinox. The mean anomaly M, listed separately in the elements of a body, has a different starting point. It begins at *perihelion*, and moves in the direction of motion of the body (which for the planets is counter-clockwise) until the mean planet is reached. The mean anomaly thus is reached by computing the movement of the body from time of perihelion to its present position. The sum of all of these distances, then, equinox to node, node to perihelion, perihelion to mean planet, is the mean heliocentric longitude, L.

Problem Explain each of the orbital elements below, and determine the mean anomaly and mean heliocentric longitude of asteroid 2010 TD54 for the time of its closest encounter with Earth at about noon on October 12, 2010 (that is, on JD 2455482).

Given The following table lists the orbital elements of 2010 TD54 as published by the Jet Propulsion Laboratory, available from its small-body database browser[7]:

Orbital elements at Epoch 2455400.5 (2010-Jul-23.0) (Heliocentric ecliptic J2000)		
Element	Value	Units
e	.6190559851793439	
a	1.785313246486003	AU
q	.6801043958288775	AU
i	5.073529889261414	deg
Node	18.76272623533118	deg
Peri	80.47725733511749	deg
M	306.6646133153061	deg
t_p	2455529.587150799390	JED
	(2010-Nov-29.08715080)	
Period	871.3047223695602	d
	2.39	yr
n	.413173475085686	deg/d
Q	2.890522097143128	AU

---

[7] Available at http://ssd.jpl.nasa.gov/sbdb.cgi#top.

Calculating the Mean Anomaly and Heliocentric Longitude of the Near Earth...   325

Assumptions We will round off the data from the table to three or four places for the purposes of our approximations. We assume for the sake of simplicity that closet encounter with Earth occurred at noon. We will ignore the inclination of the asteroid's orbit (5°) and take the Earth's orbit to be circular. These assumptions will simplify our calculations and enhance the heuristic value of the problem.

Method The first job is to determine the mean anomaly, the distance from perihelion to the asteroid's position on October 12, 2010. The mean anomaly for the "epoch date" of July 23, 2010 is given in the chart. (We will call this the "July epoch.") It is about 306.6646°. This means that the asteroid has come around about 85 % of a full orbit counterclockwise from its last perihelion, on its way to an end-of-November 2010 perihelion. This doesn't tell us where the asteroid is on the encounter date in October, 2010 (on JD 2455482). Determining the mean anomaly for the date we seek can be done by taking the days from the July epoch to the October encounter and multiplying them by the mean daily motion n. This result, when added to the mean anomaly of epoch, will yield the mean anomaly for the encounter. Then it is possible to take the next step and solve for mean longitude. To find the mean heliocentric longitude, we simply fill in the values from the table for the expression $L = \Omega + \omega + M$. As before, we will need to correct the result if it is over 360°.

Calculations *Finding the mean anomaly*:

1. First find the number of days from the July epoch (JD 2455400.5) to the October encounter (JD 2455482).

$$2455400.5 - 2455482 = 81.5 \, \text{days}$$

2. Since the mean daily motion n is .413173475° per day, the total number of degrees traversed since the July epoch is,

$$81.5 \times .413173475 = 33.6736°$$

3. Add the above angular travel distance to the July epoch given in the table of 306.66°:

$$M = M_{epoch} + 33.6736°$$

$$M = 340.34°$$

This is the mean anomaly for the encounter date. It is the angular distance going counter-clockwise from the asteroid's own perihelion to its location on a fictitious circle.

Finding the heliocentric longitude:
Now we have enough information to calculate mean longitude:

$$L = \Omega + \omega + M$$

$$L = 18.7627 + 80.477 + 340.34$$

Adding the result and subtracting 360 we have,

$$L = 79.58°$$

Observations

1. This is the mean, eastward longitude of the asteroid from the vernal equinox at closest approach to Earth on October 12, 2010, on the fictitious circle, not its true location on the ellipse. This is its total angular distance east from the vernal equinox, and is good for that 1 day/moment only.
2. We can compute the longitude of the asteroid's perihelion, which is not given in the table. The perihelion longitude is,

$$\Pi = \Omega + \omega$$

$$\Pi = 18.7627 + 80.477$$

$$\Pi = 99.237°$$

3. Comparison of the mean longitude (from the vernal equinox) of ~ 80° and the perihelion longitude (from the vernal equinox) of nearly ~100° shows that the *mean* asteroid on the fictitious circle is almost 20° from perihelion at closest approach to Earth.
4. We can check this result against the time of the asteroid's November, 2010 perihelion from the elements. The number of days to perihelion after encounter with the Earth is the angular distance to be travelled by the asteroid since October 12th times the mean daily motion in degrees:

$$\text{Days to perihelion} = \frac{\text{Angular distance to perihelion}}{\text{Mean daily motion } n}$$

$$\text{Days to perihelion} = \frac{\Pi - L}{n}$$

$$\text{Days to asteroid perihelion} = \frac{19.66°}{0.413173475}$$

$$\text{Days to asteroid perihelion} = 47.582919$$

These are days from closet encounter with Earth to the asteroid's perihelion. Is the perihelion date from the table, November 29th, in fact 47 days from October 12th? Let us check and see what the Julian Day of perihelion is:

$$\text{JD of perihelion} = \text{JD of closest encounter} + 47.582919 \text{ days}$$

$$\text{JD of perihelion} = 2455482 + 47.582919$$

$$\text{JD of asteroid perihelion} = 2455529.58$$

Compare this with the date of perihelion from the table above. From the USNO site mentioned earlier, http://aa.usno.navy.mil/data/docs/JulianDate.php, this slightly rounded result corresponds to Nov. 29, 2010, at almost 02 h UT, which is correct for this asteroid.

5. Here is the asteroid's location at closest approach in schematic form:

Schematic Diagram of Orbit of Asteroid 2010 TD54
Showing Asteroid Mean Position on October 12, 2010

Mean angular distance M from perihelion (mean anomaly) = 340.34 degrees

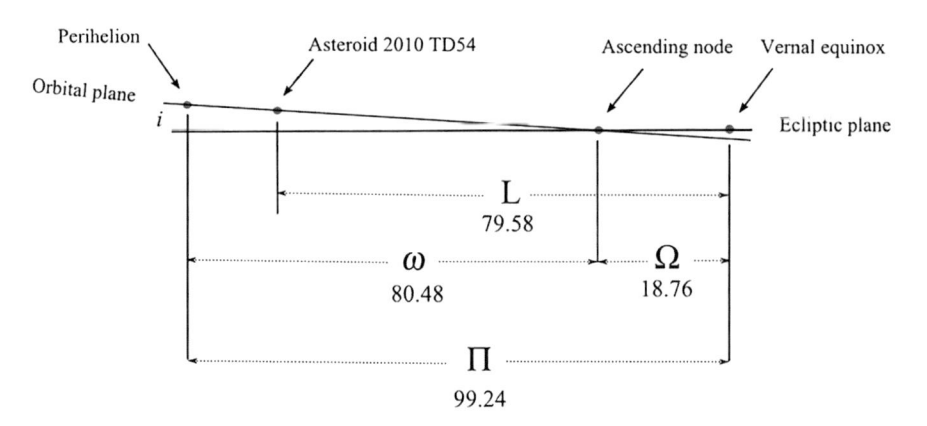

The asteroid on October 12th was as we saw only about 20° from its November perihelion, moving east toward it. Note that the ~340° mean anomaly, being measured east from the perihelion, essentially wraps around our diagram to the left and terminates at the asteroid at the location shown. This, as you can see, ends up being about 20° short of a full circle.

Graphing the Inner Orbit of Comet West C/1975 V1-A

One of the longest-period comets known is Comet West, which appeared brightly and dramatically to the naked eye in the northern hemisphere spring of 1976. This problem will take us through the steps on how to align the comet orbit in space, locate the Earth at the comet's perihelion, and visualize the comet's path on its inbound and outbound journey. As you see from the elements, comet West's orbit is highly eccentric, almost parabolic. The graph of the innermost portion of the orbit will be based upon elements provided by the NASA/Jet Propulsion Laboratory website.[8]

Problem The orbital elements of comet West as presented by the JPL site are as follows:

q	e	i	w	Node	Tp
0.19662600	0.99997100	43.06640	358.42700	118.92400	2,442,833.7216

From this information, calculate the longitude of the perihelion, semi-major axis, and the period, and graph or sketch the inner orbit of the comet at perihelion, within 2 or 3 AU from the Sun. Include the orbit of the Earth and the ecliptic plane, and locate the line of nodes and the position of Earth at the time of Comet West's perihelion. Attempt to visualize the position of the comet in the sky as seen from Earth at the time of perihelion passage.

Given

$\Pi = \Omega + \omega$	Longitude of the perihelion, from node + ω
$a = q/(1 - e)$	Semi-major axis where perihelion distance and eccentricity are known
$b = a\sqrt{1 - e^2}$	Semi-minor axis of the elliptical orbit
$P = a^{3/2}$	Kepler's Third Law, for determining the orbital period around the Sun when the semi-major axis is known
T = (JD − 2451545.0)/36525	Time in centuries from epoch J2000.0 to JD
L = 100.46457166 + 35999.37244981T	Mean heliocentric longitude of Earth at time T

Assumptions We will round off the element data to three or four places for our final approximations. Note that the inclination is 43°. We will only approximate this factor in the depiction of the comet's orbit. The Tp term means time of perihelion. Given the nearly parabolic orbit of this comet, its period will be somewhat speculative, as mentioned below. Finally, the comet broke into four pieces as it passed perihelion. The long term uncertainty in the orbit must be taken for granted.

Method Calculate each of the unknowns from the equations given. Then attempt to visualize the alignment of the orbit in space relative to the ecliptic plane, laying the vernal equinox again along the positive x axis, and placing the longitude of the perihelion counter-clockwise from that. Then the perihelion point of the orbit is placed at distance to perihelion q along the perihelion line from the Sun. From the elements we see that this q distance is a little less than .2 AU. Using the any of equations for an ellipse, graph or sketch the elliptical orbit for the comet. It may be

---

[8] From http://ssd.jpl.nasa.gov/dat/ELEMENTS.COMET.

easiest to use the Cartesian coordinate system, and for this purpose the semi-minor axis must be determined from the equation given. Here is the parametric form of the ellipse equations:

$$x = a\cos(\theta) \quad y = b\sin(\theta)$$

This creates an ellipse with the center of the ellipse at the origin of the graph (0, 0). But since we want the graph to be in the inner portion of its orbit, we will want it centered with the Sun at the origin. To offset such a graph by a certain constant, c, the equation for the x parameter would be:

$$x = c + a\cos(\theta)$$

The offset for this ellipse would be the semi-major axis minus the perihelion distance, $a - q$:

$$x = (a - q) + a\cos(\theta)$$

For the position of the Earth at comet West's perihelion, determine the Julian date of the event from the U.S. Naval Observatory site mentioned earlier.[9] Then use the given equations to determine the Earth's heliocentric longitude, and locate Earth's position on the graph counterclockwise from the vernal equinox. Try then to visualize the location of the comet in the sky from your home as the comet recedes from Earth.

Calculations We will proceed step-by-step in finding each of the needed elements:
Step 1: Find the longitude of the perihelion:

$$\Pi = \Omega + \omega$$

$$\Pi = 118.924 + 358.427$$

Subtracting 360 since the result is more than a full circle:

$$\Pi = 117.351°$$

From this result it appears that the perihelion will be in the northwest or second quadrant of our graph, about 27° left of vertical.

Step 2: Calculate the semi-major axis of the orbit:

$$a = \frac{q}{1 - e}$$

---

[9] http://aa.usno.navy.mil/data/docs/JulianDate.php.

$$a = \frac{.196626}{1 - 0.999971}$$

$$a = 6,780.2 \text{ AU}$$

This is indeed an enormous orbit! It can be seen that this comet must have originated in the Oort cloud, the primordial sphere of icy bodies surrounding our solar system and home of countless millions of such objects.

Step 3. As the graph is in the Cartesian coordinate system, it is desirable to know the semi-minor axis distance:

$$b = a\sqrt{1 - e^2}$$

$$b = 6780.2\sqrt{1 - .999971^2}$$

$$b = 51.64 \text{ AU}$$

Thus the *width* of this ellipse on its minor axis is about 100 AU, easily sufficient to encompass the entire solar system! The length-to-width ratio of this orbit is just over 130:1.

Step 4. To calculate the period of this comet, we employ the old, reliable Kepler's Third Law:

$$P = a^{3/2}$$

$$P = (6780.2)^{3/2}$$

$$P = 558,296 \text{ years}$$

The last passage of this comet by perihelion occurred before *homo sapiens* had evolved. This is the *calculated* period; however, the orbit's eccentricity is nearly parabolic, and its path is highly sensitive to the subtle perturbing conditions of the other planets, most of which disturbances occur in the inner one-thousandth portion of its orbit. Thus calculated period should be viewed with a good deal of skepticism. Moreover, the comet split into fragments in its voyaging around the Sun, and our exercise ignores that fact.

Step 5: Generate the graph. This is one of various possible ways of depicting the ecliptic plane in relation to the orbit. What is important is to fix the perihelion longitude at about 117° from the vernal equinox along the *x* axis, and the ascending node at about 119°. The little line from the Sun to the perihelion point is the distance *q* of just under .2 AU. Just to the left of the perihelion point, on the comet orbit, is the ascending node; the descending node would actually be far on the other side of the orbit, not shown here. The comet's path was inbound under the ecliptic plane, rising up at the ascending node just after perihelion passage.

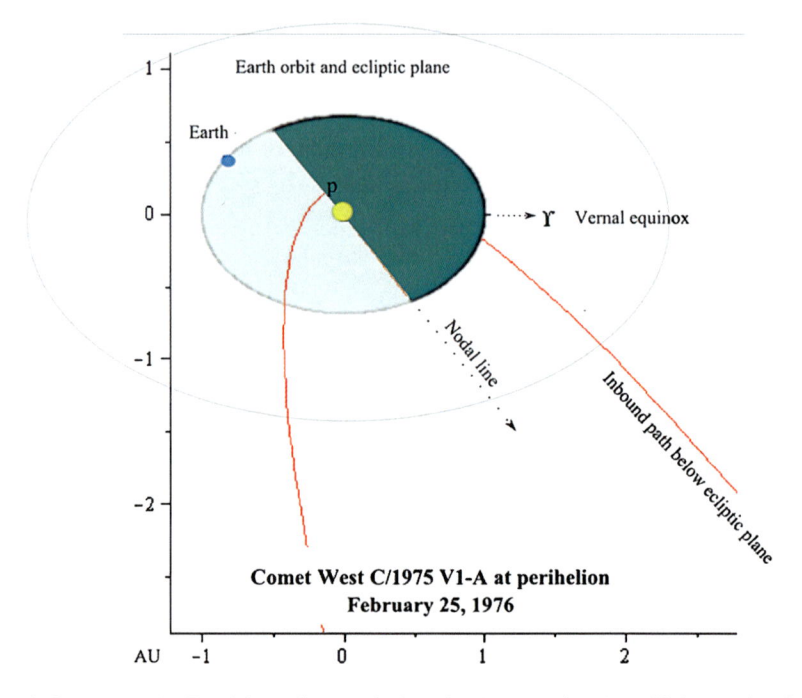

**Comet West C/1975 V1-A at perihelion
February 25, 1976**

Step 6. Compute the Earth's heliocentric longitude at perihelion. This can be done by using the given equation, after first determining the Julian Date of perihelion. For this we will consult the elements mentioned above, which tells us that February 25, 1976 is JD 2442833.5 (which is 1976: Feb 25.22160000). Use the same procedure as we used earlier.

$$T = \frac{JD - 2451545.0}{36525}$$

$$T = \frac{2442833.5 - 2451545.0}{36525}$$

$$T = -.23850787$$

Because the date is earlier than epoch 2000, the sign is negative. This value is inserted for the time variable in the longitude equation:

$$L = 100.46457166 + 35999.37244981T$$

$$L = 100.46457166 + 35999.37244981(-.23850787)$$

$$L = -8485.669072$$

Again, as this is more than 360°, it is necessary to divide by 360 and truncate the integers to find the remainder portion of the orbit:

$$L = \frac{-8485.669072}{360} = -23.57130298 = -.57130298 \text{ of a circle}$$

$$L = 360(-.57130298)$$

$$L = -205.6690728$$

It is necessary to add 360° to get a positive result:

$$L = 154.33°$$

Observations

1. Hence we arrived at a heliocentric longitude for the Earth of about 154°. It is important always to check the results to see if they may intuitive sense. The Earth's mean daily motion is about .9856°, and there are about 156 days from equinox to perihelion; their product is about 154°, confirming the result.
2. Locating the Earth on the graph clearly shows that it was well-placed for a glorious morning view of the comet in the northern hemisphere after it passed perihelion.

3. The graph does not adequately show the 43° inclination of the orbit to the ecliptic plane. If you drew a perpendicular to the line of nodes along the plane of the comet's orbit, that line would be elevated 43° from the ecliptic. Thus the part of the orbit on the left of the nodal line on the diagram is above the ecliptic plane. From the perspective of an observer at Earth's location on the plane, the comet after perihelion would appear to be rising upward and to the right as it receded. This means it would appear to be moving to continually higher declinations, toward the northern constellations, in the morning sky. As the comet moved farther from Earth, and as the Earth continued to advance in its orbit counter-clockwise, the comet would also appear to rise earlier and earlier in the morning as it became fainter with distance.

4. Comets with a high inclinations and long periods most likely originated from the Oort cloud. Comet Hale-Bopp, that miraculously bright visitor of the spring of 1997, has an inclination of 89° and a period of about 4,200 years. Comet C/1969 Y1 Bennett, whose ghostly tail hung beautifully over the morning sky in the early spring of 1970, has an inclination of about 90° and a period of almost 1,700 years. The perturbations of the planets over the millennia will subtly or not so subtly shift these orbits over time, sometimes pulling them closer into a more circular orbit, or sometimes bringing them so close to the Sun that they become "sungrazing" comets or even collide with it, ending its existence. Close encounters with Jupiter or Saturn will of course have a profound effect.

5. The orbital inclination is greater than 90° the comet's direction appears retrograde in relation to the ecliptic. A famous example of high inclination and the effects of long-term influences of the planets is comet Halley, whose inclination is about 162°, and whose period over time has become about 75 years.

6. We might be curious to determine the apsidal velocities of comet West using the simplified velocity equations:

$$\text{Perihelion velocity} \quad \text{Aphelion velocity}$$
$$v_P = \kappa\sqrt{\frac{1+e}{q}} \quad v_A = \kappa\sqrt{\frac{1-e}{Q}}$$

The constant k is 29.785 derived from the Gaussian constant, and as discussed in Chap. 12, allows distances to be inputted in astronomical units to yield outputs in units of kilometers per second. The eccentricity and perihelion distance q are given by the elements as 0.99997100 and .196626 AU respectively. The aphelion can be calculated by: $Q = a(1 + e)$. For comet West, Q is about 13,560.2 AU. From this, the velocity calculations become,

$$v_P = 29.785\sqrt{\frac{1 + .999971}{.196626}} \quad v_A = 29.785\sqrt{\frac{1 - .999971}{13560}}$$

$$v_P \simeq 95 \text{ km/s} \quad v_A \simeq .0014 \text{ km/s}$$

This comet's velocity variation from perihelion to aphelion is about 69,000 to 1! The enormous perihelion velocity puts great centrifugal stresses on the comet, and helps to explain why comet West broke into pieces during its last perihelion passage.

Exercises: The Beauty of Mysterious Conjunctions A close conjunction of bright planets in a clear sky is often a sublimely beautiful sight. Like eclipses, conjunctions of planets often have been associated with omens, both good and bad, and appear to be of persistent interest to those who perceive such events as influential on human affairs. But they are of course as predictable as are the motions of the planets. Because their orbital motion is more rapid, conjunctions of the inferior planets are far more common than conjunctions of the superior planets. Close conjunctions among the bright planets Jupiter and Saturn are the most noteworthy. Conjunctions among these giant planets are often referred to as "great conjunctions." A recent such conjunction among the giants took place on May 31, 2000. One can, of course, figure out how frequently conjunctions should occur by analysis of the approximately 12 year orbital period of Jupiter and the approximately 30 year orbital period of Saturn. Rarer of course are conjunctions of Mars, Jupiter, and Saturn, as are the so-called great "triple conjunctions" of Jupiter, Saturn, and a star. The latter event two millennia ago was conjectured to be the Star of Bethlehem. Because the motions of the planets are precisely predictable, tables of such conjunctions can be generated by computers. The rarity of the conjunctions is a function of the selective filter to be applied as to how close the planets, or planets and a star, need to be to qualify as conjoined. Conjunctions of objects separated in the sky by only minutes of arc are far scarcer than those of one or two degree separations. These events differ too from planetary line-ups where a group of planets appear to be strung in a line from the Sun, but may or may not result in tight conjunctions among some of them. One can get a sense of the possibility of a conjunction by comparing heliocentric longitudes of planets. Reliable determination of a conjunction, however, requires more precision, and specifically, knowledge of the true anomaly of the conjoining planets which will be treated in the next chapter.

Problems

1. Calculate and plot on a graph the heliocentric longitudes of Mercury, Venus, and Saturn as of December 21, 2012. Interpret your results.
2. Determine the heliocentric longitudes of Jupiter and Saturn for the date December 21, 2020.
3. Saturn's period is 10,757 days. Calculate the mean daily motion of Saturn in degrees per day. How many degrees does Saturn move across the sky in one Earth year?
4. Pluto was first seen on a photographic search plate taken of the region near the star Delta Geminorum on January 21, 1930, very close to the ecliptic plane. What was the heliocentric longitude of Pluto at the time of its discovery?

5. Dwarf planet Sedna is immensely far away and was difficult to discover because of its very slow motion against the sky. From the data available on the NASA/JPL Solar System Dynamics website discussed earlier, find its mean daily motion and the values of Ω, ω, Π, L, and M as of November 14, 2003, the date of its discovery, and plot a schematic diagram of same. (*Hint*: In the NASA/JPL HORIZONS Web-Interface, under "Type of Ephemeris," select "Elements." Use the body itself and not its barycenter on all problems here. Round to hundredths.)
6. Create a schematic diagram of the orbit of Neptune at the time of Neptune's discovery in September 23, 1846, showing longitude of the ascending node, longitude of the perihelion and the mean longitude of the planet on that date.
7. Use the NASA/JPL HORIZONS web-interface system to find the heliocentric mean longitude of the asteroid Ceres at the date of its discovery by Giuseppe Piazzi in January 1, 1801. Calculate the longitude of its perihelion, Π.
8. Referring to the previous problem, what was the heliocentric longitude of Earth as reported on the NASA/JPL HORIZONS web-interface system on January 1, 1801? Compare that with your own computation of Earth's heliocentric longitude for that date. Plot the relative heliocentric longitudes of Ceres (using the data from the previous problem) and Earth as of the date of discovery.
9. After its initial discovery and observations in the early months of 1801, Ceres was lost. A young Carl Friedrich Gauss developed an innovative new mathematical method, however, that enabled the asteroid's recovery on December 31, 1801, giving Gauss international fame. Use the NASA/JPL HORIZONS web-interface system to find the difference in the heliocentric longitudes of Ceres between discovery and recovery in 1801. Is the difference commensurate with its mean daily motion over that interval? Update the plot you did for Problem 8, now showing the relative heliocentric longitudes of Ceres and Earth as of the recovery date.
10. On March 25, 1996, spectacular comet Hyakutake C/1996 B2 passed extremely close to the Earth and was dramatically visible to the naked eye, showing visible movement even over relatively small increments of time. Using data from the from the NASA/JPL HORIZONS web-interface system: (a) Calculate the semi-major axis, and the period in years, and plot the inner orbit of the comet as of March 25, 1996. (b) Find the Julian date of periapsis (perihelion) and use the USNO site and find the calendar date and time of perihelion for the comet. How many days before perihelion was the close approach to the Earth?

References

1. Meeus J (1991) Astronomical algorithms. Willmann-Bell, Richmond, Chapter 30
2. Meeus J (1988) Astronomical formulae for calculators. Willmann-Bell, Richmond

Chapter 16
An Introduction to Kepler's Problem: Finding the True Anomaly of an Orbiting Body

We have seen how to orient the orbit of a body in space, using the Sun and the ecliptic plane as a reference, anchored externally by the position of the vernal equinox among the stars. On the ecliptic itself, we saw how to compute the heliocentric longitude of the Earth, as it moves in its almost circular orbit. It is not circular, but nearly enough, with its modest eccentricity, for us to get a sense of the Earth's location relative to the vernal equinox. Similarly aligning the orbit of a comet or an asteroid in space, as we did in the last chapter, will tell where that body is at perihelion, but it is usually necessary to predict where it will be at other times too, to trace its path anywhere along the orbit. When we discussed asteroid 2010 TD54, we knew when it passed closest to the Earth. We computed where the Earth was in its own orbit at that time, so we could determine approximately where the asteroid was too, but the two numbers did not really match up. This is because we were comparing positions for each body on their fictitious circles to find mean anomaly, M, from which the heliocentric longitudes were determined. As we will see, this provides useful information, but it does not get us to an appropriate level of accuracy when we want to compare where they are in their actual, elliptical orbits. Let us take the next step and venture to calculate more precisely where the asteroid was and is in its orbit, either at that time or any other time. That requires knowing the so-called *true anomaly* of the object.

The True Anomaly Is the Body's Actual Position in Orbit

The great goal of the astronomer trying to understand the path and motion of an orbiting body in space is to find the true anomaly of the object for any given time. This is simply the angular position (often denoted by the Greek letter Nu, v) of the asteroid in its orbit measured in the direction of its motion from perihelion, with the Sun as the focus of the elliptical orbit:

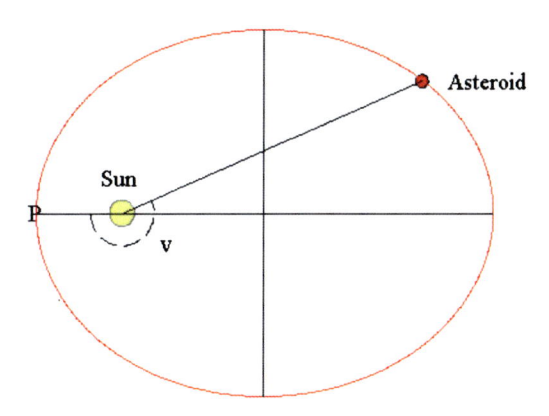

The line between the Sun and the asteroid is called the *radius vector* of the asteroid, with its usual symbol r. Its magnitude is the distance between them; its direction is always from the asteroid to one focus of the ellipse, the Sun. It is apparent from the diagram that at perihelion the true anomaly will be zero and the radius vector will have the value of the perihelion distance. Knowledge of the true anomaly tells us at any time where the body is, *anywhere* in its elliptical orbit. This is valuable information! It will inform us when a comet or asteroid, or spacecraft, will pass by Earth's orbit, or by the orbits of any other planet, as it traces its path inward and outward. For example, we saw in the last chapter, we could determine the relative positions of the Earth and comet West when the comet was at perihelion. But in the months and days before and after perihelion, we had no way of knowing this. Knowledge of the true anomaly and radius vector is crucial in plotting the path of the object in relation to the Sun and Earth. By using computers to determine the true anomaly backwards in time, and matching the comet's path with the heliocentric longitudes of the other planets, it will tell when in a comet's history it may have passed sufficiently close to any of the other planets to have had its orbit shifted and been gravitationally perturbed. Knowledge of the true anomaly is essential too for reducing the object's position to geocentric coordinates of right ascension and declination, so that it can be detected and followed by telescopes.

The true anomaly is plainly a fundamentally important parameter in celestial mechanics. How can we go about finding the true anomaly of an object in space? We need to go beyond finding mean anomaly: it is not enough to estimate an object's position by taking its mean daily motion from the time of perihelion, then working forwards or backwards the appropriate number of days to find its approximate present position on a circle. Its use in orbits of even modest ellipticity will show how inaccurate it can be.

The Limitations of Using Mean Daily Motion to Find Position in an Elliptical Orbit

In a circular orbit, we can calculate the mean daily motion from some point, and deduce where the object will be a later time. The mean daily motion of the Earth, in degrees per day, will provide an approximate check on our calculations of its heliocentric longitude. But for accurate positioning of bodies in eccentric orbits, this does not work. Velocity differences, and thus the mean daily motions, vary greatly at different places on the orbit. And the greater the eccentricity, the greater the deviation from whatever could be calculated as its mean or average motion.

Think how meaningless it would be to compute the mean daily motion of comet West. Its *average* velocity is about 47.5 km per second, but this number is in no way helpful, because the range of its orbital velocities varies from 95 km per second at perihelion to 1.4 m per second at aphelion. Its mean daily motion is 69,000 times greater at one end of its orbit than the other. Granted, comet West is an extreme example, but the principle holds for all elliptical orbits, that plotting an object's mean daily motion on an elliptical path will not yield a true prediction of its position. So the question remains, how can we find the actual position of a comet, asteroid, or spacecraft in a heliocentric orbit?

Kepler's Auxiliary Circle: Determining the True Anomaly from the Eccentric Anomaly

Johannes Kepler was keenly aware of the limitation of using mean daily motion to accurately predict the actual position of a planet or other body. Indeed, if Mars's orbit were circular, predicting its path would be easy, and Kepler's life would have been much simpler! Although in that case the world would likely not have been rewarded with his wonderful discoveries, the fruits of his great struggles to make sense of elliptical motion. He found that orbiting bodies move not in circles but in ellipses, some with large, some with small eccentricities. He needed a method that would apply to orbits of any eccentricity. His mathematical imagination came to his aid. He concluded that it would be useful to begin by picturing the orbit inscribed within a larger circle, whose center would be the point of reference for finding the angular position of the object from perihelion. In other words, Kepler began with the simpler geometry of the circle, with the center of the ellipse, rather than the Sun, as the pivot point for calculating angular distance from perihelion. The angular position of the body determined with reference to this outer *auxiliary* or *fictitious* circle is called the *eccentric anomaly* and is typically symbolized by the letter E. You can see from the diagram below, representing as an example asteroid 2010 TD 54, that the eccentric anomaly does not directly reveal where the body is on the ellipse. Rather we find it indirectly: begin at a point on the auxiliary circle at angle E from perihelion, and drop a line perpendicular to the semi-major axis until it intersects the body in its orbit.

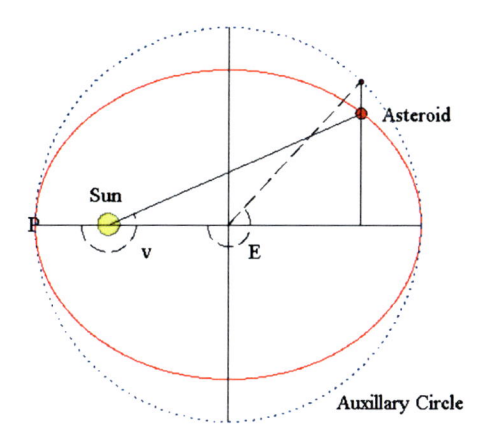

Once E is known, we can use a rather peculiar trigonometric equation, easily solved on a pocket calculator or computer, to find the true anomaly:

$$\tan\frac{v}{2} = \sqrt{\frac{1+e}{1-e}} \tan\frac{E}{2}$$

This means that twice the angle whose tangent is $v/2$ will yield the true anomaly. Finding the angle whose tangent is something is possible with the arctan function on your calculator. We could therefore write the equation for finding the true anomaly this way:

$$v = 2 \arctan\left(\sqrt{\frac{1+e}{1-e}} \tan\frac{E}{2}\right)$$

The magnitude of the radius vector can be determined from this equation:

$$r = a(1 - e\cos E)$$

In these equations, e and a are, as always, the eccentricity of the ellipse and the semi-major axis, respectively. Eccentricity has no units, a is typically in AU, and the angles are in radians (which, again, is degrees times $\pi/180$). The first equation links the eccentric anomaly to the true anomaly, so with it we can return to our familiar heliocentric reference frame.

This seems like excellent progress, but how do we determine the key unknown in each of these equations, namely, the *eccentric* anomaly?

Kepler's Equation: Finding the Eccentric Anomaly from the Mean Anomaly

Kepler worked with these odd bits of plane geometry and was finally led to a momentous breakthrough, and an equally odd equation, difficult to solve, known as *Kepler's equation*. This equation forges the next link in the chain, by relating the eccentric anomaly to the *mean* anomaly, M, which we encountered earlier and can easily calculate.

$$E = M + e \sin E$$

This is the famous Kepler's equation as it is usually written, and though it may appear simple at first glance, appearances can be deceptive. It looks easy until we notice that E appears on both sides of it. Can't we just isolate the E and make a straightforward equation out of it? The answer is no, we cannot. It is a kind of equation mathematicians call a transcendental equation, and since Kepler's day it has been the focus of hundreds of papers on ways and means to solve it. Generally, the value of E can be approximated by iterative methods. It works this way: insert an initial value of E (say, M) on the right side of the equation, and then compute what E emerges as the solution on the left hand side. We then take that value of E and plug it in on the right side and obtain a new value for E, and so on, until we get results in each iteration that differ only insignificantly from the one just before, to any degree of accuracy we choose. There are many refinements of this method, and picking the initial value of E can be something of an art. There are equations for getting at an initial value of E, so the solution to Kepler's equation converges rapidly. But with modern calculators we can do it fairly easily and quickly by these iterative methods.

The problem below will illustrate finding the asteroid's eccentric anomaly. The problem that follows after that will use the eccentric anomaly to enable computation of the asteroid's true anomaly, its place in the elliptical orbit measured in the direction of its motion, counter-clockwise from perihelion.

Determining the Eccentric Anomaly of Near Earth Asteroid 2010 TD54 at the Time of Its Closest Approach to Earth

Let us pull all of these steps together into a problem, so their workings can be better understood. We will continue to use as our case study the near Earth asteroid 2010 TD54 since we have already built a foundation of familiarity with this asteroid.

Problem Use the orbital elements given in the previous chapter for asteroid 2010 TD54 to find the mean anomaly of asteroid 2010 TD54 in its orbit on October 12, 2010.

Given

M	Mean anomaly of asteroid 2010 TD54 on October 12, 2010: 340.34° (from Problem 15.2)
$E = M + e \sin E$	Kepler's equation for finding eccentric anomaly when mean anomaly M is known
e	Eccentricity of the asteroid orbit of .6190559 given by its elements

Assumptions The same assumptions apply as we used before including particularly the circularity of Earth's orbit. We will also ignore the asteroid's 5° inclination and for our purposes be content with healthy approximations.

Method Begin the iteration of Kepler's equation by using M as the initial value of E and taking it through 15 or more iterations. Convert all degrees to radians by multiplying by $\pi/180$, or 0.01745329252. The iteration can be done by hand but access to a calculator or mathematical software that makes the iteration far easier!

Calculations The mean anomaly we found in the previous chapter for the asteroid on October 12, 2010 was 340.34° degrees. We can pick this as the initial value for E. First it is necessary to convert this value to radians, by multiplying by $\pi/180$. The result is,

$$M = 5.94 \text{ radians}$$

Now we insert this value as the initial value for E_0 in the first iteration of Kepler's equation, using for e the eccentricity value given above:

$$E_1 = M + e \sin E_0$$

$$E_1 = 5.94 + .619056 \sin(5.94)$$

The value obtained for *E1* is then plugged into the equation on the right-hand side, to obtain a new value which we label *E2*:

$$E_2 = M + e \sin E_1$$

Using a *Maple* program to repeat these steps, the following result emerged for the first 12 iterations:

Eccentric anomaly on each iteration (rad)	Degrees
E[1] = 5.731724417	328.4036183
E[2] = 5.615675350	321.7544966
E[3] = 5.556803002	318.3813594
E[4] = 5.528860579	316.7803766
E[5] = 5.516091145	316.0487419
E[6] = 5.510365224	315.7206708
E[7] = 5.507820368	315.5748613
E[8] = 5.506693864	315.5103174
E[9] = 5.506196102	315.4817977
E[10] = 5.505976334	315.4692059
E[11] = 5.505879338	315.4636485
E[12] = 5.505836535	315.4611961

Further iterations yield refinements in the result. After 20 iterations the result is 5.505802781 rad, or 315.4592621°, with a difference between each iteration of about 6.2×10^{-8} rad, which is far more accuracy than needed for our purposes. For the next steps in our problem, we will settle on $E = 5.5058$ rad, which is,

$$E = 315.46°$$

This is the eccentric anomaly of the asteroid as of October 12, 2010.

Observation

1. If the orbit were centered as in the figure above, the asteroid will have completed about 315/360 or 7/8ths of its counterclockwise orbit from perihelion.
2. Where the eccentricity of the orbit is high (above .4 or .5), one can experiment with the use of this equation for faster iterations to the result:

$$E_1 = E_0 + \frac{M + e \sin E_0 - E_0}{1 - e \cos E_0}$$

where the initial value of *E0* is again M.[1]

Finding the True Anomaly of Near Earth Asteroid 2010 TD54 at the Time of Its Closest Approach to Earth

Now we have found the eccentric anomaly of the asteroid on the encounter date, the next step is to find the asteroid's actual place in its elliptical orbit.

---

[1] See Jean Meeus [1] and Laurence G. Taff [2].

Problem Use the elements given in the previous problem to find the true anomaly of asteroid 2010 TD54 in its orbit on October 12, 2010, and plot its position on a graph, with the perihelion of the orbit lying along the negative x axis, and showing the correct angles for eccentric anomaly and true anomaly.

Given

E	Eccentric anomaly of asteroid 2010 TD54 on October 12, 2010, from the previous problem: 315.46°
$v = 2\arctan\left(\sqrt{\frac{1+e}{1-e}}\tan\frac{E}{2}\right)$	Equation for determining true anomaly when E is known
e	Eccentricity of the asteroid orbit of .6190559 given by its elements

Assumptions The same assumptions will apply here as in the previous problem, including again the circularity of Earth's orbit, the effect of which we will address later.

Method With the value for E thus determined, we simply solve the equation for true anomaly. For the plot, the angles will begin at the perihelion of the asteroid orbit, in the manner used in the graphs above.

Calculations To determine the true anomaly of the asteroid on October 12, 2010, we insert the appropriate values into the equation:

$$v = 2\arctan\left(\sqrt{\frac{1+e}{1-e}}\tan\frac{E}{2}\right)$$

$$v = 2\arctan\left(\sqrt{\frac{1+.619056}{1-.619056}}\tan\frac{5.5058}{2}\right)$$

$$v = -1.40232 \text{ radians}$$

$$v = -80.347°$$

Since the result is negative, we add 360°, and the true anomaly of asteroid 2010 TD54 is:

$$v = 279.65°$$

From these results, we now plot the position of the asteroid in its orbit at close encounter:

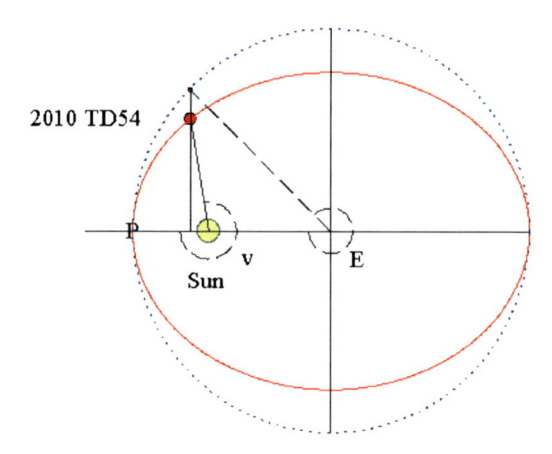

Kepler's Equation: Finding the Eccentric Anomaly from the Mean Anomaly

Asteroid Position on October 12, 2010

Observations

1. The logical next step is to compare the position of the asteroid on October 12, 2010 with that of the Earth on that date. How can we place the Earth on this diagram? Recall that the equinox is the standard external reference among the stars. From the first problem in the previous chapter we know the Earth's heliocentric longitude L at conjunction with the asteroid, as measured from the vernal equinox. We also can compute where the vernal equinox is with respect to the asteroid:

   (a) Since the asteroid's longitude of the perihelion Π is 99.24°, the vernal equinox is 260.76° counter-clockwise from perihelion point P (since 360 − 99.24 = 260.76).

   (b) Since the Earth's mean longitude L (from the vernal equinox) at encounter (from the discussion in the last chapter) is 20.81°, the angular distance of Earth from asteroid perihelion is the sum of these. We'll call this M_{Earth} or the Earth's mean anomaly, but measured here from the perihelion angle of the asteroid.

$$M_{Earth} = 260.76° + 20.81°$$

$$M_{Earth} = 281.57°$$

   (c) Compare this with the true anomaly of the asteroid of 279.65° on that date and time. Now we can plot the orbits and positions of Earth and asteroid as of October 12, 2010 when the asteroid passed close to the Earth. Removing the mathematical scaffolding of the lines, the diagram looks like this, though it is definitely not to scale.

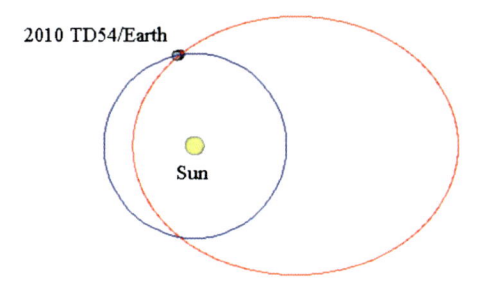

This shows quite roughly the lineup of the Earth and 2010 TD54 at conjunction. A closeup view generated again by *Maple* mathematical software shows them a little better. The assumptions we made likely account for the fact that they are not even closer together:

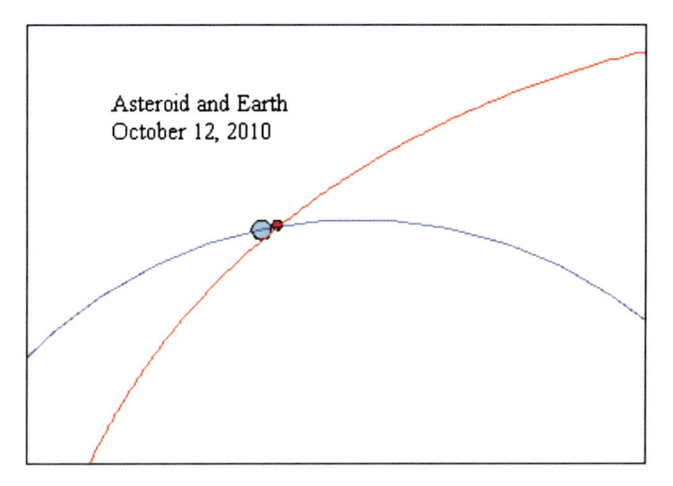

2. One of the important assumptions we made was that the Earth's orbit is circular, which it almost is, but not quite. Its actual eccentricity is small, being only .01671. But it is not insignificant when we are trying to locate an asteroid passing close to the Earth. It can make a truly big difference if the issue is whether the asteroid poses a risk of impacting the Earth! If we want to improve accuracy of the calculations, we need to take into account the true elliptical shape of our Earth's orbit. This means beginning with its mean anomaly, then finding its eccentric and true anomalies. We know from the discussion in the preceding paragraph that Earth's mean anomaly is 281.57° (from the asteroid's perihelion). This equates to about 4.9143 rad. We now adjust this circular measure for the ellipse by computing the eccentric anomaly, whose equation, again, is:

$$E_1 = M + e \sin E_0$$

$$E_1 = 4.9143 + .01671 \sin(4.9143)$$

$$E_4 = 280.629$$

$$E \simeq 280.63$$

Because the eccentricity of the Earth's orbit is so small, the result converges after only five iterations exactly to the above angle for the eccentric anomaly. This converts to 4.8979 rad. This "adjustment for eccentricity" shaved a full degree off the 281.57° mean anomaly we used for the previous problem. Let's see how that change affects the true anomaly of the Earth:

$$v = 2 \arctan\left(\sqrt{\frac{1+e}{1-e}} \tan \frac{E}{2}\right)$$

$$v = 2 \arctan\left(\sqrt{\frac{1+.01671}{1-.01671}} \tan \frac{4.8979}{2}\right)$$

The result in radians (-1.4) is converted to degrees (-80.3) and the solution in degrees is:

$$v = 279.69°$$

Compare this solution with the result of the previous problem, where we were content to use only the Earth's mean anomaly of 281.57°. How does this true anomaly of Earth compare with the true anomaly of the asteroid?

True anomalies of Earth and asteroid 2010 TD54 on October 12, 2010	
Asteroid	279.65°
Earth	279.69°

Though the results still incorporate certain approximations of time of closest encounter and various roundings-off along the way, it is evident that we have removed probably the largest source of error and greatly improved our accuracy by taking the ellipticity of the Earth's orbit into account.

Asteroid 2007 WD5's Passed by Mars: How Close Did It Come?

In late 2007 it was announced that an asteroid would come quite close to Mars early in the next year. Low inclination Asteroid 2007 WD5 was to make a close pass to Mars on January 30, 2008. This problem explores ways to find out how close it might have come.

Problem Given the orbital elements below, approximate the apparent separation of Mars and asteroid 2007 WD5 on January 30, 2008.

Given

JD 2454453.5	Date of epoch: December 19, 2007
JD 2454495.5	Date of asteroid encounter with Mars: January 30 2008
M_{epoch}	Mean anomaly of the asteroid at epoch: 14.89803°
n	Mean daily motion of asteroid: .242811°/day
Ω	Asteroid's longitude of the ascending node at epoch: 67.423964°
ω	Asteroid's argument of the perihelion for the epoch: 312.82278°
$E = M + e \sin E$	Kepler's equation for finding eccentric anomaly when mean anomaly M is known
e	Eccentricity of the asteroid orbit: .5981
e_{mars}	Eccentricity of the Martian orbit: .0933941
$v = 2 \arctan\left(\sqrt{\frac{1+e}{1-e}} \tan \frac{E}{2}\right)$	Equation for true anomaly when E is known

Assumptions As before, we assume the reasonable accuracy of the data and, for the sake of clarity of instruction, are tolerant of rounding. We will work with an assumed close encounter time of noon and ignore the attractive pull of Mars which undoubtedly would affect a precise result. We are again ignoring inclination for this asteroid (and Mars), which are quite close to the ecliptic plane.

Method The first step is to reduce the mean anomaly for the epoch to the date of encounter, which means finding the days elapsed, by subtracting the relevant Julian dates, and multiplying by mean daily motion. The mean anomaly for the epoch is then corrected by that factor. Once we are confident that we have a good mean anomaly for the encounter date, the next step is to find the eccentric anomaly E by iteration in the same way as was done in the previous problem. With the eccentric anomaly in hand, the true anomaly v emerges just by grinding through the equation for the true anomaly. After all that is done, we repeat the process for Mars, and see how close the two bodies came.

Calculations First we do the analysis on the asteroid, then the planet Mars.

Asteroid 2007 WD5:
The first step is to find the degrees traversed by the asteroid from December epoch to late January encounter. The numbers in parenthesis are the given Julian dates:

$$\Delta M = n(2454495.5 - 2454453.5)$$

$$\Delta M = .242811(42)$$

$$\Delta M = 10.198062°$$

This must be added to the mean anomaly at epoch:

$$M = M_{epoch} + \Delta M$$

$$M = 14.89803 + 10.198062$$

$$M = 25.09609°$$

This is the asteroid's mean anomaly at the time of encounter. The next step is to find the eccentric anomaly. Converting to radians,

$$M_{radians} = M\left(\frac{\pi}{180}\right)$$

$$M_{radians} = 25.09609\left(\frac{\pi}{180}\right)$$

$$M_{radians} = .43801$$

We insert this value as the initial value for E_0 in the first iteration of Kepler's equation, using for e the eccentricity value given above:

$$E_1 = M + e \sin E_0$$

$$E_1 = .43801 + .5981 \sin(.43801)$$

The value obtained for E_1 is then plugged into the equation on the right-hand side, to obtain a new value which we label E_2 and so on for the full set iterations, to generate the eccentric anomaly for the asteroid on the encounter date:

$$E = .9103 \text{ radians}$$

$$E = 52.16°$$

To find the true anomaly, we roll out this equation and insert the appropriate values into the equation for e and E (in radians!):

$$v = 2 \arctan\left(\sqrt{\frac{1+e}{1-e}} \tan\frac{E}{2}\right)$$

$$v = 2 \arctan\left(\sqrt{\frac{1+.5981}{1-.5981}} \tan\frac{.9103}{2}\right)$$

$$v = 1.5465 \text{ radians}$$

$$v = 88.61°$$

This is the angle on the asteroid orbit, counterclockwise from its own orbit's perihelion, to the asteroid. To find the distance from the vernal equinox, so we can have a standard basis of comparison with Mars, we need to add that value to the longitude of its perihelion, Π, which is:

$$\Pi = \Omega + \omega$$

$$\Pi = 67.42484 + 312.82278$$

$$\Pi = 20.25$$

Asteroid longitude from equinox as of January 30, 2008:

$$\Pi + v = 88.61 + 20.25 = 108.86°$$

Mars:
We can find where Mars was in several ways. Using the techniques mentioned in this chapter, we can find the heliocentric longitude of the planet. But then to get to mean anomaly, let us try something a little different. We can find the longitude of the perihelion Π and take the difference between that and L to find the mean anomaly, since from what we have said it follows that,

$$M = L - \Pi$$

The mean longitude we can arrive at from the polynomial formula and the information from the table. The JPL information is as of epoch J2000.0, so we again need to correct for the encounter date, JD 2454495.5:

$$T = \frac{JD - 2451545.0}{36525}$$

$$T = \frac{2454495.5 - 2451545.0}{36525}$$

$$T = 0.08078029$$

Then we find heliocentric longitude of Mars for that date:

$$L = a_0 + a_1 T$$

$$L = 355.4465680 + 0.44441088(.08078029)$$

Kepler's Equation: Finding the Eccentric Anomaly from the Mean Anomaly

After dividing the result by 360 and finding the remaining part of the circle that is left, we have,

$$L = 101.60577°$$

We can do essentially the same process for finding the Martian longitude of the perihelion. From the tables on the JPL site[2] referred to before, the equation and coefficients are:

$$\Pi = a_0 + a_1 T$$

where $a_0 = -23.94362959$ and $a_1 = .44441088$. Working these into the equation,

$$\Pi = -23.94362959 + 0.44441088(.08078029)$$

$$\Pi = -23.9077°$$

$$\Pi = 336.09°$$

Subtracting this from the heliocentric longitude gives,

$$M = L - \Pi$$

$$M = 101.60577 - (-23.9077)$$

$$M = 125.5135°$$

$$M = 2.1906 \text{ radians}$$

Again, we insert this value as the initial value for E_0 in the first iteration of Kepler's equation, using for e for Mars eccentricity value given above:

$$E_1 = M + e_{mars} \sin E_0$$

$$E_1 = 2.1906 + .0933941 \sin(2.1906)$$

After iteration, we have,

$$E = 2.26255 \text{ radians}$$

$$E = 129.6345°$$

---

[2] See http://ssd.jpl.nasa.gov/txt/p_elem_t1.txt.

To find the true anomaly, we again insert the appropriate values into the following equation for e and E

$$v = 2\arctan\left(\sqrt{\frac{1+e}{1-e}}\tan\frac{E}{2}\right)$$

$$v = 2\arctan\left(\sqrt{\frac{1+.0933941}{1-.0933941}}\tan\frac{2.26255}{2}\right)$$

$$v = 2.3325 \text{ radians}$$

$$v = 133.64°$$

To find the distance from the vernal equinox, to enable comparison with the asteroid, again we need to add that value to the longitude of its perihelion, Π, which we already found to be 336.09°. Hence,

Mars longitude from equinox as of January 30, 2008:

$$\Pi + v = 336.09 + 133.64 = 109.73°$$

Comparison of true anomalies of Mars and asteroid 2007 WD5 on January 30, 2008 (angular distances from vernal equinox)	
Asteroid	108.86°
Mars	109.73°

Observation

1. If one compares the value of Mars' true anomaly with the precise JPL HORIZONS' output for that planet for January 30, 2008, one will notice that the reported true anomaly is 133.593°, which compares reasonably well with our calculated true anomaly for Mars of 133.64°. Similarly, the HORIZONS' readout for the asteroid's true anomaly for that date is 89.491° which compares reasonably well with our 88.61°.
2. The attentive reader may have noticed a possible source for even this small discrepancy, apart from the rounding off of certain numbers. The Mars value was determined by taking the JPD/SSD Keplerian Elements data and correcting it forward from epoch 2000.0 to the encounter date. The asteroid value, on the other hand, was found by using the published elements for the asteroid as of December 19, 2007, and adjusting only the mean anomaly forward (by the periodic daily motion times the days elapsed to encounter). The asteroid values are not as well established as those of Mars, of course, and refinements in the elements were being made as more observations were analyzed.
3. The reader is encouraged to explore and become familiar with the JPL HORIZONS site. The "Small-Body Browser"[3] is a good place to start, and orbit diagrams can be automatically generated for almost any object desired.

---

[3] See http://ssd.jpl.nasa.gov/sbdb.cgi.

Kepler's Equation: Finding the Eccentric Anomaly from the Mean Anomaly

Exercises: Look Out Above! Now that robotic optics of all sorts are continually scanning the skies, scores of new near earth objects (NEOs) are being detected, some of them passing discomforting close to Earth. For example, in mid-2012, the popular website spaceweather.com reported the sighting of a new object, 2012 KT42 that was among the half-dozen closest NEO objects ever spotted. Likely several meters wide, it passed only 14,000 km above the Earth's surface, well under the altitude of geosynchronous satellites. Similarly, in February 2011, NEO 2011 CQ1 passed within a mere 11,855 km above our planet. The large number of these objects and the evident frequency of impacts into planets and moons in the life of our solar system make watching out for these objects, and calculating their orbits, definitely a worthwhile endeavor!

Problems

1. Draw separate diagrams of the orbits of Mars and asteroid 2007 WD5 as of January 30, 2008, with the vernal equinox aligned along the positive x axis. For each one, draw and label arcs of Ω, ω, υ in different colors, with lengths in degrees approximating their values. Draw the nodes, semi-major axes, perihelion and aphelion points, and place the planet and asteroid on their respective diagrams.

2. Recalculate the true anomaly for the asteroid 2007 WD5 using data from the NASA/JPL HORIZONS web-interface system for epoch January 30, 2008. Compare and interpret your result with the value determined above.

3. Use the radius vector equation $r = a(1 - e \cos E)$ to find the radius vector of asteroid 2007 WD5. Does your result make intuitive sense?

4. Comet Halley's last perihelion passage was on February 6, 1986, JD 2446467.395317051. Its orbital eccentricity is .9673 and its mean anomaly M on that date was 359.953°. Using the iterative equation for high eccentricity orbits given at the end of the first problem of this chapter, find the eccentric anomaly of Comet Halley at that date and time.

5. Using the eccentric anomaly you found from the last problem, find the true anomaly of Comet Halley at that date and time. Express your answer in radians and degrees. Confirm your answer using the NASA/JPL HORIZONS web-interface system.

6. Asteroid 2005 YU55 was reached perihelion on September 9, 2011 then came quite close to the Earth on November 7, 2011. At the time of its encounter its mean anomaly was 46.886°. Its semi-major axis was 1.143333 AU and the eccentricity of its orbit was .4294. Find its true anomaly on that date and its distance from the Sun.

7. Use the Orbit Diagram section of the NASA/JPL Solar System Dynamics website to locate asteroid 2005 YU55 as of November 7, 2011. Identify the Earth's location in relation to the asteroid on that date.

8. Comet 21/P Giacobini-Zinner (2012 epoch) has an eccentric orbit of $e = .71$ that takes it just past Jupiter's orbit before it returns to the inner solar system in a 6.6 year period. Using the iterative equation for high eccentricity orbits given at the end of the first problem of this chapter, and given that its mean anomaly

M as of September 23, 2018 will be 1.90779°, find its true anomaly as of that date. Express your answer in radians and degrees.
9. The Earth's mean anomaly on September 23, 2018 was 256.139. Calculate the Earth's approximate angular distance from Comet Giacobini-Zinner on that date.
10. On June 1, 1981 Jupiter was about 5.44 AU from the Sun in the vicinity of Comet Giacobini-Zinner. Use the radius vector equation $r = a(1 - e\cos E)$ to find the magnitude of the heliocentric radius vector of the comet on that date. Assume a semi-major axis for the comet of 3.505 AU and a mean anomaly of 126.3577°.

References

1. Meeus J (1991) Astronomical algorithms. Willmann-Bell, Richmond, p 187
2. Taff LG (1985) Celestial mechanics, a computational guide for the practitioner. Wiley, New York, p 34

Chapter 17
What Causes the Tides?

The phenomena of the ocean tides is one of the most visible and dramatic manifestations of gravity. It is a vivid demonstration of the "action at a distance" which characterizes its deep mystery. We know that the force of gravity is determined by two things: the existence of a mass and its distance from us. Any particle of mass will attract every other particle of mass, and particles twice as far away will exert their attraction one fourth as strongly. Understanding this *inverse square* relationship is the key to understanding the tides. As Rachel Carson put it in her classic, *The Sea Around Us,*

> The tides are a response of the mobile waters of the ocean to the pull of the moon and the more distant sun. In theory, there is a gravitational attraction between every drop of sea water and even the outermost star of the universe. In practice, however, the pull of the remote stars is so slight as to be obliterated in the vaster movements by which the ocean yields to the moon and sun. Anyone who has lived near tidewater knows that the moon, far more than the sun, controls the tides. He has noticed that, just as the moon rises later each day by fifty minutes, on the average, than the day before, so, in most places, the time of high tide is correspondingly later each day. And as the moon waxes and wanes in its monthly cycle, so the height of the tide varies. ***
>
> That the sun, with a mass 27 million times that of the moon, should have less influence over the tides than a small satellite of the earth is a first surprising. But in the mechanics of the universe, nearness counts for more than distant mass, and when all the mathematical calculations have been made, we find that the moon's power over the tides is more than twice that of the sun.[1]

Let us therefore begin with the Moon's effect on the oceans. The Moon pulls on the Earth from an average distance from the center of the Earth of 384,400 km. But the oceans on the near side of the Earth are closer than this to the Moon. Being closer, they experience a greater pull or lift toward the Moon than is felt by the mass of the Earth under them. If we now go around to the far side of the Earth, we see that it is the Earth that is closer to the Moon than the oceans are. It tends to pull away from them – to be closer to the Moon than they are. Because the oceans are

---
[1] Carson [1].

fluid and mobile, and the Earth is relatively more rigid, the oceans move and fill in, causing a heaping up or bulge, in the areas on the near and far side of the Earth where these differences in forces are the greatest. To a lesser extent, the Sun's more remote attraction has the same effect, and the varying height of the tides is due to the varying configurations of the Moon and the Sun together. Put most simply, the cause of the tides is the difference in the gravitational attraction of the Moon and the Sun on the oceans as compared to their attraction on other parts of the Earth.

Tidal forces are thus due to the *variation* of gravitational forces acting over different parts of an extended body. They apply to any object, not just the Earth and its oceans. Great and small tidal forces are found everywhere in the universe. Strong tidal forces can tear a body apart, deform stars, prevent planetary formation and give rise to rings around a planet. They are *deforming* forces. Our oceans, in fact, are not the only manifestation of the variation of gravitational forces affecting Earth over its volume. The Earth's crust is somewhat elastic and is itself deformed by the differential forces of the Moon and Sun. But the rise and fall of the fluid oceans as the Moon and Sun pass by are the visible manifestations of the tugs of these bodies.

Newton dealt with the tides in Book III of the *Principia*. He described the timing of the tides and the blended forces of the Sun and Moon in this way:

PROPOSITION XXIV. THEOREM XIX

That the flux and reflux of the sea arise from the actions of the sun and moon

... [I]t appears that the waters of the sea ought twice to rise and fall every day, as well lunar as solar; and that the greatest height of the waters in the open and deep seas ought to follow the approach of the luminaries to the meridian of the place by a less interval than six hours; as happens in all that eastern tract of the *Atlantic* and *Ethiopic* [South Atlantic] seas between *France* and the *Cape of Good Hope*; and on the coasts of *Chile* and *Peru* in the *South Sea* [Pacific ocean]; in all which shores the flood falls out about the second, third or fourth hour, unless where the motion propagated from the deep ocean is by the shallowness of the channels, through which it passes to some particular places, retarded to the fifth, sixth, or seventh hour, and even later.

... The two luminaries excite two motions, which will not appear distinctly, but between them will arise one mixed motion compounded out of both. In the conjunction or opposition of the luminaries their forces will be conjoined, and bring on the greatest flood and ebb. In the quadratures [where the Moon is at right angles to the Earth–Sun line] the sun will raise the waters which the moon depresses, and depress the waters which the moon raises, and from the difference the smallest of all tides will follow.

...But the effects of the luminaries depend upon their distances from the earth; for when they are less distant, their effects are greater, and when more distant, their effects are less, and that as the cube of their apparent diameter. Therefore it is that the sun, in the winter time, being then in its perigee, has a greater effect, and makes the tides in the syzygies [when all three bodies are in a line] somewhat greater, and those in the quadratures somewhat less than in the summer season; and every month the moon, while in perigee, raises greater tides than at the distance of fifteen days before or after, when it is in its apogee.

Newton went on to discuss the influences of the varying declination of the Moon, which is its distance from the equatorial plane of the Earth, and the effect of one's

17 What Causes the Tides?

latitude on Earth, the inertia of tidal waters, and the spin of the Earth.[2] The tides are not caused by the rotation of the Earth, though their timing and local extent is influenced by the Earth's rotation. Newton noted this in the Corollary to Proposition XXXVI of Book III of the *Principia*, referring to the "centrifugal force of the parts of the earth" having different effects as one goes from the equator to the poles, as we discussed in Chap. 9.[3] Recall that the Earth's rotation creates a uniform equatorial bulge on the planet, so that it is more an "oblate spheroid" rather than a uniform sphere. While this rotation deforms the Earth to some extent, the bulge around its middle is symmetrical with respect to the Earth's axis, and does not cause the periodic tides.

In the first problem, we will look simply at the differences in the pull of the Moon on the Earth at three points. Our frame of reference will be the center of the Earth, point b, at distance r_b from the Moon. We will compare the Moon's gravitation-induced acceleration on the center-Earth distance with its pull on the sub-lunar point a, at distance r_a and the anti-lunar point c, at distance r_c. The points are illustrated by the accompanying figure.

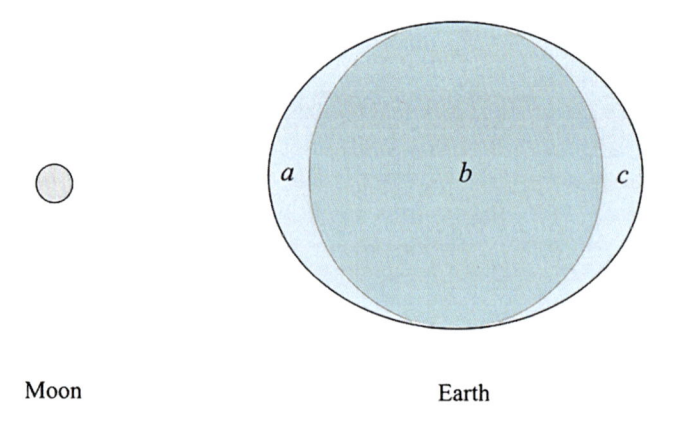

Moon   Earth

Why do we mention a frame of reference? Taking just the case of the Moon and Earth, the Earth is attracted to the Moon just as the Moon is attracted to the Earth. Hence the Earth, too, is in accelerated "free fall" toward the Moon, just as the Moon is in free fall toward the Earth. (This free fall does not result in collision with the Moon because the fall is continually off-set by the Earth and Moon each "falling around" the center of mass of the Earth–Moon system.) With respect to the fixed stars – the so-called *inertial reference frame* (that is, a frame of reference that is not rotating or otherwise accelerating), we can say that our non-inertial, *geocentric* reference frame is in free fall. The free-fall acceleration on the Earth (determined at

---

[2] Sir Isaac Newton, *Mathematical Principles of Natural Philosophy*, Book III, Proposition XXIV, Theorem XIX, pp 395–337 (Translation by Andrew Motte, 1729, revised by F. Cajori). University of California Press, Berkeley, 1949. (Short title) *Principia*.

[3] *Principia*, Book III, Proposition XXXVI, Corollary, 478.

its center) induced by the Moon is Gm/r_b^2. All points within this reference frame are accelerated in this free fall, affected equally and in the same direction. To isolate the effect of the Moon on different parts of the Earth, as distinguished from the free-falling Earth itself, we need to mathematically subtract out the accelerated free-fall motion of this geocentric reference frame. This will enable it to be seen as it would appear from an inertial frame of reference. After subtracting this overall acceleration of the Earth itself, we can then imagine the Earth as an enormous spaceship, and ourselves as astronauts weightless and floating at its center. As we move away from the center within the spaceship in the direction of the Moon, we feel an increasing tug from the Moon pulling us toward it. As we move in the opposite direction, we perceive the spaceship increasingly moving toward the Moon away from us.[4]

Another way to describe this weightless situation is to see it as the perfect equilibrium of accelerations. The acceleration induced by the Moon's gravity at the center of the Earth acts as if it is balanced by an acceleration acting in the opposite direction, in the amount of $-Gm/r_b^2$ that acts on the Earth as a whole. Whether the method is seen either as a netting out of forces and accelerations or as a shifting of the frame of reference from an accelerating (non-inertial) reference frame to a non-accelerating (inertial) reference frame, the mathematical result is the same. We may from either perspective examine the resulting forces (i.e., those not determined by the free-falling Earth) that act on points a and c.

Calculating the Differential Gravitational Forces Exerted by the Moon on the Earth

Problem Calculate the Moon's gravitational acceleration on three points on the Earth: the sub-lunar point in the oceans nearest the Moon, at distance r_a, the center of the Earth, at distance r_b, and the anti-lunar point in the oceans on the far side of the Earth, at distance r_c. Find the differences (the tidal effects) between the

---

[4] We have avoided the terminology of "pseudo forces" and "fictitious forces" which may be confusing to the student. These are forces that emerge in accelerating reference frames and are not physically real in the sense of the four fundamental forces of nature. For example, a ball placed on a moving merry-go-round will move to the outer rim. But no "force" has pushed or pulled it – it is the rotation of the reference frame itself that has displaced its location due to the ball's inertia, its tendency not to move with the rotating frame. If we say that the gravitational force (or acceleration) of the Moon acting at the center of the Earth is exactly balanced by the fictitious force of inertia $F = -mf$ (or acceleration, $F/m = -f$) acting in the opposite direction (a so-called "centrifugal force"), we get exactly the same result as described in the text. For a useful discussion of frames of reference and the confusion they can cause when explaining tides, see P. Sirtoli, *Tides and Centrifugal Forces* (2005), at http://www.vialattea.net/maree/eng/index.htm. An interesting but considerably more technical treatment is given by E.I. Butikov, *A Dynamical Picture of the Oceanic Tides* (1989), http://faculty.ifmo.ru/butikov/Planets/Tides.pdf. The image of the spaceship Earth was drawn from this paper.

Calculating the Differential Gravitational Forces Exerted by the Moon... 359

accelerations at points a and c compared to the acceleration and force at point b, the center of the Earth.

Given

$f_n = Gm/r_n^2$	Gravitional acceleration f at distance r_n between Earth and Moon
G	Gravitational constant: 6.674×10^{-11} N·m²/kg
m	Mass of the Moon: 7.349×10^{22} kg
R_E	Equatorial radius of the Earth: 6.378×10^6 m
r_b	Mean center-to-center Earth-Moon distance: 3.844×10^8 m

Assumptions The center of the Earth will be the frame of reference. We'll assume that the Earth is spherically symmetric, and will for the moment ignore the effects of the Sun.

Method We need to find the value of the acceleration due to the Moon's gravity at three different distances from the Moon, which correspond to the points a and c on Earth. To do this we just use acceleration equation above with the applicable values of r for each distance. Then we subtract the forces and accelerations at points r_a and r_c from r_b. This will yield the differential accelerations caused by the Moon's pull at opposite points on Earth.

Calculations First find the distances from the Moon to Earth-points a and c, remembering that R_E is the Earth's radius in meters. To do this, we simply subtract and add the Earth's radius to the mean lunar distance of r_b which is 384,400 km. The results in meters are:

$$r_a = 3.7802 \times 10^8 \text{ meters}$$
$$r_b = 3.8440 \times 10^8 \text{ meters}$$
$$r_c = 3.9078 \times 10^8 \text{ meters}$$

To calculate the applicable accelerations, we use the given acceleration equation with the applicable value for r each time. For example, we do the calculation for gravitational acceleration at each applicable point, a, b or c:

$$f_n = \frac{Gm}{r_n^2} \quad \text{where } n = a, b \text{ or } c$$

Using for Gm the value $4.90471 0^{12}$ and the radius values determined above, the lunar-induced accelerations at each of the three points a, b, and c on Earth are:

At point a on Earth (near side): $f_a = 3.43 \times 10^{-5}$ m/s²
At point b on Earth (center): $f_b = 3.32 \times 10^{-5}$ m/s²
At point c on Earth (far side): $f_c = 3.21 \times 10^{-5}$ m/s²

It is apparent that the accelerative pull of the Moon on Earth is stronger on the near side of Earth relative to the center, and correspondingly weaker on the far side: $f_a > f_b > f_c$.

At the distance from the Moon to the center of the Earth, the accelerative pull of lunar gravity/s² is thus quite small. Since the Earth weighs about 5.97×10^{24} kg, the total *force* (from Newton's Second Law, $F = mf$) exerted by the Moon on Earth, however, is that mass times the 3.32×10^{-5} (the acceleration at the center of the Earth) or almost 2×10^{20} newtons. The pull on the near side is slightly more; on the far side, slightly less. However small these accelerations are, their differences, acting upon the great mass of the oceans, cause the lunar component of the tides we see.

It is now necessary to compare these accelerations against those exerted by the Moon on the center of the Earth, at r_b, to ascertain the *tidal* differences between the near and far sides of the Earth relative to the Moon:

Tidal accelerations:

$f_a - f_b = 3.43 \times 10^{-5} - 3.32 \times 10^{-5} \quad = 1.1 \times 10^{-6}\,\text{m/sec}^2 \quad \leftarrow$ Near side

$f_c - f_b = 3.21 \times 10^{-5} - 3.32 \times 10^{-5} \quad = -1.1 \times 10^{-6}\,\text{m/sec}^2 \quad \leftarrow$ Far side

Observations

1. If we considered a giant mass, say an aircraft supercarrier, whose mass we will assume is about 9.5×10^6 kg, and compared its weight one side of the Earth with the Moon overhead at a certain phase, with its weight on the other side of the world with the Moon in the same phase (with all our given assumptions), the difference in force on this mass from the Moon's pull would be[5]:

$F_{a-b} = 10.45$ newtons (near side of Earth to Moon)

$F_{c-b} = -10.45$ newtons (far side of Earth to Moon)

2. As noted in Chap. 10, we could derive these same accelerations from the velocities of the Earth and Moon in their orbits around the center of mass using the equation for centripetal acceleration discussed earlier: $f = 4\pi r/P^2$.

Deriving a General Equation for Determining Tidal Forces

Our approach with the Moon was a numerical one. That background helps in undertaking the next problem, which is to derive a general analytical expression for calculating tidal forces for either the Sun or Moon, or potentially any other celestial object.

---

[5] Using Newton's Second Law, $F = mf$. See Eric M. Rodgers, *Physics for the Inquiring Mind*, p. 328 (Princeton, 1960) for a similar treatment of this problem which inspired the approach.

Deriving a General Equation for Determining Tidal Forces

Problem Derive a general equation for determining the tidal effects of one orbiting mass on another. Starting with Newton's law of gravitation and the Earth and Moon as our example, show the tidal effects as a change of force, and also the rates of change of force and acceleration. Do the calculation for a distance Δr closer than r to the body exerting the tidal force.

Given

$F = GMm/r^2$	Newton's Law of Gravitation, where in our example M is the mass of the Earth and m as the mass of the Moon
r_b	The mean, center-to-center distance between the two masses
Δr	The radius of the Earth, though it could be any small, arbitrary distance from r_b

Assumptions Again we assume ideal circular orbits and spherical masses of uniform density, and the absence of all other perturbing effects.

Method Examining the steps in our numerical solution above, we could have expressed the steps mathematically this way, where m is the mass of the Moon, R_E the radius of the Earth, and f_a and f_c are the tidal accelerations at the near and far sides of the Earth relative to the Moon, which are just statements of Newton's gravitational accelerations at each point:

$$f_a = \frac{Gm}{(r_b - R_E)^2} - \frac{Gm}{r_b^2} \qquad f_c = \frac{Gm}{(r_b + R_E)^2} - \frac{Gm}{r_b^2}$$

$\underbrace{}$ Tidal acceleration caused by Moon at point a on Earth $\qquad$ Tidal acceleration caused by Moon at point c on Earth

We want an even more general expression that applies to the tidal forces between any two points separated by any distance Δr, where the main bodies are r meters distant from each other. We can call the difference between the two points Δr, where $\Delta r \ll r$. Moving from r to Δr means a change in acceleration. In our Earth and Moon example, the Δr is the radius of the Earth. Here we will take the first of the equations and generalize the denominator. The tidal force ΔF between the two points will then be:

$$\Delta F = \frac{GMm}{(r - \Delta r)^2} - \frac{GMm}{r^2}$$

The task is to find a simplified expression for ΔF.

Calculations Beginning with the general relation just discussed, we do some algebra heavy-lifting to simplify it:

$$\Delta F = \frac{GMm}{(r - \Delta r)^2} - \frac{GMm}{r^2}$$

$$\Delta F = \frac{GMm}{r^2}\left(\frac{r^2}{(r-\Delta r)^2} - 1\right)$$

$$\Delta F = \frac{GMm}{r^2}\left(\frac{r^2 - r^2 + 2r\Delta r - \Delta r^2}{r^2 - 2r\Delta r + \Delta r^2}\right)$$

Now after some cancelling, given that $\Delta r \ll r$ and $\Delta r^2 \ll r^2$, we may achieve an acceptable approximation by eliminating the isolated Δr and Δr^2 terms:

$$\Delta F \simeq \frac{GMm}{r^2}\left(\frac{2r\Delta r}{r(r - 2\Delta r)}\right)$$

$$\Delta F \simeq \frac{2GMm}{r^3}\Delta r$$

This is the change of the *force* in going from r to $r - \Delta r$.

If we divide through by m (since by Newton's Second Law $f = F/m$ and continuing to use f as our notation for acceleration), then the change of *acceleration* in going from r to $r - \Delta r$ is

$$\Delta f \simeq \frac{2GM}{r^3}\Delta r$$

Observations

1. Whereas the gravitational force diminishes with the square of the distance, the tidal effects diminish more rapidly, with the *cube* of the distance, a result discovered by Newton.
2. We could have chosen to analyze the equation by going from r to $r + \Delta r$, that is, to move outward from the center of m. As an exercise, rework the above problem, and satisfy yourself that the result is the same but with a negative sign, appropriately indicating a diminution of acceleration as one moves farther away from M.
3. To actually calculate the change in acceleration induced by the Moon upon the Earth by this equation, using the mass of the Moon m and the center-to-center distance between Earth and Moon:

$$\Delta f \simeq \frac{2Gm}{r^3}\Delta r$$

Here again we are interested in knowing the acceleration differences between the center of the Earth and point a or point c on the Earth. So Δr will be the radius of the Earth. The calculation to find the tidal acceleration at either point is,

$$\Delta f \simeq \frac{2(6.674 \times 10^{-11})(7.349 \times 10^{22})}{(3.844 \times 10^8)^3}(6.378 \times 10^6)$$

$$\Delta f \simeq 1.1 \times 10^{-6} \, \text{m/s}^2$$

which could be plus (for near-side point *a*) or minus (for far-side point *c*) and is the same value we derived above for the difference in accelerations.

4. This is a job perfectly suited for calculus, which Newton used to determine the effects of infinitesimally small changes. Those familiar with it will see that an easy approach could be to find the rate of change of force in Newton's gravitation equation, $F = GMm/r^2$. The rate of change of force may with respect to radius may be expressed as dF/dr where dF and dr have the same function as ΔF and Δr. That is, an infinitesimal change in radius being dr with the resulting force differential being ΔF. We would need to "take the derivative of the function of F with respect to r." This means determining how the force changes with small changes in orbital radius. One finds the derivative of a function like this by reducing the exponent of the variable by one, and multiplying the function by the old exponent, keeping the constants intact as before. The variable is one over radius squared, or r^{-2}. The new variable is therefore $-2r^{-3}$. The derivative of the equation (the change of force with respect to distance along r) becomes:

$$F = \frac{GMm}{r^2}$$

$$\frac{dF}{dr} = -\frac{2GMm}{r^3}$$

Thus, the minute change in force over a very small distance is,

$$\frac{dF}{dr}$$

Over the distance Δr, the change of force would be,

$$\Delta F = \left(\frac{dF}{dr}\right)\Delta r$$

which is the same as,

$$\Delta F = -\frac{2GMm}{r^3}\Delta r$$

This equation is the same as what we found just above by non-calculus means.

5. The solar versus lunar tidal effects on Earth may easily be compared with this equation. Call the respective tidal accelerations induced on Earth by the Sun and Moon f_s and f_m. If the mass of the Sun is M_s and the mass of Moon is M_m, and their respective mean distances from Earth are r_s and r_m, it will be apparent that, taking the ratio of solar to lunar accelerations,

$$\frac{\Delta f_s}{\Delta f_m} = \frac{\frac{2GM_s}{r_s^3}}{\frac{2GM_m}{r_m^3}}$$

$$\frac{\Delta f_s}{\Delta f_m} = \left(\frac{M_s}{M_m}\right)\left(\frac{r_m}{r_s}\right)^3$$

The ratio of the accelerations of the Sun to the Moon is as the ratio of their masses and the inverse ratio of the cubes of their distances from Earth. Working through the numbers, we have,

$$\frac{\Delta f_s}{\Delta f_m} = \left(\frac{2 \times 10^{30}}{7.3 \times 10^{22}}\right)\left(\frac{3.84 \times 10^8}{149.6 \times 10^9}\right)^3$$

$$\frac{\Delta f_s}{\Delta f_m} = .46$$

The cubing of the great distance from the Sun, as compared with the nearby Moon, offsets its vastly greater mass. The resulting solar tidal effects are only about half that of the Moon. The lunar tidal effects on Earth are thus compounded by the pull of the Sun, and their motions together primarily determine the tidal cycles.

6. We can determine the actual tidal forces induced by the Sun. A simple way would be to take .46 of the lunar tidal value of 1.1×10^{-6} m/s², which is about 5.07×10^{-7} m/s². Let's confirm this by using the tidal acceleration equation:

$$\Delta f \simeq -\frac{2GM}{r^3}\Delta r$$

The mass of the Sun is 1.989×10^{30} kg, the gravitational constant G is 6.674×10^{-11} N m²/s, and the mean Earth-Sun distance (the astronomical unit) r is 1.496×10^{11} m. Again, we can use the Earth's radius as $\Delta r = 6.378 \times 10^6$ m. Inserting these values into the tidal equation yields the following result:

$$\Delta f_s = \frac{2(6.674 \times 10^{-11})(1.989 \times 10^{30})}{(1.496 \times 10^{11})^3}(6.378 \times 10^6)$$

$$\Delta f_s \simeq 5.07 \times 10^{-7} \text{ m/s}^2$$

This again is the *tidal* acceleration (difference in gravitational acceleration) induced by the Sun over the radius of the Earth, which also contributes to the tides.

7. The solar-induced *gravitational* (not tidal) acceleration on Earth at point r_b may be found from,

$$f_{sun} = \frac{GM_{sun}}{r^2}$$

$$f_{sun} = \frac{(6.674 \times 10^{-11})(1.989 \times 10^{30})}{(1.496 \times 10^{11})^2}$$

$$f_{sun} \simeq 5.93 \times 10^{-3} \, \text{m/s}^2$$

This is about 179 times more powerful than the gravitational acceleration induced by the Moon of 3.32×10^{-5} m/s². But at its great distance and the fact that tidal forces diminish as the cube of the distance, the Sun's tidal force is as we saw less than half that of the Moon.

8. The *timing* of the tides and their actual height at different locations, are far beyond the scope of this problem. As noted in Newton's description of the tides, they depend on many complex physical factors, including the contours of the oceans and bays, the declination of the Moon, the latitude on Earth, and the other characteristics and variables of the lunar orbit. One can see this immediately by inspection of any tidal table.

Visualizing Tidal Pull by Comparing Theoretical and Actual Orbital Velocities at Opposite Points on Earth

A helpful way of visualizing the tides is to look further at the gravitational dynamics of a two-body system. Imagine a satellite orbiting the center of mass of the system. If its actual velocity were to slightly exceed that just needed at its center for circular orbital motion (to keep it in perpetual "free fall" around the center of mass of the system), it would "want to" migrate outward to make a larger orbit, if it could; it will tend to drift outward—demonstrating a kind of gravitational "buoyancy." This would occur either with an increase in velocity or a decrease in the accelerative pull of the primary: the projectile or tangential vector of its motion would be slightly increased relative to the inward centripetal/gravitational component. This is so just as a spacecraft in circular orbit will move into a broader elliptical orbit if it is given a tangential, propulsive "kick" to increase its velocity. On the other hand, an increase in the gravitational component acting on it, or a decrease in the tangential velocity, will have the opposite tendency. The gravitational component there is stronger, and its actual speed is less than that needed for circular orbital motion at such distance, so it will find that the gravity of the attracting mass will tend to pull it inward, and it will tend to "sink" toward the primary. Now think of that satellite as the Earth orbiting the center of mass of the Earth–Moon system. Parts of it are accelerated differently. If we could take three free-floating particles and place them at points a, b and c on the Earth, as in the figure, we would see them tending toward different orbits, because their accelerations differ. Since centripetal acceleration is $f = v^2/r$, then the applicable velocity is $v = \sqrt{fr}$. As acceleration differs among points a, b and c, the velocity of the particles differ. We will see how this applies to the oceanic tides here.

Problem The center of the Earth (point b in the figure) is orbiting the center of mass of the Earth–Moon system, and the radius of that orbit is about 4,671.6 km. Imagine three particles located at ponits a, b and c, respectively, as shown in the figure of the Earth at the beginning of this chapter. Compute the orbital velocity of Earth around the center of mass of the two-body system and compare it with theoretical velocities that unattached particles at points a and c, not held by the Earth, would have.

Given

$v = \sqrt{fr}$	Circular orbital velocity at radius r and acceleration f
r'_b	4.6716×10^6 m, the distance from the center of the Earth to the center of mass of the Earth–Moon system (the radius of the Earth's small orbit)

Assumptions We assume again ideal circular orbits and spherical masses of uniform density, and the absence of all other perturbing effects. We neglect the rotation of the Earth. Since we assume the Earth has no rotational motion as to any point with respect to the center of mass of the system, we must assume that every point must move in a circle of radius 4,671.6 km from the center of mass. To emphasize this important point, since the Earth's revolution about the Earth–Moon center of mass is non-rotational, the actual velocities at any points within the Earth will be the same. In other words, there is no rotational component to the Earth's revolution around the Earth–Moon center of mass.

Method Our task is to compare the circular orbital velocity of the center of the Earth around the center of mass of the Earth–Moon induced by the Moon's gravitation (from the equation $v_b = \sqrt{fr'_b}$) with the velocities which hypothetical particles at points a and c on the Earth would have. Since from our assumption the radius of revolution in each equation $v = \sqrt{fr}$ will be the same in this hypothetical, velocity must vary as the square root of the gravitational acceleration, increasing and decreasing with it.

Calculations The lunar-induced acceleration f at point b, as found above, is $f_b = 3.32 \times 10^{-5}$ m/s^2. Using equation $v = \sqrt{fr}$, we can determine the velocity of the center of the Earth, at distance r'_b from the center of mass of the Earth–Moon system:

$$v_b = \sqrt{f_b r'_b} \quad \rightarrow \quad v = \sqrt{(3.32 \times 10^{-5})(4.6716 \times 10^6)} \cong 12.45 \text{ m/sec}$$

This is the Earth's velocity of revolution in its orbit around the center of mass of the Earth–Moon system. Importantly, it is the velocity of *every part* of the Earth, since the Earth's motion is translational, not rotational. We now compare the Earth's orbital velocity just found with theoretical velocities of particles located at points *a* and *c*, which are also in orbit around the center of mass of the Earth–Moon system. We know from above the accelerations at those two points:

Lunar – induced acceleration at *a* : $f_a = 3.43 \times 10^{-5} \text{ m/s}^2$

Lunar – induced acceleration at *c* : $f_c = 3.21 \times 10^{-5} \text{ m/s}^2$

Hence, the hypothetical velocity of particle at *a* is:

$$v_a = \sqrt{f_a r'_b}$$
$$v_a = \sqrt{(3.43 \times 10^{-5})(4.671 \times 10^6)}$$
$$a \simeq 12.66 \text{ m/s}$$

By a like calcuation using for f_c the value 3.21×10^{-5} m/s², and the same value for r'_b, we obtain for the velocity at point *c*:

$$v_c \simeq 12.25 \text{ m/s}$$

Observation These velocities may be regarded as the "natural" circular orbital velocities of unattached particles at points *a* and *c*, not held by the Earth. But in fact they are bound to it (albieit the oceans more loosely), and are all moving at 12.45 m/s. Hence the inward (sub-lunar) particles at *a* are actually moving *slower* (at 12.45 m/s) than the natural circular orbital velocity would dictate at that distance (12.66 m/s), and so the inclination is for the oceans on the near side is to go inward. The outward (anti-lunar) particles at *c* are moving *faster* (at 12.45 m/s) than the natural circular orbital velocity would dictate at their distance (12.25 m/s), and so the inclination is for the oceans on the far side to go outward. This is another way of visualizing the outward trends on the oceans at the two points.

Estimating the Lifting Force of the Moon

An interesting mathematical approach to the analysis of tidal forces is to imagine a mass of any small size on one of the bodies and compare the forces that compete with it, to hold it fast or to rip it off. The parent body with its own gravity wants to hang onto it with a gravitational force; the competing body will be tempting it away with its tidal force. Even if the parent body, like the Earth, has a vastly stronger hold on the mass (which could be an element of the ocean, for example) the competing body (such as the Moon) may nevertheless have measurable, even dramatic effects. By looking at the ratio of tidal to gravitational forces, we can quantify the effects, at least within the idealized limits of our governing assumptions.

Overall, we are looking to see the magnitude of tidal distortion. What do we mean by *tidal distortion*? It is the consequence of the pulling of the respective masses toward each other, in effect warping and elongating them. If the other mass is close enough to overcome the gravitational pull of the parent, it can ultimately rip mass away from the parent mass. Just before this happens, we can expect to see significant distortion. Tidal distortion, in one degree or another, occurs not just in close double stars, but also in the gravitational interaction between planets and their satellites. Even the Earth's Moon pulls on the crust of the Earth, and we see the warping most visibly in the tides. Its effects are geologically apparent in some of the other satellites in the solar system, such as Jupiter's moon Io, which we'll explore shortly.

Problem Derive an equation for the ratio of tidal to gravitational forces acting on any mass on the surface of the Earth, calculating actual values for each force; then find the amount by which the Earth's gravitational acceleration g is offset by the passage of the Moon overhead; finally, use the equation to roughly approximate the maximum lift of the oceans on an idealized Earth-sphere caused by each of the Moon and the Sun.

Given

$F_g = GM_1 m/r^2$	Newton's gravition equation
$F_t = GM_2 m \Delta r/r^3$	Tidal force equation
G	Gravitational constant: 6.674×10^{-11} N·m^2/kg
M_1	Mass of the Earth: 5.972×10^{24} kg
M_2	Mass of the Moon: 7.349×10^{22} kg
m	An arbitrarily small mass on the Earth's surface
Δr	Equatorial radius of the Earth: 6.378×10^6 m
r	Mean center-to-center Earth-Moon distance: 3.844×10^8 m

Assumptions We will assume a circular lunar orbit and that the Earth and Moon are perfectly symmetric spheres with uniform densities, with the oceans evenly covering the Earth at a uniform average depth. We will ignore all other factors, such as the eccentricity of the lunar and solar orbits, their inclination with respect to the equatorial plane of the Earth, tidal friction, latitude on Earth, the ocean's variable depth, contours of the sea floor and coastal land, ocean currents, and other physical determinants of the actual tides in a given location.

Method For this problem, M_1 is the Earth and M_2 is the Moon. A small mass element m is on the surface of the Earth at radius distance Δr from the center our idealized spherical Earth. The small mass m is attracted by the Earth's (M_1's) own gravitational pull, F_g, represented by Newton's gravitation equation, as if all the mass of M_1 were at its center. The Moon, M_2, will exert a tidal force F_t, on the small mass, quantified by the tidal equation. The ratio of the two equations, F_t/F_g, when

Estimating the Lifting Force of the Moon

simplified algebraically, gives us the first result. Their difference $F_g - F_t$ will yield the diminution of g caused by the Moon's passage overhead, which "new" g we can call g'. To find the lifting force, we will imagine a surface of even gravitational potential on the Earth, and see the degree to which the Moon's pull affects it. The gravitational potential energy at the Earth's surface (our assumed reference level) is given by the equation $PE = -mgh$. The work done in lifting a mass m to height h is ΔPE. Since the variations in g near the (assumed spherical) Earth's surface, are negligible, ΔPE from one location to another will be linearly related to the height h. The *equipotential surface* we assume for the Earth is an equal-energy surface, such that ΔPE per unit mass is constant. Hence gh is a constant. If we let the reference height be the Earth's surface, $h = R$ (choosing here for simplicity to represent Earth's radius as R) then the equation $C = gR$ will describe that equipotential surface, where C is a constant.[6] The task then is noting that for any g, such as g', a "new" R can be found, which we will call R'. For any such surface, $C = gR = g'R'$. The lift, L will be the difference between R' and R. Let us see how this all works in practice.

Calculations For the first part of the problem, we need to find the ratios of the applicable equations. The *gravitational* force acting on the little mass m by Earth's own gravity is given by the Newtonian gravity equation:

$$F_g = \frac{GM_1 m}{(\Delta r)^2}$$

The gravitational acceleration on a little mass on the surface of the Earth is,

$$f_g = \frac{GM_1}{(\Delta r)^2}$$

Inserting the given values, we have,

$$f_g = \frac{(6.674 \times 10^{-11})(5.972 \times 10^{24})}{(6.378 \times 10^6)^2}$$

$$f_g \simeq 9.8 \,\mathrm{m/s^2}$$

which is the expected result, here rounded to a "standard" g for simplicity of reference. The *tidal* force acting on the little mass m on the surface of the Earth

---

[6] A succinct way of looking at it is this: if R is the radius of the Earth, the gravitational potential on the surface is $U = -GM_1 m/R$. Since $g = GM_1 m/R^2$, then $GM_1 m = R^2 g$. Substituting this last expression into the first equation yields $U = -gR$. We can ignore the sign if we choose in this context, since the difference in U is what will concern us.

caused by the Moon, M_2, is expressed by this equation, where r is the mean separation between the centers of the Earth and Moon:

$$F_t \simeq -\frac{2GM_2 m}{r^3}\Delta r$$

The tidal acceleration is,

$$f_t \simeq -\frac{2GM_2}{r^3}\Delta r$$

Plugging in the right values tells us the tidal acceleration:

$$f_t \simeq -\frac{2(6.674 \times 10^{-11})(7.349 \times 10^{22})}{(3.844 \times 10^8)^3}(6.378 \times 10^6)$$

$$f_t \simeq -1.1015 \times 10^{-6} \text{m/s}^2$$

which is the acceleration we found earlier in this chapter using different means. The proportion of the tidal to the gravitational acceleration (and thus force per unit mass) is therefore,

$$\frac{f_t}{f_g} \simeq \frac{-\frac{2GM_2}{r^3}\Delta r}{\frac{GM_1}{(\Delta r)^2}}$$

Simplifying this equation yields the ratio of tidal to gravitational force on a small element of mass m on the Earth, though for other applications it could be any planet or star we designate as M_1 pulled by an M_2:

$$\frac{f_t}{f_g} \simeq \frac{2M_2}{M_1}\left(\frac{\Delta r}{r}\right)^3$$

Considering any two objects in a binary system, as the ratio of satellite (or star) radius to distance to the primary diminishes (i.e., as the fraction $\Delta r/r$ grows larger), the greater are the tidal effects. Let's call this ratio by the Greek letter τ (for "tidal"). To quantify this ratio for the Earth–Moon system, we run the numbers through it:

$$\tau \simeq \frac{2M_2}{M_1}\left(\frac{\Delta r}{r}\right)^3$$

$$\tau \simeq \frac{2(7.349 \times 10^{22})}{(5.972 \times 10^{24})}\left(\frac{6.378 \times 10^6}{3.844 \times 10^8}\right)^3$$

$$\tau \simeq 1.12416 \times 10^{-7}$$

Estimating the Lifting Force of the Moon

The reader may confirm this by taking the ratio of the tidal and g values just computed. This equation may be interpreted as the fractional lessening (or strengthening) of g over the distance Δr, just as we saw in the first part of this chapter, when we calculated the actual lunar-induced acceleration differences on each side of the Earth, one positive (near side) and the other negative(far side). In our example it amounts to about 1/87 millionth the Earth's surface gravity.

For the second part of the problem, we need to find the reduction in Earth's surface g caused by the overhead Moon:

$$g' = f_g - f_t$$

$$g' = 9.8 - 1.1015 \times 10^{-6}$$

$$g' = 9.799998 \text{ m/s}^2$$

This very slight reduction in g may seem small, but remember, its effects are felt on every kilogram of ocean water.

Finally, we need to determine the lift caused by the Moon on our idealized spherical Earth as the result of tidal distortion.

When the gravitational acceleration of the Earth at its surface is slightly reduced (or augmented) by the lunar gravitational acceleration to become g' the "distorted" Earth's "new" radius (reflected in the lift of the oceans along the Earth – Moon syzygy line in our model) is R'. Since we said that any product of g and R will describe a surface of constant potential, such that $C = gR = g'R'$, then,

$$\frac{R'}{R} = \frac{g}{g'}$$

Calling the lift L_{moon}, then $L_{moon} = R' - R$. From this, substitution for R' may be made in the above equation, since $R' = L_{moon} + R$:

$$\frac{L_{moon} + R}{R} = \frac{g}{g'}$$

Solving for L readily yields,

$$L_{moon} = R\left(\frac{g}{g'} - 1\right)$$

The "lift" of the lunar tide from this equation is thus:

$$L_{moon} = (6.378 \times 10^6)\left(\frac{9.8}{9.799998} - 1\right)$$

$$L_{moon} \simeq .71 \text{ m}$$

Or 71 cm. The lift induced by the Sun would be as we saw earlier, 46 % of this, or

$$L_{sun} \simeq .33 \text{ m}$$

The combined lunar and solar lifts distort the liquid part of the planet to create our tides, as the Earth rotates beneath the Earth – Moon line.[7] The combined solar and lunar forces at new and full moons, assuming our idealized spherical Earth with even ocean depths, would raise it about a meter. And these augment and diminish at near and far approaches of the Earth to the Sun and the Moon to the Earth in their eccentric orbits.

Observations

1. To avoid calculating the g difference, we could convert the lift equation to one using just the tidal force ratio τ we developed above. We know that

$$g' = g - f_t$$

And since $f_t = \tau g$, then,

$$g' = g(1 - \tau)$$

Substituting this for the g' in the above lift equation, we have,

$$L_{moon} = R\left(\frac{\tau}{1-\tau}\right)$$

The reader should feel free to confirm that this generates the same lift from the Moon as found before.

2. We have found the approximate lunar contribution to the tidal rise along the Earth – Moon line, in our idealized sphere of even ocean depth. More generally, for any radius and where the denominator (as above) substantially approaches unity, the lift may be approximated by the following expression:

$$L = \tau R$$

3. If the Earth's net gravitational attraction on a body is reduced by the Moon passing overhead, what will its effects on you be? Let's suppose you weigh 50 kg

---

[7] We again emphasize that the actual tides and their timing are as noted earlier affected of course by many factors, such as the eccentricity of the lunar and solar orbits, inclination of the orbits, friction, latitude on the Earth, ocean depth, contours of the land, currents, etc. The point to remember is that water *flows* and the oceans tend to move and hump up under the areas of least earthly gravitational potential, along the Earth – Moon line. Owing to these differences, tides can be many meters in some places on Earth and barely detectable in others.

when the Moon is in quadrature (that is, when the Moon is at right angles to the Sun-Earth line). How can the Moon affect your weight? Ignoring the effects of the Sun, what will you weigh when the Moon stands high overhead? Using the same principles we explored above, we know that the g will be less by the factor $g\tau$. And, since weight is nothing more than gravitational acceleration acting on your mass (by Newton's Second Law), then if the gravitational acceleration is less you will weigh less. So, since $F = w = mg$, your "new" weight will be $w' = mg'$. Using the above information, it is apparent that the difference in your weight will be:

$$\Delta w = w\tau$$

Thus, $\Delta w = 50(1.2416 \times 10^{-7}) = 5.62 \times 10^{-6}$ kg, or about 5.6 mg! Your new weight will be $w' = w - \Delta w$ or 49.99999438 kg. This difference will not significantly improve your performance in the high jump, but it is interesting to know that your weight changes twice a day![8]

The same principles of tidal distortion that that cause the tides we see, and that cause you to weigh less when the Moon is at syzygies, also cause distortions – sometimes extreme distortions, in all close systems of planets, stars and galaxies, where tidal forces are significant.

4. The moon raises tides on the Earth, but also the Earth exerts enormous tidal pull on the Moon, producing a tidal bulge on the Moon. Over hundreds of millions of years, the Earth's gravitational pull on that bulge has caused the Moon to lock itself into a *synchronous orbit* with the Earth. This means that the Moon's rotation rate is precisely the same as its orbital period, so that it always faces in the same direction to the Earth. The Moon is thus *tidally locked* to the Earth. This phenomenon is found throughout the solar system.

Exercises: The Prevalence of Tidal Forces in the Solar System While we are most familiar with the tides here on Earth, and how they are manifest in the rise and fall of the oceans, there are significant tides of all sorts elsewhere in the solar system, and among other star systems and even galaxies. Take for example Jupiter's large inner moon Io. The first Voyager spacecraft mission detected dramatic, copiously erupting volcanoes on Io. What is the cause of this activity? It is certainly close to the giant planet, about 5.9 Jupiter radii, and we would therefore expect it to experience strong tidal forces from Jupiter; but many moons experience strong tidal forces (our own Moon, for example, which is quite similar to Io in size and density) but are not volcanically active. Io is too small to have any geologic activity of its own—it should be dead like our Moon. The answer lies in the eccentric shape of its orbit, and how its uneven velocity causes the Jovian-induced tides on Io to stir up its

---

[8] I credit Moaz [2] for this problem. This difference in weight is measurable and it would be an enjoyable experiment to try over the course of a lunar month.

interior. The gravitational influence of Io's next-door neighbor Europa, at 9.4 Jovian radii has rendered Io's orbit eccentric (*e* about .004). Io's velocity thus varies as it orbits, and it cannot always present the same face to Jupiter. As a result, the moon cannot rest comfortably in a synchronous orbit. Io's tidal bulge, however, always faces Jupiter. The bulge on Io thus moves back and forth quite rapidly across Io's surface as the moon orbits the planet, just as our oceans move back and forth under the influence of the Moon, only on Io it is crust, not fluid water, and it moves far more rapidly. These tides within Io's interior thus create enormous stresses and heat, resulting in the intense volcanic activity we see on Io which is unlike anything else in the solar system.

Closer in, it is not surprising that Mercury, orbiting so close to the sun, should also be subject to great tidal stresses. We know this from examination of its surface in photographs relayed to us by the Messenger spacecraft. All of these effects are the result of the strong gravity of one body tugging at the mass of another which is struggling by its own gravity to hold itself together.

Problems

1. Referring to the diagram of the Earth–Moon system at the beginning of the chapter, calculate the *solar* induced accelerations at points a and c on the Earth, and compute their approximate differences from the solar induced acceleration at point b. Take the mass of the Sun to be 1.189×10^{30} kg and its mean distance from the Earth to be 1 AU or 1.49598×10^8 km. The equatorial radius of the Earth is 6.378×10^6 m.
2. Find the solar tidal acceleration on Earth using the tidal equation.
3. Using the idealized assumptions about the Earth and Moon in this chapter, calculate the combined tidal accelerations of the Sun and Moon on the Earth's oceans at new moon and full moon.
4. What is the tidal acceleration on the Moon induced by Earth? The Earth's mass is 5.97×10^{24} kg. The Moon's radius is 1,737.5 km.
5. Jupiter's innermost Galilean moon Io is slightly larger than our Moon and slightly farther from its planet than is our Moon: its radius is 1821.6 km, and it is 421,800 km from the center of Jupiter. But the mass of the mother planet, however, is vastly larger than Earth, being 1.89813×10^{27} kg. What is the tidal acceleration induced by Jupiter on its moon Io?
6. How many times greater is the Jovian-induced tidal force on Io than the tidal force induced by Earth on our Moon?
7. Jupiter's very small inner moon Metis is only 128,000 km from the Jovian center and spins around it once in only 7 h. Its mean radius is only about 21.5 km. What is the tidal acceleration induced by Jupiter on Metis?
8. Identify any inapplicable answers: The Jovian-induced tidal force on Metis in relation to the tidal force on Io is (a) almost two and a half times greater because it is closer to Jupiter; (b) less than half that of Io partly because it is smaller than Io; (c) almost 3.6 mm/s^2 less than Io.

9. What is the tidal acceleration induced on Mercury as a result of the Sun's close pull? Mercury's mean distance from the Sun is only .3871 AU, and is its radius is 2439.7 km.
10. Mercury's orbit is quite eccentric, e being equal to 0.20563593. What are the tidal accelerations at the apsides, and their difference?

References

1. Carson R (1961) The sea around us, Rev edn. New American Library, New York, pp 142–143
2. Moaz D (2007) Astrophysics in a nutshell. Princeton University Press, Princeton, p 112

Chapter 18
Moons, Rings, and the Ripping Force of Tides

In the last chapter we calculated the lifting force of the Moon by imagining a mass of any small size resting on the Earth's surface. We looked at the *ratio* of the Earth's gravitational force holding that mass fast to the surface to the competing tidal force of the Moon (and Sun) tending pull it away. The result was the proportionate tidal pull of the Moon (and Sun) on the mass. The greater was that ratio, which we called τ, the greater was the tidal pull on that mass. Multiply the pulling effects on the little mass by the great quantity of the Earth's substance at its surface and the result is seen in the tides and (less evidently) a rise in the Earth's crust.

We thus examined the effect of a satellite, the Moon, on its parent Earth. But tidal forces are found in many regions of the solar system and space where the effect on the satellite, being the less massive object, is the more noteworthy, and where the satellite is so close to the primary that it is decidedly warped by the parent's gravitational pull. There are close double star systems where the stars are distended into a kind of tear-drop shape by their companions. What if the bodies are extremely close; will one rip apart the other? Suppose we take our mathematical model and put the little mass on the satellite instead of the planet. Then, we might ask, is there ever a case where a satellite could orbit so close as to endanger its own integrity, so the little mass on its surface (or elsewhere in the satellite) could be ripped off by the parent's gravity?

When an object is exposed to unequal forces it may bend or warp, deforming it from its original shape. Even rigid objects, such as the hull of a submarine, the wing of a plane, or the span of a bridge, may buckle and break if the uneven distribution of forces is greater than its design tolerance. In everyday experience, it is the strength of the object's materials – originating from chemical forces, combined with its structural design, distributing the loads of water, wind, traffic, gravity, etc., as the case may be, that keeps it stable. The relatively thin hull of a submarine has integrity from the enormous forces of deep water because the external pressure is distributed around the hull, and there is a compensating, even, internal pressure. There is a balance of forces and an equilibrium of pressures. In fact many structures, such as fuselages, wings, balloons and eardrums, are themselves often relatively fragile, and are only maintained intact because forces are distributed more or less evenly around it. These same principles apply in space, too, to planets, satellites, comets, stars and even whole galaxies. In space, tidal forces are as we saw gravitational forces that are unevenly distributed across a mass, causing a tendency to deform the mass. The internal tensile strength of a planet's satellite may resist gravitational tidal deformity from its parent planet.

In the end, the predominant large-scale disruptive force in the universe is unquestionably gravity.Where a planet, star, or galaxy is sufficiently massive, it can create an enormous gravitational gradient (difference of force) across a neighboring satellite, star or galaxy. If the tidal force of the parent is almost as great as the gravitational forces that hold the other together, the latter may be strongly deformed. If the tidal force exceeds the internal forces that bind the sibling together, it will be ripped apart. Tidal forces may prevent a planet, satellite or star from forming in the first place. The evidence of ancient and not-so-ancient tidal rippings can be seen in the rings of some planets.

The French mathematician Edouard Roche (1820–1843) determined mathematically how close a satellite could come to a planet before being ripped apart. He found this by comparing the gravitational forces holding a body together with the tidal forces tending to break it up. If a satellite is too close to the planet, the tidal forces will overwhelm the internal gravitational forces of the satellite, and it will begin to disintegrate. This forbidden zone is known as the *Roche limit*. If the body has its own internal tensile strength, it may resist and stay whole a little within the limit. If the body is rather fluid, it may distend and begin to pull apart before the limit. It may elongate and stretch closer, which in turn increases the tidal differences in the distended bulge. Thus the Roche limit of a planet or star is relative to the object affected. The Roche limit of Jupiter with respect to an icy, low density comet passing near will be much farther out from the planet than the Roche limit for a relatively rigid satellite.

Deriving an Equation for the Roche Limit and Applying It to the Earth–Moon System

Problem Imagine two small spheres, each of mass m and radius Δr. They are just in contact with each other and orbit the parent planet. Using this model, derive an equation for finding the Roche limit of a planet that exerts a tidal force on a satellite, simplify that equation by putting it in terms of the densities of the two bodies, and use the equation to compute the Roche limit of the Earth in the Earth–Moon system.

Given

$F_g = GM_1 m/r^2$	Gravition equation
$F_t = GM_2 m \Delta r / r^3$	Tidal force equation
G	Gravitational constant
m	Mass of two small spheres orbiting at distance r
Δr	Radii of each of the spheres

(continued)

Deriving an Equation for the Roche Limit and Applying It to the Earth–Moon...   379

R	Mean radius of the Earth: 6,371 km
r	Mean center-to-center Earth-Moon distance: 384,400 km
ρp	Planetary density. Earth's mean density[a] is 5,513 kg/m^3
ρs	Satellite density. The Moon's mean density is 3,344 kg/m^3

[a]Density values are drawn from the NASA/JPL websites, http://ssd.jpl.nasa.gov/?planet_phys_par for the Earth and http://ssd.jpl.nasa.gov/?sat_phys_par for the Moon

Assumptions We assume that the Earth and Moon are spherical, rigid bodies with uniform densities.

Method First we imagine two small spheres, each of mass m and radius Δr. They are just in contact with each other and orbit a planet of mass M. The pair is aligned along the radius, so that a line connecting their centers passes through the center of the planet. Hence the gravitational pull of M will be slightly stronger on the inner sphere than on the outer sphere, tending to pull them apart. On the other hand, they have their mutual gravitational attraction toward each other resisting separation. The question is, which force wins? If their distances are such that the two forces, gravitational and tidal, have exactly the same magnitude, then we have found their "Roche limit." If there is a venturing of the little masses past this limit closer to the planet, then tidal forces will prevail and cleave the twins from each other. It is that difference that we want to explore here. We can do it more simply using radial notation, analyzing the forces on each sphere. We may substitute the expressions for the masses of the bodies with the expressions for their densities, then see how this simplifies the result. Putting the equation into that form will also enable the calculation of Roche limits for objects, such as satellites of other planets or comets, whose masses may not be known but whose densities in some cases may be approximated.

Calculations First we need to find the gravitational attraction of the two spheres toward each other. Since the total center-to-center distance between them is twice their radii, their mutual gravitational attraction is,

$$F_{grav} = \frac{Gm^2}{4\Delta r^2}$$

This we will put aside for later use.

The next step is to develop the force equations for the orbiting spheres. Since r is the distance to the point of contact of the two spheres, the *gravitational* force from the planet on inner sphere is:

$$F_{g\ inner} = \frac{GMm}{(r - \Delta r)^2}$$

The gravitational force from the planet on outer sphere is:

$$F_{g\ outer} = \frac{GMm}{(r + \Delta r)^2}$$

Since the spheres are in orbit, they experience an inertial, centrifugal force that keeps them from falling into the planet. Letting ω be the angular speed of the twin spheres, the *centrifugal* force on inner sphere is:

$$F_{c\ inner} = m(r - \Delta r)\omega^2$$

The centrifugal force on outer sphere is:

$$F_{c\ outer} = m(r + \Delta r)\omega^2$$

The respective net dynamical forces on the spheres are the balances of these two respective forces:

$$F_{inner} = \frac{GMm}{(r-\Delta r)^2} - m(r-\Delta r)\omega^2 \bigg\} \text{Dynamical force on inner sphere}$$

$$F_{outer} = \frac{GMm}{(r+\Delta r)^2} - m(r+\Delta r)\omega^2 \bigg\} \text{Dynamical force on outer sphere}$$

Ultimately, we will want to equate each of these dynamical forces with the forces of gravitational attraction of the two spheres *toward each other*. We want to know the point at which their own attraction for each other is out-competed by the differential force of the primary. This point is the Roche limit for the two small spheres. We can simplify each of these equations by using Kepler's Third Law in radial notation. Since $r \gg \Delta r$, the orbital motion of the pair may be represented by that equation, where again ω is the angular speed of the twin spheres:

$$\omega^2 = \frac{GM}{r^3}$$

The GM terms in the dynamical equations can be replaced by $\omega^2 r^3$. A rather furious amount of algebra is required to simplify this,[1] but the result is, for each equation, the *tidal* force[2]:

$$F_{tidal} \simeq 3m\omega^2 \Delta r$$

---

[1] The key to simplifying what results in a cubic equation is approximating. Because $\Delta r \ll r$ the terms Δr^2 and Δr^3 can be eliminated, and the resulting fraction is quite close to unity to a reasonable approximation.

[2] By substituting GMm/r^3 for ω^2 in this equation, you will recognize it as the same tidal force equation encountered before, but in radial notation and with the coefficient 3 instead of 2: $F_{tidal} = 3GMm\Delta r/r^3$. The coefficient is different because we used a different mathematical model (two small spheres) to create it. This "twin sphere" analysis of dynamical forces is found in many texts, but that given by James Van Allen, presented here, is one of the clearest. Van Allen [4].

The Roche limit is the point where this differential tidal force just equals the spheres' mutual gravitational attraction:

$$\frac{Gm^2}{4\Delta r^2} \simeq 3m\omega^2 \Delta r$$

Remembering that $\omega^2 = GM/r^3$, the equation can be simplified and solved for r, which we'll now call r_{Roche}:

$$Gm \simeq 12\omega^2 \Delta r^3$$

$$r_{Roche}^3 \simeq 12\left(\frac{M}{m}\right)\Delta r^3$$

$$r_{Roche} \simeq 2.3\Delta r \left(\frac{M}{m}\right)^{1/3}$$

Note that we may not know the mass of smaller body, so it is often convenient to replace the mass terms with those for density, where ρ is density and mass is equal to $4/3\pi r^3 \rho$.[3] In each equation, subscripts p and s denote the densities of the planet and satellite spheres, respectively. Denoting R as the radius of the planet, we now substitute the density-based expressions into the above equation, eliminating the mass terms:

$$r_{Roche} \simeq 2.3\Delta r \left(\frac{\frac{4}{3}\pi R^3 \rho_p}{\frac{4}{3}\pi \Delta r^3 \rho_s}\right)^{1/3}$$

$$r_{Roche} \simeq 2.3R \left(\frac{\rho_p}{\rho_s}\right)^{1/3}$$

A more complete mathematical analysis would yield this same form, but a slightly larger number for the limit of a fluid body:

$$r_{Roche} \simeq 2.44R \left(\frac{\rho_p}{\rho_s}\right)^{1/3}$$

The units will be in whatever units R is in. The limit in terms of planetary radii is r_{Roche}/R.

---

[3] This is so since density is mass per unit volume ($\rho = m/V$) then $m = \rho V$. Volume is found by the traditional formula, $V = 4/3\pi r^3$.

To find the Roche limit for the Earth-Moon system, under the assumptions above, we insert into the equation the given values for radius and density of each body:

$$r_{Roche} \simeq 2.44R \left(\frac{\rho_p}{\rho_s}\right)^{1/3}$$

$$r_{Roche} \simeq 2.44(6371)\left(\frac{5513}{3344}\right)^{1/3}$$

$$r_{Roche} \simeq 18,364 \text{ km}$$

This is almost 2.9 Earth radii.

Observations

1. This result means that a loose body less than eighteen and half thousand kilometers from the Earth's center would be torn apart, unless other forces held it together. Why don't we get torn apart? It is because chemical forces hold us together, and those forces are stronger than the gravitational differences. For the same reason orbiting satellites, bolted together with strong materials, don't get deformed and destroyed by the tidal force of Earth's gravity.
2. Using the technique of two small spheres helps in the intuitive understanding of planet and satellite formation. If the spheres are imagined to be reduced to the size of small particles, then it is apparent that they will not cohere into a single body within the Roche limit of the other attracting body. On the other hand, if a body is formed more distantly and later captured by a planet, and is orbits within the 2.44 radii Roche limit, then it suggests that its composition is more rigid, with greater internal tensile strength. The likely minimum Roche limit even for rigid bodies is about 1.26 radii.
3. A variation on the above analysis can show the derivation of the 1.26 radii Roche limit for solid bodies. Instead of visualizing two small spheres orbiting a primary, we imagine one body with a small mass element m on its surface, at distance Δr from the center, as we did in the analysis of tidal forces of the Moon acting to lift tides on the Earth. The gravitational force acting on the little mass m by the satellite's own gravity is given by the gravity equation:

$$F_g = \frac{GM_2 m}{(\Delta r)^2}$$

The tidal force acting on the little mass m on the satellite, M_2, is expressed by the tidal equation, where r is the mean separation between the centers of the planet, M_1, and satellite:

$$F_t \simeq -\frac{2GM_1 m}{r^3}\Delta r$$

Equating these forces, and isolating r will give the distance between them that would mean the little mass gets pulled off if the bodies got any closer:

$$\frac{GM_2 m}{(\Delta r)^2} = \frac{2GM_1 m}{r^3}\Delta r$$

$$r^3 = \frac{2M_1}{M_2}(\Delta r)^3$$

Solving as before for r, we find,

$$r_{Roche\ rigid} = 1.26\Delta r \left(\frac{M_1}{M_2}\right)^{1/3}$$

Again it is useful to replace the mass with density terms to simplify it, where ρ is density and mass is proportional to $r^3 \rho$. The radius of M_1 is denoted by R, and again terms cancel to yield this:

$$r_{Roche\ rigid} = 1.26R \left(\frac{\rho_1}{\rho_2}\right)^{1/3}$$

Where the densities are the same, for rigid bodies the Roche limit is a little more than one and a quarter times the radius of the planet. What does this mean? In the last chapter we found the *ratio* of tidal to gravitational forces (which we called τ) as well as the difference in g entailed by the pull of a neighboring mass, which in that chapter was the Moon acting on the Earth. The "new" g resulting from the other mass was $g' = f_g - f_t$. Here, by *equating* the gravitational and tidal forces, we are examining a situation whereby decreasing the distance between the masses eventually causes the new g to become zero. The net g is null. That is, a situation where the mass of the secondary, however firm to begin with, is now completely inadequate to hold anything to it, even itself, and disintegration ensues, unless other factors, such as its internal tensile strength, act beyond mere gravitational force to "glue" it together or otherwise to prevent its disruption.

4. Some bodies, like comets, being mere porous collections of ice and dust, are loosely held together, with a likely Roche coefficient near 1.8.[4] The Roche limit range for rigid to deformable bodies thus appears to be between 1.26 and 2.44 times the radius of the larger body, respectively, where densities are equal, with

---

[4] See the reference cited in footnote 14, at page 1.

comets perhaps in between. Of course, densities are not typically equal: moons are often less dense than their mother planet. In the case of the Earth-Moon system, the ratio of satellite-to-planet density is about 3:5. For Saturn and its innermost moons, it broadly ranges from about 1:2 to 5:8.

Finding the Roche Limit for Saturn and Its Innermost Satellite Pan

Saturn's famously beautiful rings and many moons provide a good arena for exploring the Roche limit concepts. Its largest moon is Titan, and at least five other major moons are visible through small telescopes.

Less well-known are its numerous tiny inner moons, which hover near and gravitationally interact with the rings. The location of Saturn's rings and its smaller moons is shown in the table below.[5] It does not include some of the more recently discovered fine rings that appear to be associated with the ejecta of some of Saturn's satellites.[6] It is evident that the main rings are all within the 2.44 Roche limit. Through even a small telescope on Earth, one can clearly discern (in order of increasing distance from the planet) the C, B and A rings, with the dark Cassini division sharply dividing the broad B ring from the A ring. The other rings, and the

---

[5] The distance data is derived from http://nssdc.gsfc.nasa.gov/planetary/factsheet/saturnfact.html. The density information comes from http://ssd.jpl.nasa.gov/?sat_phys_par.

[6] See generally, http://en.wikipedia.org/wiki/Rings_of_Saturn#G_Ring. There are rings that extend far beyond the Roche limit for Saturn. Some rings seem to be associated with particles being blasted off the surface of moons by meteorid or micrometeorid impacts, or ejecta from the moons themselves. Saturn's E ring, for example, appears to be fed by *cryovolcanic* plumes – ice geysers – from the surface of Enceladus.

[7] The Cassini-Huygens mission has added enormously to our knowledge of the Saturn system. See http://saturn.jpl.nasa.gov/index.cfm for a host of pictures and data.

smallest moons, were discovered by spacecraft.[7] Saturn's innermost moon, Pan, orbits about 2.217 Saturn radii from the planet. There are no moons lying inward of Pan. It and Daphnis are just inside the outer edge of the main A ring. Pan is one of the "shepherd moons" that help sculpt and contour the rings. Pan sweeps a gap in the A ring called the Encke gap, and Daphnis clears the Keeler gap.

Location of Saturn's rings and approximate densities and distances of some of Saturn's inner moons (within 10 radii)

Moon/ring	Approximate density of moon (kg/m³)	Distance in Saturnian radii	Distance from Saturn in kilometers
Saturn equator		1	60,268
D Ring inner edge		1.11	66,900
C Ring inner edge		1.239	74,658
B Ring inner edge		1.527	92,000
B Ring outer edge		1.951	117,580
Cassini division			
A Ring inner edge		2.027	122,170
Pan (and Encke Gap)	360	2.2165	133,583
Daphnis (and Keeler Gap)	560	2.26	136,500
A Ring outer edge		2.269	136,775
Atlas	437	2.282	137,670
Prometheus	475	2.2843	139,353
F Ring center		2.326	140,180
Pandora	500	2.3512	141,700
Epimetheus	689	2.5099	151,422
Janus	634	2.5125	151,472
G Ring center		2.86	172,500
E Ring inner edge		3	181,000
(Many small and large moons)	~250 up to 1,500	3.08–8.75	185,520–527,040
E Ring outer edge		8	483,000

The average density of the innermost moons seems in the neighborhood of 550 kg/m³ or about 4/5ths the density of the planet. The questions that present themselves here are, why are there no moons inward of Pan? What forces gave rise to Saturn's rings?

Problem We wish to find out whether Saturn's closest satellite is a rigid or a somewhat fluid body. Find the Roche limit of Saturn for Pan and interpret the results in terms of the likely nature of Pan.

Given

$R_{roche} = 2.44\, R(\rho_p/\rho_s)^{1/3}$	Equation of determining Roche limit for a fluid body
ρ_p	Mean density of Saturn 687 kg/m³
ρ_s	Estimated mean density of Pan: 360 kg/m³
R	Radius of Saturn 60,278 km
Pan's semi-major axis	133,585 km or about 2.217 Saturn radii[a]

[a]See http://ssd.jpl.nasa.gov/?sat_elem for the source of the physical data about Pan

Assumptions We assume bodies of uniform density and the reasonableness of our estimate of Pan's density. There are of course many unknowns about the true densities of these objects, and most data give a healthy range of uncertainty around them. We ignore the effects of any internal tensile strength of Pan, though our analysis may tell us something about it.

Method We will find the planet-to-satellite density ratio and use the Roche limit equation for the Saturn–Pan system and see how the limits compare with Pan's actual distance from Saturn. From this we may get an approximate idea about whether Pan is a rigid or deformable body. If we assume that Pan is on the edge of where satellites of about its density break up, we can try using the density of Pan-like moons and find an approximate practical value for the Roche limit for Saturn for such moons. Begin by using the radius of Saturn as the unit of measurement (i.e., Saturn's radius $= 1$).

Calculations We insert the given quantities in the Roche equation and see what the Roche limit of Saturn is for Pan. Using Saturn radii as our units, $R = 1$, the Roche equation is,

$$r_{Roche} \simeq 2.44 \left(\frac{\rho_p}{\rho_s}\right)^{1/3}$$

$$r_{Roche} \simeq 2.44 \left(\frac{687}{360}\right)^{1/3}$$

$$r_{Roche} \simeq 3.03 \text{ Saturn radii}$$

Since Pan is about 2.2 radii from Saturn, Pan must be held together more rigidly than a loose, semi-fluid body. Pan is the innermost moon, so it is possible that a Pan-like moon any closer to Saturn than Pan would not be internally rigid enough to survive Saturn's strong tidal forces.

Our working assumption is that Pan's orbit is more or less at the Saturn Roche limit of moons of its composition. Let us work backwards to see what the Roche numerical coefficient might be for a Pan-like moon. We can write the equation this way, where x is the factor we seek:

$$2.217 \simeq x \left(\frac{\rho_p}{\rho_s}\right)^{1/3}$$

$$x = \frac{2.217}{\left(\frac{\rho_p}{\rho_s}\right)^{1/3}}$$

$$x = 1.8 \text{ Saturn radii}$$

Finding the Roche Limit for Saturn and Its Innermost Satellite Pan   387

Thus, a Roche limit factor of moons of Pan's composition is more likely about 1.8. Altogether, it suggests that Pan is more rigid than fluid.

Observations

1. The rings are most likely the result of moons that might-have-been but were unable to form inside the Roche limit, or those that were captured by Saturn's gravity then disintegrated after passing within the Roche limit. This could have been a larger moon that collided or simply fragmented in the distant past, or smaller moons whose internal gravity could not overcome the competing tidal forces from the mother planet, or a combination of such events. Saturn, Jupiter, Uranus and Neptune all have rings and numerous, small, low-density satellites both in and beyond them, in addition to their main entourage of higher density, larger and farther-out moons. The main rings of all these planets mostly lie within the innermost moons. See the accompanying figure of the main rings and selected innermost moons of the gas giant planets. The densities of these tiny moons, often at the limit of visibility, must be assumed from orbital parameters and magnitudes and assumed albedos, or simply given a best guess. The graph includes the selected moons and rings of Saturn, Jupiter, Uranus and Neptune, in the order of nearest to the planet outward.

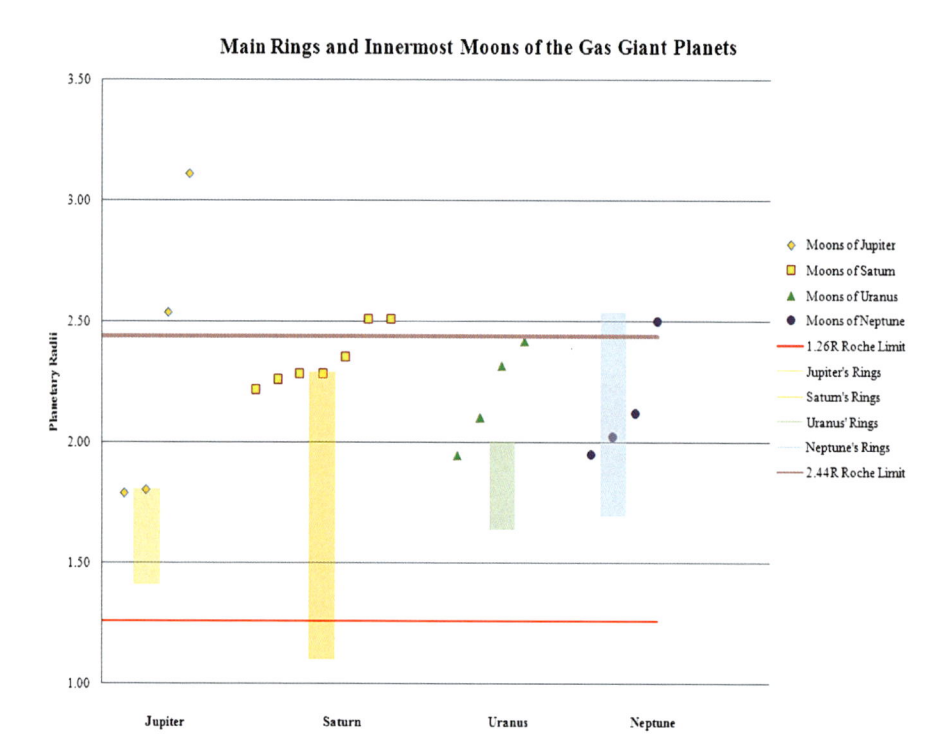

The dozens of other, farther-out moons are not included. Here are the inner moons referred to in the graph[8]:

Planet	Selected inner moons	Rings
Jupiter	Metis, Adrastea, Amalthea, Thebe	Halo, Main
Saturn	Pan, Daphnis, Atlas, Prometheus, Pandora, Epimetheus, Jamus	D,C,B,A,F,G
Uranus	Cordelia, Ophelia, Bianca, Cressida	6,5,4, Alpha, Beta, Gamma, Delta, Lambda, Epsilon
Neptune	Naiad, Thalassa, Despina, Galatea	LeVerrier, Lassell, Arago, Unnamed, Adams

As noted earlier, certain rings are apparently the product of dust emitted from moons. The outermost gossamer rings of Jupiter and E ring of Saturn fall into this category, and were excluded from the above graph. In general, the rings appear to lie closer in than the planet's family of moons, a conspicuous exception being Neptune's moons.[9]

2. It will be convenient as a working assumption to use a "fluid" Roche coefficient (which corresponds to the distance in parent-body radii between equal density bodies) of 2.44 for extremely loose, fluid, or deformable bodies, such as stars or small particles that would tend to coalesce under the force of mutual attraction; a "rigid" 1.26 coefficient for hard, rocky bodies, such as some meteorites and some rocky asteroids; and an "intermediate" 1.8 coefficient for those bodies somewhere in between, such as some small moons, "rubble pile" asteroids, and comets, the latter of which are usually described as rather porous and made of a conglomerate of water-ice and dust. A more cautious approach is to consider ranges for the Roche limit of bodies since their internal composition is usually not well known.

Comet Shoemaker-Levy 9's Fatal Encounter with Jupiter

When Eugene and Carolyn Shoemaker and David Levy examined a photograph of the sky taken in March 1993 with the Schmidt telescope on Palomar Mountain in California, they discovered something most unusual. It was clear there was a new comet, the ninth on their list of remarkable joint discoveries. What was astonishing

---

[8] See also the JPL website on *Planetary Satellite Physical Parameters*, http://ssd.jpl.nasa.gov/?sat_phys_par#legend

[9] Many moons of each of these planets thus lie beyond the ones shown in the graph. But there are no moons within the inner Roche limit, and most hover close to or are beyond 1.8 radii. Wide density and orbital differences indicate that some moons may have formed at the time their parent planet coalesced, and that others were no doubt captured in passing. A *retrograde* orbit (moving in reverse direction from the revolution of the planet in its orbit) is good evidence of capture. Jupiter, for example, at the outer edge of the asteroid belt, has captured many rocky, high density moons.

was that the comet had appeared somehow "squashed" on the image. Photographs from other observatories later confirmed that Comet Shoemaker-Levy 9 had actually broken into pieces, evidently torn apart by a prior close encounter with Jupiter. More surprises were to come. Indeed, it was learned that their comet had actually been *captured* by that planet several decades earlier and was orbiting the planet with an approximately 2 year period. More incredibly, calculations showed that the fragments would soon impact the planet. This revelation set off a world-wide effort to witness the impact and its aftermath with every telescope that could be brought to bear on it, including the Hubble Space Telescope, which could observe it far above the Earth's atmosphere. In the words of one scientist, "The disruption of a comet into multiple fragments is an unusual event, the capture of a comet into an orbit about Jupiter is even more unusual, and the collision of a large comet with a planet is an extraordinary, millennial event."[10] The comet fragments finally collided with Jupiter's southern hemisphere between July 16 and July 22, 1994. The actual impact occurred on the far side of the planet, but Jupiter's rapid rotation quickly brought the dramatic impact scars into view.

According to NASA, "From July 16 through July 22, 1994, pieces of an object designated as Comet P/Shoemaker-Levy 9 collided with Jupiter. This is the first collision of two solar system bodies ever to be observed, and the effects of the comet impacts on Jupiter's atmosphere have been simply spectacular and beyond expectations. Comet Shoemaker-Levy 9 consisted of at least 21 discernable fragments with diameters estimated at up to 2 km."[11]

Problem Orbital calculations of Comet Shoemaker-Levy 9 showed that the comet was closing in on Jupiter with each pass, and was destined to collide with the planet in 1994. This trend can be seen in the following table, calculated before the impact. The date of each perijovial pass is given, together with the distance in astronomical units:

Comet Shoemaker-Levy 9's pre-impact passes by Jupiter		
Year	Date	Distance AU
1971	April 26	0.08963
1975	April 26.8	0.06864
1977	May 7.0	0.07000
1980	February 1.8	0.11896
1982	May 26.0	0.12453
1984	October 4.5	0.11937
1987	July 12.4	0.07031
1989	August 2.5	0.06090
1992	July 8.0	0.00072
1994	July 16.8	Impact

---

[10] Rob Landis [1]. His paper includes a fascinating description of the detailed preparations for the event.

[11] http://www2.jpl.nasa.gov/sl9/sl9.html

Using a Roche coefficient of 1.8, find the Roche limit in Jovian radii and astronomical units, and determine *earliest* likely date of break-up of the comet.

Given

$R_{roche} = 1.8\,R(\rho_p/\rho_s)^{1/3}$	Equation for determining Roche limit with a coefficient of 1.8, estimated as reasonable for a comet
ρ_p	Mean density of Jupiter 1,326.2 kg/m^3
ρ_c	Assumed mean density of comet nucleus: 500 kg/m^3
R	Radius of Jupiter 71,492 km
1.496×10^8	1 AU, in kilometers

Assumptions Though unknown, we assume the given mean density of the comet, based on best-guess estimates.[12] The nucleus of a comet is likely a porous mixture of water ice and dust, neither liquid nor rocky, so we assume a Roche coefficient of about 1.8. We assume no other factors affecting its disintegration, such as the internal tensile strength of the materials within the cometary nucleus, outgassing from the nucleus due to solar heat (which in any event would be minimal at 5 AU), or even the size of the comet.

Method Here we use the Roche limit equation, assuming the comet is a more or less loose collection of rock and ice, and find the limit in Jovian radii. To convert to astronomical units, simply multiply the Roche Jovian radii by the radius of Jupiter, which yields the distance in kilometers, then divide by the number of kilometers in an AU.

Calculations First we find the Roche limit for Jupiter in Jovian radii using a coefficient of 1.8:

$$r_{Roche} \simeq 1.8 \left(\frac{\rho_p}{\rho_c}\right)^{1/3}$$

$$r_{Roche} \simeq 1.8 \left(\frac{1326.2}{500}\right)^{1/3}$$

$$r_{Roche} \simeq 2.49 \text{ Jovian radii}$$

This many Jovian radii is 178,132 km. In AU it is,

$$r_{Roche} \simeq \frac{178,132}{1.496 \times 10^8}$$

---

[12] E.g., D.A. Crawford (Sandia National Laboratories), *Comet Shoemaker-Levy 9 Fragment Size and Mass Estimates from Light Flux Observations*, http://www.lpi.usra.edu/meetings/lpsc97/pdf/1351.PDF. See also, http://www.nature.com/nature/journal/v370/n6485/abs/370120a0.html.

$$r_{Roche} \simeq .0012 \text{ AU}$$

Thus it appears that the comet did not likely break up until the very last pass near Jupiter in July, 1992. Only then did it enter the critical Roche limit for the planet.

Observations

1. The comet's final pass was .00072 AU from Jupiter, or 107,712 km, which is only about one and a half Jovian radii and well within the Roche limit (.001) even for rigid bodies. Comet Shoemaker-Levy 9 was a vivid demonstration of the tidal forces that will destroy any body sojourning too close to a planet.
2. The closest perijovial distance of the comet was 107,712 km. For comparison, the mean radius of Jupiter's moon Amalthea is 181,400 km. Jupiter's moons Metis and Adrastea are respectively 128,000 and 129,000 km from Jupiter.[13]

How Comet Lovejoy C/2011 W3 Barely Survived the Solar Furnace

In late November, 2011 Australian amateur astronomer Terry Lovejoy identified a small, unknown smudge in a star field on a CCD image with his small 8″elescope. He could not know that the smudge would turn out to be one of the most beautiful and interesting comets in recent years. It soon became evident that Comet Lovejoy C/2011 W3 would be a "sungrazer" that would reach perihelion December 16th, coming within a hair's-breadth 200,000 km of the Sun's scorching surface. The comet was a member of the so-called Kreutz group, comets that round the Sun within only a few solar radii.[14] According to long-time amateur astronomer and comet expert John Bortle: "Among the most extraordinary of all comets, the family of sungrazers is thought to be the remnants of a huge comet that broke apart millennia ago, and it might even have been the one Aristotle wrote about in the year 371 BC that had a tail that spanned a third of the heavens. Progressive fragmentation at each subsequent swing by the Sun broke up the major fragments of this progenitor to spawn a host of ever smaller pieces."[15]

Images of the comet disappearing behind the Sun were visible on the coronagraph of the orbiting Solar and Heliospheric Observatory (SOHO). According to

---

[13] Data was taken from http://ssd.jpl.nasa.gov/?sat_elem.

[14] The comets of the Kreutz group all have similar, though not identical orbital elements. Their orbits are highly inclined and eccentric, with semi- major axes of ~100 AU and periods of ~ 1,000 years. The original parent comet of this group is estimated to have had a diameter of perhaps 100 km. The majority of the Kreutz group comets have diameters in the range of a meter to 10 m that are only detectable when they are close to the sun. Gundlach et al. [2].

[15] Bortle [3].

Bortle, the images "showed the comet's tail being violently distorted by the solar corona, looking like cigarette smoke in a turbulent breeze, just before the comet disappeared behind the solar limb. Incredibly, however, the same camera soon captured images of the comet emerging from behind the Sun's opposite limb. Comet Lovejoy had somehow survived its hellish passage through the solar atmosphere."[16]

When, a few days later, the comet finally became dramatically visible to telescopes in the southern hemisphere, it appeared to be all tail. Indeed, the head of the comet had transformed into a "bright tailward-pointing ray instead of a compact central condensation. This ray grew longer each day as the comet had become weaker and more diffuse. Comet expert Zdenek Sekanina at JPL proposed that Lovejoy's nucleus completely disrupted about December 17.6 UT, and that the stream of resulting debris was moving rapidly outward into the tail, forming the ray-like feature."[17] Let us see what might have happened to the head of that comet.

Problem Using a Roche coefficient of 1.8, determine the Roche radius of the Sun in solar radii and kilometers, and evaluate whether such a comet at a perihelion that close could have survived without fragmentation.

Given

$R_{roche} = 1.8\, R(\rho_p/\rho_s)^{1/3}$	Equation for determining comet Roche limit using a coefficient of 1.8, estimated as reasonable for a comet
P_{sun}	Mean density of the Sun: 1,408 kg/m^3
ρ_c	Assumed mean density of comet nucleus: 500 kg/m^3
R	Radius of the Sun: 696,000 km
.00554	Comet Lovejoy's perihelion distance in AU[a]
1.496×10^8	1 AU, in kilometers

[a] Data taken from: http://ssd.jpl.nasa.gov/sbdb.cgi#top

Assumptions We again assume a likely mean density of the comet, and a Roche coefficient of about 1.8. We assume no other factors affecting its disintegration, such as the size of the comet, internal tensile strength of the materials within the cometary nucleus, or the likely significant outgassing from the nucleus due to solar heat.

Method Here we use the Roche limit equation. To convert to astronomical units, simply multiply the Roche Jovian radii by the radius of Jupiter, which yields the distance in kilometers, then divide by the number of kilometers in an AU.

---

[16] Ibid., 39.
[17] Ibid., 40.

Calculations To find the Roche limit for the Sun and the comet in solar radii, we are asked to use a coefficient of 1.8:

$$r_{Roche} \simeq 1.8 \left(\frac{\rho_{sun}}{\rho_c}\right)^{1/3}$$

$$r_{Roche} \simeq 1.8 \left(\frac{1408}{500}\right)^{1/3}$$

$$r_{Roche} \simeq 2.39 \text{ solar radii}$$

In kilometers this is 696,000 × 2.39 or only 1,663,440 km. In AU it is,

$$r_{Roche} \simeq \frac{1,663,440}{1.496 \times 10^8}$$

$$r_{Roche} \simeq .01 \text{ AU}$$

Lovejoy, at a perihelion distance of .00554 AU, was well within this radius and should have fragmented.

Observation The real question is, why did Comet Lovejoy survive at all? Both Shoemaker-Levy 9 and Lovejoy penetrated deep into their parent bodies Roche radius; the first died, the other lived to see another day, albeit in likely much reduced form. The answer may lie in the outgassing that Lovejoy experienced as it ventured well into the solar heat, which may have kept the nucleus more intact and resistant to fragmentation, and its size, which may have been far larger than the usual Kreutz-group comet.[18]

Exercises: How Close Encounters Can Be Perilous We have emphasized the effects of tidal disintegration of bodies close at hand, such as comets and moons of our solar system. But they are everywhere in the universe. Astronomers see vivid evidence of tidal forces pulling passing galaxies apart. Our own Milky Way has "cannibalized" nearby dwarf galaxies, and the stringy remnants of their former structures have been discerned among the clouds of other stars in our galaxy.

---

[18] This idea was put forth in Gundlach et al. [2]. The authors concluded that the radius of the comet must have been between .2 and 11 km before perihelion. Matthew Knight of the Lowell Observatory and Johns Hopkins Applied Physics Lab estimated that the comet's core must have been at least 500 m in diameter; otherwise it couldn't have survived so much solar heating. http://science.nasa.gov/science-news/science-at-nasa/2011/16dec_cometlovejoy/.

Common too are "contact" binary stars. They are in such close proximity that their Roche limits touch, creating merging "Roche lobes" which allow star-matter to overflow from one star to the other (or from star to black hole) over a kind of interstellar bridge. Algol in the constellation of Perseus is a famous example of such a star. Returning to our own solar system, the recent and dramatic comet disintegrations, the *Voyager* discovery of rings around the outer planets, pictures of active extra-terrestrial volcanoes, the still-puzzling mysteries of Saturn's complex moon-ring dynamics, and even the inward-spiral of the Martian moon Phobos, give plenty of examples of the effects of gravitational tides close to home. And of course our moving oceans are our ever-present reminder of this mysterious action-at-a-distance force which so deeply connects us to the Moon and Sun.

Problems

1. How close could a comet come to Mars before it disintegrated? Assume a Martian density of 3,934 kg/m^3. For this and the problems that follow, assume a Roche limit coefficient of 1.8 unless otherwise indicated.
2. Phobos, a likely captured asteroid, is revolving rapidly in a low orbit around Mars, at about 2.76 Martian radii, and is said to be gradually falling inward toward the planet. Assuming the satellite's density is 1,872 kg/m^3, tell how close to the Roche limit, in Martian radii and kilometers, Phobos is currently orbiting.
3. If comet Shoemaker-Levy were an asteroid of density 2,500 kg/m^3, how close to Jupiter (in radii) could it come before disintegrating?
4. Modify the Roche equation to use instead of densities, just the mass of the primary and the density of the secondary. What units would result from using the modified equation?
5. The densities of Jupiter's most inward satellite Metis orbits at 128,000 km from Jupiter's center, and is estimated to be about 3,000 kg/m^3. Find the likely Roche coefficient for that satellite. Jupiter's equatorial radius is 71,492 km.
6. Find an approximate minimum perihelion distance within which a comet would likely be torn apart within the Sun's Roche limit. Assume the Sun's mean density is 1,408 kg/m^3, and has a mean radius of 696,000 km. Confirm your result by using the equation you developed in Problem 4, assuming the Sun's mass to be 1.9891×10^{30} kg.
7. How close could a comet come to the Earth without being disrupted by the Earth's Roche limit? Assume a mean density for the Earth of 5,513.4 kg/m^3.
8. What do you estimate could be the closest orbit of a small body around the asteroid Vesta? Assume its and its visitor's density to be about 5,290 kg/m^3.
9. Neptune's little moon Naiad is in a very circular orbit about 1.948 Neptunian radii distant from the planet (48,227 km). Its density is estimated to be 1,300 kg/m^3. Neptune's density is about 1,638 kg/m^3. Why hasn't that moon disintegrated? What does its distance tell you about its composition?
10. Explain how different Roche coefficients may come into play when talking about the formation of planets and satellites in contrast to the breakup of already existing bodies.

References

1. Landis R (1994) Comet P/Shoemaker-Levy's collision with Jupiter: covering HST's planned observations from your planetarium. http://www2.jpl.nasa.gov/sl9/hst1.html. International Planetarium Society Conference, Space Telescope Science Institute
2. Gundlach B, Blum J, Skorov Yu V, Keller HU (2012) A note on the survival of the sungrazing comet C/2011 W3 (Lovejoy) within the Roche limit (preprint submitted to Icarus) 9 Mar 2012. http://arxiv.org/pdf/1203.1808.pdf
3. Bortle JE (2012) The remarkable case of comet Lovejoy. Sky Telescope 123(5):36
4. Van Allen JA (1993) Elementary problems and answers in solar system astronomy, University of Iowa Press, Ames, pp. 213–214

Chapter 19
Hovering in Space: Those Mysterious Lagrangian Points

The James Webb Space Telescope will hover in an orbit not around the Earth, as does the Hubble Space Telescope, but around the Sun. It will mimic Earth's orbit, but be farther out, orbiting at the so-called *second Lagrangian point*, or the *L2 point*. The telescope will maintain a stable temperature, unaffected by passing in and out of the Earth's shadow. Yet the reader may ask: how may an object orbiting the Sun farther out from the Earth, in a larger orbit, keep up with us? Bodies in more distant orbits move more slowly; we continually overtake the slower, outer planets as we whirl around the Sun, and the inner planets outpace us. Won't the JWST fall behind? The answer is no, and to know why requires understanding the Lagrangian points.

What does it mean when we say the new space telescope will be in the L2 point? Interestingly, a study of the Lagrangian points, as the "L" points are called, yields useful insights on how bodies orbit and interact in gravitational fields. While the interactions of two bodies are straightforwardly analyzed with Newtonian mechanics (the so-called *two body problem*), things become exceedingly difficult when three or more bodies are involved (the *three body problem*). Many such problems cannot be solved by analysis, but require computer iterations to approximate solutions. The Lagrangian points restrict the inquiry to certain spots in front and behind and to each side of a planet's (or moon's) orbit, where solutions are much simpler. Thus it is an interesting form of what is known as a *restricted three body problem*. To investigate this, we need to step back and take a look as some basic principles of gravitation.

What do we know about the gravitational effects of the Sun on the Earth? We know that while the Sun is massive (in kilograms, about two followed by 30 zeros) it is also far away – about 149.6 million kilometers distant. And because gravity's force diminishes with the square of the distance away, the actual acceleration perceived at our earthly outpost is rather modest. The Sun pulls on the Earth with a gravitational acceleration of only about .00593 m/s^2. That is, about 6 mm per second, every second.[1] That is Earth's "free fall" acceleration toward the Sun. How can such a tiny pull keep our big Earth from travelling off into space? While this seems like a tenuous hold on our precious home, the Sun's attraction is felt by every particle of the Earth, so the Sun's net pull on the globe as a whole is about 3.54×10^{22} N. It is this attraction that (by Kepler's Third Law) keeps us in an orbital period of about 365¼ days. This solar acceleration is also fairly slight compared to the Earth's own gravity at its surface. The downward pull felt by you and me on the ground is about 9.8 m/s^2. This is about 1,650 times more than the 6 mm/s^2 gravitational acceleration from the Sun at Earth's distance.

But now let's try a thought experiment. First, any object, such as a spacecraft, at Earth's distance from the Sun will also have Earth's 1 year period. (By Kepler's Third Law, the period is proportional to the 3/2 power of the distance from the Sun. If the distance is the same, the period is the same.) What if we launch a solar-orbiting spacecraft in a smaller, more inward circular orbit? Venus moves with a quicker period than Earth, and Mercury is quicker still. The spacecraft's orbital period would, like the inner planets, also be less than a year, its increased velocity counteracting the increased pull of the Sun tending to draw it inward. The actual period would of course be determined by Kepler's Third Law.

Now let's go a step farther: What if the gravitational attraction of the Sun perceived by our space-faring craft were somehow made less? This could happen if the spacecraft is still close enough to Earth to feel its gravity pulling it outward, offsetting the inward pull of the Sun. If the ship is poised at some as-yet unknown distance between Earth and Sun, the *net* gravitational acceleration acting on it could be just right so that the period of the spacecraft would not be as short as it otherwise would be that distance. In fact, if its distance from Earth were just so, its period could in theory exactly *match* the Earth's – it would "track" the Earth exactly. Likewise, a spacecraft orbiting the Sun at a larger orbit than Earth would normally have a longer period than Earth (as in the case of the outer planets). But if it were positioned just close enough to Earth so that the Earth's gravitation supplemented the Sun's pull, then it would have a shorter orbital period than it otherwise would at that distance, were it experiencing the Sun's gravity alone. At some particular point, its period too would just match the Earth's period. It would hover in a solar orbit just outside the Earth's orbit, and stay there. This is similar to the way a geosynchronous Earth-orbiting satellite stays over one spot of the Earth, its orbital period in perfect synch with the Earth's rotation.

These *Lagrangian points* were discovered by a brilliant Italian-born mathematician and astronomer, Joseph-Louis Lagrange (1736–1813), after whom they were named. Lagrange mastered and extended Newtonian mechanics, and in 1788 published his most famous work, *Mécanique Analytique* (Analytical Mechanics). Consider three cases: where the spacecraft is between the Earth and Sun it is at the L1 point. Where it is outward of the Earth, it is at the L2 point. Where the spacecraft is opposite the Sun from us it is at the L3 point. The L1, L2, and L3 points are all on the Earth-Sun axis. Lagrange

---

[1] This can be readily calculated from the Newtonian equation for gravitational acceleration, $f = GM/r^2$.

also found two other points of stability on the planet's orbital path itself: the L4 point, leading at a 60° angle from the Earth-sun axis, and the L5 point, following at 60°.[2]

The Lagrangian points can be found in almost any three-body system. A well-known example is the Trojan asteroids clustering at the L4 and L5 points in Jupiter's orbit. The Lagrangian points are of tremendous benefit to space exploration. The Solar and Heliospheric Observatory (SOHO) is stationed in solar orbit at Earth's L1 point, allowing SOHO continuous observation of the Sun without the interference of Earth that would be occasioned if it were in an Earth orbit. And we already mentioned that the James Webb Space Telescope is scheduled to be placed in Earth's L2 point, where it will have an uninterrupted view of the heavens, and join the Herschel and Planck missions. We will discuss later the potential benefits of the lunar Lagrangian points for exploring the Moon. Now let us now mathematically work through how the Earth's L1 and L2 points work.

Deriving Equations for Finding the L1 and L2 Points in the Earth–Sun System

Problem You are asked to help plan launches of next-generation solar and astronomical telescopes to the L1 and L2 points respectively. Use the concept of gravitational acceleration toward the Sun being offset (for L1) or supplemented (for L2) by the Earth's gravitational acceleration, to derive the mathematical relationship between period and distance from Earth R for the respective L1 and L2 points. (For now, do not try to solve for R).

Given

$f_s = GM_1/r^2$	Gravitational acceleration f_s induced by the Sun on an object at distance r from the Sun where M_1 and G are the mass of the Sun and the gravitational constant, respectively. At the distance of the Earth, r is one astronomical unit
$f_e = GM_2/R^2$	Gravitational acceleration f_e induced by the Earth on an object at distance R from the center of the Earth, where M_2 and G are the mass of the Earth and the gravitational constant, respectively
$P = 2\pi\sqrt{r/f}$	Period a circularly orbiting object experiencing centripetal acceleration f at r distance from the center of force[a]
R_{L1}	The distance from the center of the Earth to the L1 point; the difference $r - R_{L1}$ is thus the distance of the L1 point from the Sun
R_{L2}	The distance from the center of the Earth to the L2 point; the difference $r + R_{L2}$ is thus the distance of the L2 point from the Sun
$\omega = 2\pi/P$	Angular velocity of a circularly revolving object, where P is period. Units are radians per second if period is in seconds

[a]This equation was derived in Chap. 9

---

[2] An animation of the Lagrangian points can be found at the European Space Agency website, at http://www.esa.int/esaSC/SEMM17XJD1E_index_0.html.

Assumptions We will simplify our task by assuming the Earth is a sphere of uniform density, and that the Earth and the spacecraft move in circular, planar orbits. We ignore the Moon, the Earth's motion about the Earth–Moon barycenter, and any perturbing forces of other planets. The spacecraft will be assumed to have negligible mass relative to the Earth's and Sun's masses.

Method At L1, the gravitational acceleration of the Sun on orbiting telescope is offset by the *opposing* pull of the Earth, just enough to slow it to Earth's period. And at L2, the Earth's gravity on the object *supplements* that of the Sun enough to make it match Earth's period. Things that would otherwise orbit faster are slowed in L1 things that would orbit slower are speeded up at L2. Since we need to quantify this and find the distance of L1 and L2 from the Earth, the question to ask is: at what distance will the period of the Earth and the period of the object be the same? We are not inquiring where in space the gravitational pulls of the Sun and Earth are equal; only where the *periods* of Earth and the spacecraft are equal.

If in each case we combine the equation for solar acceleration with the given period equation, we can find out how the period depends on the gravitational acceleration. The goal is to be sure the end result has R in final equation, which is the distance to the applicable Lagrangian point. To summarize: start with the appropriate expressions for net the Sun-induced gravitational acceleration at the L1 and L2 points (reduced by Earth's pull at L1, augmented at L2). Plug these into the period equation with the appropriate expressions for the distances, R_{L1} and R_{L2}. Each of the resulting equations for the period of an object is cast in terms of the gravitational acceleration it experiences. Each equation will then have only one unknown, the distance to the Lagrangian point.

Calculations

Step 1: *Find equations for Sun's and Earth's gravitational accelerations (forces per unit mass) on the spacecraft at L1 and L2 points.*

From the equations given, we already have the gravitational accelerations induced by (a) the Sun, f_s and (b) the Earth, f_e at distances r and R, respectively. Here we modify the equations to take the distances to the L1 and L2 points into account:

Equation	L1 Point	L2 Point
(a) $f_s = \frac{GM_1}{r^2}$	$\to f_s = \frac{GM_1}{(r-R_{L1})^2}$	$f_s = \frac{GM_1}{(r+R_{L2})^2}$
(b) $f_e = \frac{GM_2}{R^2}$	$\to f_e = \frac{GM_2}{R_{L1}^2}$	$f_e = \frac{GM_2}{R_{L2}^2}$

Step 2: *Express the values of solar acceleration f_s as reduced or supplemented by Earth's acceleration, as appropriate to the respective L1 and L2 Lagrangian points.*

We noted that at L1, the gravitational acceleration of the Sun on an object is slightly canceled or offset by the opposing pull of the Earth, just enough to slow it to Earth's period, while at L2, the Earth's gravity on the object supplements that of the

Deriving Equations for Finding the L1 and L2 Points in the Earth–Sun System

Sun just enough to make it match Earth's period. In this step we express these ideas algebraically in terms of f_s being (a) reduced or (b) supplemented by Earth's gravitationally-induced acceleration, f_e, to yield the centripetal accelerations f_{L1} and f_{L2} at those Lagrangian points:

$$\text{(a) } f_{L1} = f_s - f_e$$
$$\text{(b) } f_{L2} = f_s + f_e$$

Step 3: *Combine the equations from Steps 1 and 2 to show the net gravitational accelerations at the L1 and L2 points.*

(a) $f_{L1} = f_s - f_e \rightarrow f_{L1} = \frac{GM_1}{(r-R_{L1})^2} - \frac{GM_2}{R_{L1}^2}$ ← Net acceleration toward Sun at L1

(b) $f_{L2} = f_s + f_e \rightarrow f_{L2} = \frac{GM_1}{(r+R_{L2})^2} + \frac{GM_2}{R_{L2}^2}$ ← Net acceleration toward Sun at L2

Step 4: *Substitute above expressions for centripetal acceleration into the period equation*

Objects in the L1 and L2 points have the same period as the Earth. If we can recast the accelerations in terms of the orbital period, which is a known constant, it will be possible to solve for distance R in each. We saw in Chap. 9 that period can be expressed in terms of centripetal acceleration: $P = 2\pi\sqrt{r/f}$. Substituting into that equation $r = r - R_{L1}$ for the distance to the L1 point and $r = r + R_{L2}$ for the distance to the L2 point; and also for f, the acceleration expressions for each Lagrangian point found in the last step, we have:

Substitute values for distance :   Substitute values for acceleration :

(a) <u>L1 Point</u>: $P = 2\pi\sqrt{\dfrac{r - R_{L1}}{f_{L1}}}$ → $P = 2\pi\sqrt{\dfrac{r - R_{L1}}{\frac{GM_1}{(r-R_{L1})^2} - \frac{GM_2}{R_{L1}^2}}}$

(b) <u>L2 Point</u>: $P = 2\pi\sqrt{\dfrac{r + R_{L2}}{f_{L2}}}$ → $P = 2\pi\sqrt{\dfrac{r + R_{L2}}{\frac{GM_1}{(r+R_{L2})^2} + \frac{GM_2}{R_{L2}^2}}}$

These are rather ungainly equations! But they are the expressions we sought, relating the period to the distance to the L1 and L2 points. The only unknowns in each equation are the distances from Earth to the Lagrangian points. The period (by definition) P will be equal to Earth's period of 365.256 days (expressed in seconds), r is the Earth's distance from the Sun (in meters), and the gravitational constant and mass terms (in kilograms) are known.

Step 5: *Simplify the equations.*

The equations invite simplification by squaring each side and cross-multiplying. Beginning with the L1 point equation, we have,

$$P = 2\pi \sqrt{\dfrac{r - R_{L1}}{\dfrac{GM_1}{(r-R_{L1})^2} - \dfrac{GM_2}{R_{L1}^2}}}$$

$$P^2 \left(\dfrac{GM_1}{(r - R_{L1})^2} - \dfrac{GM_2}{R_{L1}^2} \right) = 4\pi^2 (r - R_{L1})$$

Now do the same thing with the L2 equation, and divide each side by P^2. You can see by inspection that the resulting constant term on the right will be $4\pi^2/P^2$. This is the square of the orbital velocity (or frequency) in radians. Replace it with ω^2:

(a) **L1 Point:** $\dfrac{GM_1}{(r-R_{L1})^2} - \dfrac{GM_2}{R_{L1}^2} = \omega^2 (r - R_{L1})$

(b) **L2 Point:** $\dfrac{GM_1}{(r+R_{L2})^2} + \dfrac{GM_2}{R_{L2}^2} = \omega^2 (r + R_{L2})$

These have the familiar form of the dynamical balance of gravitational (left side of the equations) and centripetal (right side of the equations) acceleration of orbiting objects. Even though we reduced the equations to one unknown each and simplified, it is still quite difficult algebraically to *solve* for R_{L1} and R_{L2} in these equations. Using *Maple* software program, however, we find the following approximate values:

(a) R_{L1}:1.49 × 10⁹ meters (1,490,000 km)← L1 is this far from the Earth *toward* the Sun

(b) R_{L2}:1.50 × 10⁹ meters (1,500,000 km) ← L2 is this far from the Earth *away* from the Sun

Observations

1. The above equations are still not in their simplest form, but they are intuitive. Each term represents components of the net acceleration acting upon the spacecraft at the Lagrangian point.
2. Examine the L1 equation. The first term on the left side is the solar gravitational acceleration; the second is the reduction of that quantity by the Earth's gravitational acceleration. The result on the left-hand side is thus the *net* gravitational acceleration acting on the object. The signs between the terms show that the solar pull is reduced by Earth's gravity at the L2 point and augmented at the L2 point. The term on the right-hand side of each equation is the centripetal acceleration that must equal the net gravitational acceleration.
3. Here is the L1 equation annotated:

Deriving Equations for Finding the L1 and L2 Points in the Earth–Sun System

$$\underbrace{\frac{GM_1}{(r-R_{L1})^2}}_{\substack{\text{Sun's} \\ \text{gravitational} \\ \text{acceleration at} \\ \text{distance } R_{L1} \\ \text{from Earth} \\ \text{(the L1 point)}}} - \underbrace{\frac{GM_2}{R_{L1}^2}}_{\substack{\text{Earth's} \\ \text{gravitational} \\ \text{acceleration at} \\ \text{distance } R_{L1} \\ \text{from Earth} \\ \text{(the L1 point)}}} = \underbrace{\omega^2(r-R_{L1})}_{\substack{\text{Object's} \\ \text{centripetal} \\ \text{acceleration at} \\ \text{distance } R_{L1} \\ \text{from Earth} \\ \text{(the L1 point)}}}$$

Note that by moving the centripetal acceleration term to the left-hand side of the equation, all the accelerations (and thus all of the forces acting on each unit of mass) will equal zero.

4. In the above calculations, we created equations that held the period fixed while we saw what relations the distances must have in order to fulfill that condition. We could have approached the problem from the beginning more like the physicist than the mathematician, and said: "Let's find the point where the *vector sum of all the forces on the spacecraft is zero*, so that in a rotating frame of reference a spacecraft at the Lagrangian point is weightless." Here, we imagine that the object orbiting around the Sun is the rotating frame of reference, with the Sun at the center. We are inside the spacecraft at L1. We feel the gravitational pull of the Sun in one direction opposed by the gravitational pull of the Earth in the other. It can be shown that at L1 the pull of the Sun is slightly stronger than the Earth's pull (see the next Problem below). So our craft would drift toward the Sun. But we must account for the fact that moving objects tend (by Newton's First Law) to go in a straight line; hence, the orbiting spacecraft seems to drift outward by its circular motion.[3] Let us account for this tendency. At the point of weightlessness the inertial tendency to rectilinear motion should just balance the differential pull of the Sun over that of the Earth.

---

[3] If one is sitting in a circular frame of reference, though, there appears to be no particular thing acting on the object that accounts for this outward force except the rotation of the reference frame itself. So in a merry-go-round, one would not necessarily know why coins spilled on the ground would all move outward from the center. As noted earlier, physicists often refer to the centrifugal force as a "fictitious" force. It effectively cancels the accelerative effects of the reference frame itself to make it an "inertial" reference frame – that is, without its own acceleration, and where Newton's laws remain valid.

404   19 Hovering in Space: Those Mysterious Lagrangian Points

Determining the Accelerations on a Satellite at the Sun–Earth L1 Point

Problem From the L1 equation derived above, draw a graph of the net gravitational vs. centripetal accelerations of an object at the Sun–Earth L1 point to find an approximate value for R_{L1}; and calculate each of the actual accelerations on such an object at the L1 point to confirm that the net gravitational and centripetal accelerations are equal.

Given

$\frac{GM_1}{(r-R_{L1})^2} - \frac{GM_2}{R_{L1}^2} = \omega^2(r - R_{L1})$	L1 equation derived above
$f_s = \frac{GM_1}{(r-R_{L1})^2}$	Gravitational acceleration f_s toward the Sun at L1
$f_e = \frac{GM_2}{R_{L1}^2}$	Gravitational acceleration f_e toward the Earth at L1
$f_c = \omega^2(r - R_{L1})$	Centripetal acceleration at L1
P	Period of the Earth, 365.256 days or 3.15581184×10^7 s
G	Gravitational constant: 6.674×10^{-11} N·m²/kg
M_1	The mass of the Sun: 1.981×10^{30} kg
M_2	The mass of the Earth: 5.9736×10^{24} kg
r	The mean distance from Earth to the Sun: 1.496×10^{11} m

Assumptions We will again assume circular, planar orbits, and ignore the Moon, the Earth's barycentric and elliptical orbit, and any perturbing forces of other planets. We also assume the Earth is a sphere of uniform density. The spacecraft has negligible mass relative to the Earth's and Sun's masses. All forces will be expressed as forces per unit mass (that is, as accelerations).

Method We want to find the location at distance R_{L1} from the center of the Earth toward the Sun, where all the above accelerations are in balance: where the difference between the gravitational pull of the Sun and Earth ($f_s - f_e$) should just equal the centripetal acceleration (f_c) needed to deflect it away from a rectilinear path. This location should therefore satisfy the equation $f_s - f_e = f_c$.

One we have created the equation, we will graph it to find an approximate solution. Graphing complicated equations, whether manually or on a calculator of computer, can often more easily be accomplished if each side of the equation is plotted separately. Then zoom in on the graph to see where the two lines intersect, where $f_s - f_e$ intersects with f_c. The intersection point (or points) is the solution to the equation. The second part of the problem requires simply calculating the values for $f_s - f_e$ and for f_c, to confirm numerically what the accelerations are and if they do in fact balance.

Calculations For the first part of the problem, make a plot of the *net* gravitational acceleration of Sun minus Earth, as one proceeds to go away from Earth toward Sun (that is, as R increases). Let the L1 distance from Earth be R. On the same graph plot

Determining the Accelerations on a Satellite at the Sun–Earth L1 Point

the centripetal acceleration of an object, also as R increases from Earth. Be sure the centripetal acceleration is calculated at fixed Earth-year period. Below is one such graph, at a close-in scale, where R has been chosen (by approximation or trial and error) to be near the crossing point.

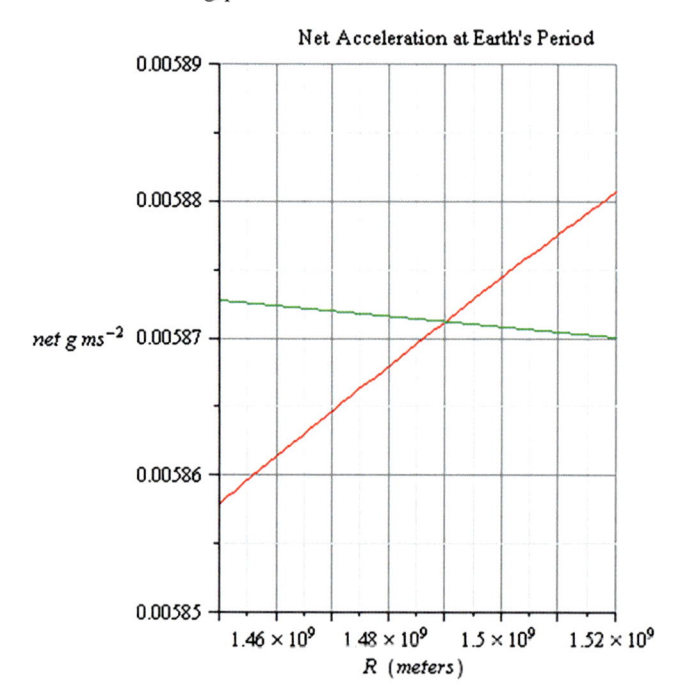

Net Acceleration at Earth's Period

net g ms⁻² vs R (meters)

The upward-sloping red line is the net gravitational acceleration (Sun minus Earth) and the downward-sloping green line is the centripetal acceleration (again, with P = Earth's period).

One can see graphically where the L1 point is. Note the intersection point is at just over 1.49×10^9 m (1,490,000 km), which is where we expected the L1 point to be. The accelerations there balance out at a little more than .00587 m/s², which is 5.87 mm/s per second acceleration.

For the second part of the problem, calculate the values for f_s and f_e and for $f_s - f_e$ acting on an object, such as a spacecraft, at the L1 point between Earth and the Sun. Then use the value 1.49×10^9 for R_{L1} and substitute that value into each equation:

(a) The Sun's gravitational pull at L1:

$$f_s = \frac{GM_1}{(r - R_{L1})^2} \rightarrow \frac{(6.674 \times 10^{-11})(1.9891 \times 10^{30})}{(1.496 \times 10^{11} - 1.49 \times 10^9)} \approx .00605 \text{ m/s}^2$$

(b) The Earth's opposite gravitational pull at L1:

$$f_e = \frac{GM_2}{R_{L1}^2} \to \frac{(6.674 \times 10^{-11})(1.9736 \times 10^{24})}{(1.49 \times 10^9)^2} \approx .0001796 \text{ m/s}^2$$

The difference, $f_s - f_e$ is the *net* gravitational pull on the spacecraft at L1:

$$f_s - f_e \to .00605 - .0001796 \approx .00587 \text{ m/s}^2$$

Note that this value of 5.87 mm per second per second, pulling toward the Sun, is about what appears on the graph. The only thing that will keep our spacecraft from falling into the Sun is its inertial tendency (by Newton's First Law) to go in a straight line, from an imparted tangential push at its original orbital insertion. The resulting circular orbital motion around the Sun is caused by the equality of the gravitational and centripetal accelerations, the latter being calculated as follows:

(c) The centripetal acceleration at L1:

$$f_c = \omega^2(r - R_{L1})$$

$$f_c = \frac{4\pi^2(r - R_{L1})}{P^2} \to \frac{4(3.14..)^2(1.496 \times 10^{11} - 1.49 \times 10^9)}{(3.15581184 \times 10^7)^2} \approx .00587 \text{ m/s}^2$$

where we again used the identity, $\omega = 2\pi/P$. Thus we have our equality of accelerations, $f_s - f_e = f_c$, since $.00587 = .00587$.

Observations

1. The calculation of accelerations for the L2 point (which we label R_{L2}) follow in the same way, with just a change in sign: the distance from the Sun to the L2 point is $r + R_{L2}$. The distance from the Sun to the L2 point is thus 1.4976×10^{11} m. So the equation for centripetal acceleration at the L2 point is just a little closer to 6 mm per second per second:

$$f_c = \frac{4\pi^2(r + R_{L2})}{P^2} \to \frac{4(3.14..)^2(1.496 \times 10^{11} + 1.49 \times 10^9)}{(3.15581184 \times 10^7)^2}$$

$$f_c \approx .00599 \text{ m/s}^2$$

2. The ledger-sheet for our L1 accelerations now appears as shown on the accompanying graph arranged as a kind of accountant's balance sheet. The direction of accelerations is shown along with the contributor to that acceleration. Note that the overwhelmingly important factors are the gravitational and centripetal accelerations, with the contribution of the Earth's own gravity being relatively small, yet crucial to make the books balance. Note that while

Determining the Accelerations on a Satellite at the Sun–Earth L1 Point   407

centripetal acceleration is always acceleration toward the center (e.g., the Sun in this case) we represent its vector as pointing away from the Sun (with a negative sign) since it is the acceleration required for an object to be deflected from its inertial tendency to go in a straight line. Some books would call this the "centrifugal acceleration" vector.

Balance of accelerations on spacecraft at the Earth–Sun L1 point

Source	Sun	Earth	Centripetal	Net
Direction of vector	Toward Sun	Away from Sun	Away from Sun	Equal
Acceleration: (mm/s^2)	6.05	−.18	−5.87	0.0
Percentage Contribution: (each direction)	100 %	3 %	97 %	0.0

3. The differences in acceleration on the above graph are on a scale of hundredths of a millimeter per second per second. It would seem that the slightest pulls and tugs from other planets could dislodge a spacecraft from that point. Is this so? For example, would the gravitational pull of Jupiter at its closest affect a spacecraft hovering in the L1 point? Let us assume that Jupiter at its closest opposition some year is about 650 million kilometers (6.5×10^{11} m) away (about 4.3 AU). Its mass is 1.8986×10^{27} kg. Its distance from Earth's L1 point will thus be about 6.5149×10^{11} m. The gravitational acceleration imparted by Jupiter at that location will be,

$$f_J = \frac{GM_J}{r^2}$$

$$f_J = \frac{(6.674 \times 10^{-11})(1.8986 \times 10^{27})}{(6.5149 \times 10^{11})^2}$$

$$f_J = 2.985 \times 10^{-7} \, \text{m/s}^2$$

which is only about .0003 mm per second per second. The pull of Jupiter on the L1 point is therefore but a fraction of the 5.87 mm/s^2 acceleration induced by the Sun and Earth — about one part in 20,000. The varying influences of other planets are small but are cumulative. Having a greater effect at L1 and L2 might be the (so far unaccounted-for) varying pull of the Moon. The new Moon (when it is closest to the L1 point) actually contributes .00221 mm/s^2 of gravitational acceleration on the L1 point, or about one part in 2,700. This is over seven times the pull of Jupiter at opposition, and these effects too accumulate to perturb the L1 orbit. In fact, if we roughly approximate the Moon's effect by simply adding the lunar mass to the Earth's mass and repeat the L1 calculation, it turns out (with all other assumptions remaining intact) that the L1 point is about 6,100 km (near one Earth radius) closer to the Sun. For the L2 point it is about 6,200 km farther out. But we should always keep our original assumptions in mind and beware of the false sense of accuracy that numbers can sometimes convey. A true accounting would show the variation in the Lagrangian points as the Moon orbits the Earth in its inclined and slightly eccentric orbit. An even more complete accounting would consider that the Earth's orbit around the Sun is slightly eccentric, so there is a variation in the solar acceleration at different times of the year. All of these and other complex perturbations reduce the stability of any object in the L1 and L2 points. Spacecraft and space telescope designers therefore realize that any spacecraft in the L1 or L2 points will need periodic adjustments to its orbit to remain in position.

Calculating the Heliocentric Orbital Velocities of a Spacecraft at the Sun–Earth L1 and L2 Points

Problem Calculate the mean orbital velocities in kilometers per second of the spacecraft as if it were orbiting the Sun at the Earth – Sun L1 point, and as if it were orbiting at the L2 point.

Given

$v_{L1} = \sqrt{f_c(r - R_{L1})}$	Velocity of an object in circular orbit at L1[a]
$v_{L2} = \sqrt{f_c(r + R_{L2})}$	Velocity of an object in circular orbit at L2
f_{cL1}	Centripetal acceleration at the L1 point, $\approx .00587$ m/s^2
f_{cL2}	Centripetal acceleration at the L2 point, $\approx .00599$ m/s^2
r	The mean distance from Earth to the Sun: 1.496×10^{11} m
R_{L1}	Distance from the Earth to the L1 point, approximated above: 1,490,000 km or 1.49×10^9 m
R_{L2}	Distance from the Earth to the L2 point, approximated above: 1,500,000 km or 1.50×10^9 m

[a]This equation was discussed in Chap. 9. It is a rearrangement to solve for v of the equation for centripetal acceleration equation $f = v^2/r$

Assumptions Same assumptions as noted in the previous problem.

Method To find the distances from the Sun to the respective Lagrangian points, it is necessary to add or subtract the respective R_{L1} and R_{L2} distances from the Earth's mean distance r from the Sun. Then insert the correct values into the velocity equations and solve for velocity.

Calculations

(a) $v_{L1} = \sqrt{f_{c_{L1}}(r - R_{L1})} \rightarrow \sqrt{.00587(1.481 \times 10^{11})} \simeq 29.49$ km/s

(b) $v_{L2} = \sqrt{f_{c_{L2}}(r - R_{L2})} \rightarrow \sqrt{.00599(1.511 \times 10^{11})} \simeq 30.08$ km/s

Observations

1. The Earth's mean orbital velocity is about 29.78 km/s. Thus, for the spacecraft to keep up with Earth at the L1 point, it must orbit about 290 m/s slower at the L1 point, and about 300 m/s faster at the L2 point. These differences are about 1% greater and lesser than Earth's velocity, respectively.
2. The equation for the L1 point,

$$\frac{GM_1}{(r-R_{L1})^2} - \frac{GM_2}{R_{L1}^2} = \omega^2(r-R_{L1})$$

is intuitive in its form since it shows the individual components of acceleration. It can be simplified. As a first step in cleaning up the equation, divide each side by $(r - R_{L1})$:

$$\frac{GM_1}{(r-R_{L1})^3} - \frac{GM_2}{R_{L1}^2(r-R_{L1})} = \omega^2$$

From the above discussion we note that since $\omega^2 = GM_1/r^3$, the ω^2 term can be put into the same general terms as the rest of the equation, which will create more opportunities for simplification:

$$\frac{GM_1}{(r-R_{L1})^3} - \frac{GM_2}{R_{L1}^2(r-R_{L1})} = \frac{GM_1}{r^3}$$

Now multiply each term by r^3/GM_1 and call the ratio of M_2/M_1 in the second term M_r. Dividing by M_1 is the same as calling the solar mass equal to one unit (that is, one solar mass), which is a typical unit of mass in astronomical calculations:

$$\frac{r^3}{(r-R_{L1})^3} - \frac{r^3 M_r}{R_{L1}^2(r-R_{L1})} = 1$$

Divide the numerator and denominator of each fraction by r^3:

$$\frac{1}{\left(\frac{r-R_{L1}}{r}\right)^3} - \frac{M_r}{\frac{R_{L1}^2(r-R_{L1})}{r^3}} = 1$$

$$\frac{1}{\left(1-\frac{R_{L1}}{r}\right)^3} - \frac{M_r}{\frac{R_{L1}^2}{r^2}\left(1-\frac{R_{L1}}{r}\right)} = 1$$

Now call the ratio of distances $R_{L1}/r = u$:

$$\frac{1}{(1-u)^3} - \frac{M_r}{u^2(1-u)} = 1$$

Where the Fictional "Counter-Earth" Would Be: The L3 Point in the Earth–Sun System

Problem Using all the same assumptions noted in the previous problems, construct an equation for the Earth–Sun L3 point, on the opposite side of the Sun from the Earth. Using calculator, spreadsheet, mathematical software, or manual trial and error, approximate the distance from the Earth to the Earth–Sun L3 point.

Given

G	Gravitational constant: 6.674×10^{-11} N·m²/kg
M_1	The mass of the Sun: 1.981×10^{30} kg
M_2	The mass of the Earth: 5.9736×10^{24} kg
r	The mean distance from Earth to the Sun: 1.496×10^{11} m

Assumptions All of the assumptions of the previous problem.

Method The R_{L3} distance we want is on the far side of the Sun. Thus the combined gravitational pull of the rather distant Earth and the Sun are additive (as in the case of the L2 point). Locating the L3 orbit again means finding the sweet spot where the centripetal acceleration will offset the combined gravitational accelerations of Sun and Earth. One thing to be aware of: the combined pull of the Earth and the Sun would make the L3 point *inside* the orbital trace of the Earth, on the far side of the Sun, because of the slightly stronger gravitational attraction of the combined bodies vs. that of the Sun alone. The equation should thus resemble the L2 equation. The L3 distance from the Earth will be R_{L3}. The distance from the Sun to the L3 point will be R_{L3} less 1 AU.

Calculations Here for reference is the L2 equation developed above:

$$\text{L2 Equation}: \quad \frac{GM_1}{(r+R_{L2})^2} + \frac{GM_2}{R_{L2}^2} = \omega^2(r + R_{L2})$$

We noted that the distance from Earth at L3 will be R_{L3}, and the distance from the Sun to R_{L3} will be that minus r. We thus construct the L3 equation in this way:

$$\text{L3 Equation}: \quad \frac{GM_1}{(R_{L3} - r)^2} + \frac{GM_2}{(R_{L3})^2} = \omega^2(R_{L3} - r)$$

Using *Maple* software, we obtain about 2.992×10^{11} m or 299.2 million km from Earth, on the Earth-Sun line. This puts it about 12,726 km from the Earth's orbit on the far side.

Observations

1. The L3 point too is relatively unstable, since any object there would be perturbed by other planets and by the uneven effects of non-circular (eccentric) orbits, just as are objects at the L1 and L2 points.
2. The L3 point has been a source of science fiction inspiration: What if there is another planet, a "counter-Earth" revolving around the Sun there, but hidden from our view? Of course, we now have the means to know that this is not true.

Space Station Parking Spot for Lunar Exploration: Determining the L1 Point in the Earth–Moon System

The Earth–Moon L1 point is an excellent place to park a space way-station for voyages to and from the Moon. Let's find out where this is.

Problem Construct equations for the distance from the Earth to the Earth–Moon L1 and L2 points, taking into account the Earth–Moon barycenter as the center of revolution of the Earth–Moon system, and find the distances from the Earth and Moon to the two Lagrangian points, R_{L1} and R_{L2}.

Given

M_1	The mass of the Earth: 5.9736×10^{24} kg
M_2	The mass of the Moon: 7.349×10^{22} kg
r_b	The distance from the Moon to the barycenter (center of mass) of the Earth–Moon system: 3.79728×10^8 m
r	Mean distance between the centers of Earth and Moon: 3.844×10^8 m
G	Gravitational constant: 6.674×10^{-11} N·m²/kg
$\omega = \sqrt{\frac{G(M_1+M_2)}{a^3}}$	Angular velocity for the binary system where here $a = r$.

Assumptions We will again assume circular, planar orbits, and ignoring the ellipticity of the Moon's orbit, and any perturbing forces of other planets. We also assume the Earth and Moon are spheres of uniform density. The spacecraft has negligible mass relative to the Earth's and Moon's masses.

Method We will start with the L1 equation derived above, doing just what we did with the Earth–Sun L1 point:

$$\frac{GM_1}{(r-R_{L1})^2} - \frac{GM_2}{R_{L1}^2} = \omega^2(r - R_{L1})$$

But in the Earth–Sun system it was safe to ignore the vastly smaller mass of the Earth and assume that the center of the Sun as the center of mass of the system. We cannot do that for the Earth–Moon system; the mass of the Moon is not insignificant in relation to the Earth's mass. The Earth and Moon revolve, as do all binaries, around their common barycenter along the line between them. The distance from the Moon to the barycenter is found as we saw in Chap. 10 using the equation:

$$r_b = r\left(\frac{m}{M+m}\right)$$

The result is 3.79728×10^8 m. The centripetal acceleration of the Moon and Earth, offsetting the gravitational attraction between them, are generated by their spins around that center of mass, so the distances in the $\omega^2 r$ term are going to have to reflect the distances from the center of mass. The distance in the last term of the L1 equation must therefore be changed to $r_b - R_{L1}$ because that is the distance of L1 from the center of mass, and therefore the center of revolution for the L1 point, from which centripetal acceleration is calculated. The angular velocity in the L1 equation, which is a function of the period, can be rather accurately determined by the given equation, which is Kepler's Third Law, remembering, however, that we are assuming circular orbits for these problems.

Calculations First find the angular velocity of the Moon around the barycenter:

$$\omega = \sqrt{\frac{G(M_1 + M_2)}{r^3}}$$

$$\omega = \sqrt{\frac{(6.674 \times 10^{-11})(5.9736 \times 10^{24} + 7.349 \times 10^{22})}{(3.844 \times 10^8)^3}}$$

$$\omega = 2.665377395 \times 10^{-6}$$

We will use this result shortly. Now construct the L1 equation as before, with a slight change described above to accommodate the center of mass:

$$\frac{GM_1}{(r-R_{L1})^2} - \frac{GM_2}{R_{L1}^2} = \omega^2(r_b - R_{L1})$$

Inserting the correct values we have:

$$\frac{(6.674 \times 10^{-11})(5.9736 \times 10^{24})}{(3.844 \times 10^8 - R_{L1})^2} - \frac{(6.674 \times 10^{-11})(7.349 \times 10^{22})}{R_{L1}^2}$$

$$= (2.665377395 \times 10^{-6})^2 (3.79728 \times 10^8 - R_{L1})$$

Using *Maple* software to solve for *RL1*, we get

$$R_{L1} = 58{,}023 \text{ kilometers}$$

This is the distance from the Moon. The distance from the Earth is thus 326,377 km.

Observation This distance is quite close to the Moon, about 85 % of the way there, yet an object there would be in orbit around the Earth, not the Moon.

The Equilateral L4 and L5 Lagrangian Points in the Earth–Moon System

The L4 and L5 Lagrangians have the unique and interesting quality of each being at one apex of an equilateral triangular relationship with the centers of the other two masses. The points are 60° ahead and behind a line joining the primary and secondary. In the relationship of the Earth and Moon, for example, the base of the equilateral triangle is the line between the centers of those two bodies. The other two sides of the triangle may be constructed from the Earth and Moon, respectively, to slant toward the L4 point in one direction, ahead of the motion of the Moon, and to the L5 point trailing it. Thus the distance to the L4 and L5 points to the Moon is the same mean distance as the Earth is from the Moon, 384,400 km. The next problem will explore how the balance of the balance of gravitational and centripetal accelerations in this three-way arrangement is maintained.

Problem Show numerically that the L4 and L5 points in the Earth–Moon system are places where the balance of the gravitational and centripetal accelerations is equal.

Given

M	The mass of the Earth: 5.9736×10^{24} kg
m	The mass of the Moon: 7.349×10^{22} kg
r_b	The distance from the Moon to the barycenter (center of mass) of the Earth–Moon system: 3.79728×10^8 m
r	Mean distance between the centers of Earth and Moon: 3.844×10^8 m
G	Gravitational constant: 6.674×10^{-11} N·m²/kg
$\omega = \sqrt{\frac{G(M+m)}{a^3}}$	Angular velocity for the binary system where here $a = r$

Assumptions We will again assume the Earth and Moon are spheres of uniform density in circular, planar orbits, and ignore the ellipticity of the Moon's orbit, and any perturbing forces of other planets.

Method We can take either L4 or L5 to make the case, and here we will illustrate with the L5 point. Unlike the L1, l2 or L3 points, the L4 and L5 points are not in a line, but are in a triangular relationship. Before looking at the specifics of the Earth and Moon, let us examine the situation of any two masses, a primary M and a secondary m. The center of mass is somewhere between them.

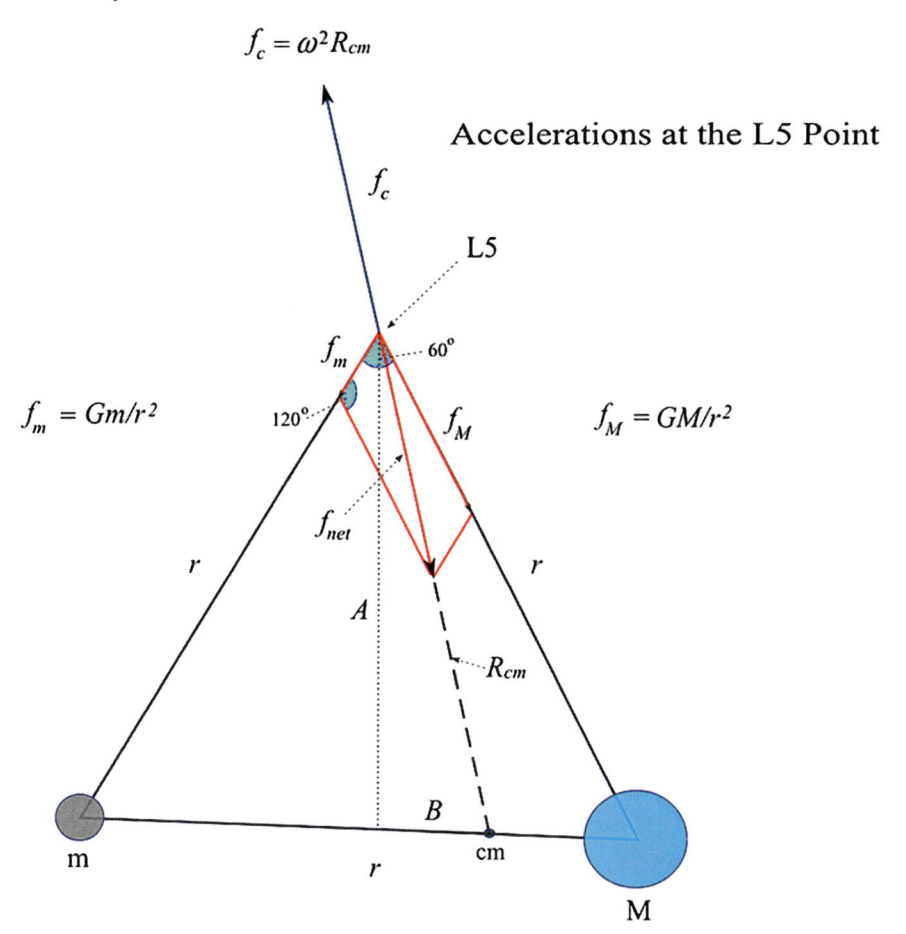

Accelerations at the L5 Point

$$f_c = \omega^2 R_{cm}$$

$$f_m = Gm/r^2$$

$$f_M = GM/r^2$$

Here, in the accompanying diagram, we let f_M be the gravitational acceleration of the primary M, and show its vector as pointing toward M along the right side of the triangle. Our task is to find the values of the vectors f_c and f_{net} to see if they are in fact equal. This we will do with the Earth and Moon, using numbers rather than a geometrical proof, although we will need to apply geometry to find the values.

The acceleration f_m represents the smaller gravitational acceleration of the lesser mass m, and its vector is slanting along the other side of the triangle, toward m. The composition of those vectors, the resultant vector, is the downward line from L5 toward the center of mass, denoted f_{net}. Exactly opposing this downward, resultant, gravitational acceleration vector is the upward-pointing vector f_c. This is the centripetal acceleration vector, here directed outward as a "centrifugal acceleration" vector which must just balance the net gravitational acceleration vector

The Equilateral L4 and L5 Lagrangian Points in the Earth–Moon System

induced by M and m to stay in the same relative position as it orbits the center of mass. Remember, everything orbits the center of mass of this system, which is marked cm on the diagram.

To find the length of the vector f_c, we need to know the length of line R_{cm} so we can multiply it by $\omega^2 R_{cm}$. This invites use of the Pythagorean Theorem, because the triangle ABR_{cm} is a right triangle. Side B is just $r/2$ minus the distance of M to the center of mass, cm, which as we found in Chap. 10 is 4,671.6 km. Since the large triangle is equilateral, we know from elementary trigonometry that side A is r times $\sqrt{3}/2$.[4] The vector f_c must balance the net gravitational acceleration vector, f_{net}. To find that from the other two gravitational vectors, we can use the law of cosines.[5] When all the accelerations have been computed, we can see if they balance, such that $f_{net} = f_c$.

Calculations Starting with the centripetal acceleration and employing the Pythagorean Theorem, we create the following equation to find the distance from the L5 point to the center of mass, R_{cm}. This is the orbital radius of an object at the L5 point:

$$R_{cm} = \sqrt{A^2 + B^2}$$

From the information given:

$$A = \frac{\sqrt{3}}{2} r \quad B = \frac{r}{2} - 4.6716 \times 10^6$$

$$A = 3.329 \times 10^8 \text{ m} \quad B = 1.8753 \times 10^8 \text{ m}$$

Inserting these values into the equation for R_{cm} we have,

$$R_{cm} = \sqrt{(3.329 \times 10^8)^2 + (1.8753 \times 10^8)^2}$$

$$R_{cm} = 3.821 \times 10^8 \text{ m}$$

Next find angular velocity from the given equation:

$$\omega = \sqrt{\frac{G(M+m)}{r^3}}$$

---

[4] Length A is $r\cos(30°)$. The cosine of 30° is equivalent to $\sqrt{3}/2$.

[5] The law of cosines is allows one to calculate the third side of a triangle when we know the other two and the angle between them. It is usually written this way: $c^2 = a^2 + b^2 - 2ab\cos\theta$. Its use is shown in the problem.

$$\omega = \sqrt{\frac{6.674 \times 10^{-11}(5.9736 \times 10^{24} + 7.349 \times 10^{22})}{(3.844 \times 10^8)^3}}$$

$$\omega = 2.6653774 \times 10^{-6}$$

From these last two values we can compute centripetal acceleration:

$$f_c = \omega^2 R_{cm}$$

$$f_c = (2.6653774 \times 10^{-6})^2 (3.821 \times 10^8)$$

$$f_c = 2.71442668 \times 10^{-3} \text{ m/s}^2$$

Let us compare this value with the net gravitational acceleration at L5. To do that, we have to use the law of cosines to find f_{net} from f_M and f_m. The equation is:

$$f_{net} = \sqrt{f_M^2 + f_m^2 - 2f_M f_m \cos(120°)}$$

Since the separate accelerations are,

$$f_M = \frac{GM}{r^2} \qquad f_m = \frac{Gm}{r^2}$$

Since the cosine of 120° is $-1/2$, the equation may be simplified to,

$$f_{net} = \frac{G}{r^2}\sqrt{M^2 + m^2 + Mm}$$

Inserting the correct values yields the following result:

$$f_{net} = \frac{6.674 \times 10^{-11}}{(3.844 \times 10^8)^2}$$

$$\times \sqrt{(5.9736 \times 10^{24})^2 + (7.349 \times 10^{22})^2 + (5.9736 \times 10^{24})(7.349 \times 10^{22})}$$

$$f_{net} = 2.71442668 \times 10^{-3} \text{ m/s}^2$$

As can be seen, the gravitational and centripetal accelerations are precisely the same at the L5 point:

$$f_{net} = f_c = 2.71442668 \times 10^3 \text{ m/s}^2$$

Observations

1. The same procedure can be repeated for the L4 point. Note that the locations of the L4 and L5 points are just the same as the mean distance from the Moon as the Earth is. L5 rides behind the Earth, and L4 leads, and each hangs in the balance, so to speak, of the joint gravitational attractions of both bodies.
2. Once can do a balance sheet of these accelerations, just as we did earlier. It turns out that the Earth contributes over 99 % of the gravitational acceleration acting on the L5 point.
3. We worked though the individual steps to show the method intuitively. The above equations could be generalized. For example, the length of side B of any other such system is,

$$B = \frac{r}{2} - r\left(\frac{M}{M+m}\right)$$

which can be simplified and integrated into the above analysis, if desired.

Exercises: The Strangeness of n-Body Gravitational Motions Things become sometimes strangely interesting when three or more bodies are in delicate equilibrium. The Trojan asteroids accompanying Jupiter are in the stable L4 and l5 points, but it is unknown why more cluster in the leading L4 point than the L5 point. Asteroids have been captured at the stable L-points of Earth, Mars and Venus. The small moon Cruithne orbits the Sun but is in resonance with the Earth's motion. Like a car on a race track, however, it will appear to switch lanes and go from outside to inside the Earth's orbit; from the Earth's perspective it traces an odd horseshoe-like pattern.[6] Similarly, the small Saturnian satellites Janus and Epimetheus seem to share an orbit: one is slightly inward and the other outward, so the inner one will overtake the other, but as they get close, their weak gravitational interaction causes them to switch positions. The moons Telesto and Calypso maintain their orbits around Saturn in the L4 and L5 points with respect to Tethys. Hyperion, orbiting between Saturn's Titan and Iapetus, seems to exhibit chaotic behavior.[7]

Sometimes the powerful force of a planet will even fling or "eject" a small body well out of its customary orbit. The near-Earth Apollo asteroids and the Kirkwood gaps in the asteroid belt appear to bear the tell-tale fingerprints of mighty Jupiter's influence. With the aid of computers and the tools of computational physics, these strange behaviors can be better understood. On a larger scale, the long-term stability of multiple interacting bodies in their gravitational fields is related to a profound

---

[6] See http://www.astro.uwo.ca/~wiegert/3753/3753.html for an interesting discussion of this moon.

[7] Chaisson and McMillan [1].

question that has been puzzling scientists for a long time: Is the solar system itself stable? Oddly, this question may depend on how one defines "stable" and, it appears, has not been fully settled.[8]

Problems

1. Just as was done above, compute a chart of the balance of net accelerations at the Earth–Sun L2 point.
2. Compute the heliocentric orbital velocity of a spacecraft stationed at the Earth–Sun L3 point in km/s.
3. If the Moon's mass were added in to the determination of the L3 point, what would be the difference in the L3 distance from that calculated above using just the Earth's mass?
4. Using calculator, spreadsheet, mathematical software, or manual trial and error, approximate the distance from the Earth and the Moon to the Earth–Moon L2 point. Take the lunar period to be 27.3217 days, and its mass to be 7.349×10^{22} kg. (*Hint*: Use distances to center of mass of the EM system in your calculations.)
5. Compute the gravitational accelerations induced by the Earth and Moon at the L2 point and create a table of the gravitational and centripetal accelerations like the one shown in the text.
6. Approximate the L1 and L2 points for Mars. Assume its mass is 6.41693×10^{23} kg and its semi-major axis is 1.52371034 AU.
7. Approximate the L1 and L2 points for Jupiter. Assume its mass is 1898.13×10^{24} kg and its semi-major axis is 5.202887 AU.
8. Compute the accelerations acting on the Trojan asteroids at the Jovian L4 or L5 points, and show that the centripetal and net gravitational accelerations induced from the Sun and Jupiter are equal.
9. In the above problem, what approximate percentage of the net gravitational acceleration acting on the Trojan L4 or L5 points is due to the Sun, and what percentage is due to Jupiter?
10. Compute the accelerations on a spacecraft at the Earth–Sun L5 point and show that the centripetal and net gravitational accelerations induced from the Sun and Earth are equal.

References

1. Chaisson E, McMillan S (2005) Astronomy today, 5th edn. Pearson Prentice Hall, Upper Saddle River, pp 324–324
2. Szebehely VG, Mark H (1998) Adventures in celestial mechanics, 2nd edn. Wiley, New York, Chapter 13

---

[8] For a readable discussion of this problem with references to further reading, see Szebehely and Mark [2].

Appendix: Solutions to Problems

Chapter 2

1. 30 m/s^2. 13,005 m.
2. 4,410 m.
3. 4,591.8 m.
4. PE is 9,800 J; (a) KE is 9,800 J; v is 140 m/s; (b) 14.286 s
5. The distances would be reduced proportionally by 0.3786. For the given times, distance is proportional to acceleration.
6. The distances would be increased proportionally by 1.138, by the same reasoning.
7. (a).
8. By virtue of the independence of forces, both hit the ground at the same time.
9. (c).
10. Since the mass is 1 kg, the force will be equal to the acceleration. Acceleration is constant at 9.8 m/s^2. Velocities each second in meters per second: 9.8, 1.96, 2.93, 3.92; velocity at bottom is 4.427 m/s, which continues.

Chapter 3

1. 0.00365 m.
2. 2.512 m.
3. 0.3759 m; period is 2 s.
4. 1.855 m.
5. 0.19 m/s^2.
6. 0.0195 m.
7. 1.9; same for density.
8. Same reason that unequal masses hit the ground at the same time, as explained in Chap. 1.

9. $P = 2\pi\sqrt{lr^2/GM}$; $P = 2\pi r^{3/2}/\sqrt{GM}$; $P = kr^{3/2}$.
10. About 27.3 days, due to the greatly reduced value of the Earth's surface gravity at the lunar distance. The actual lunar period is about 27.3 days.

Chapter 4

1. e is 0.9962; a is 141.214 AU; b is 12.31 AU.
2. 5.458 years.
3. 541.436 AU.
4. 3.295 years.
5. q is 15.202 radii; Q is 15.269 radii.
6. (c).
7. P is 1.76 years; b is 1.42 AU; e is 0.22253.
8. 1.0549 to 1; 0.46126 to 1.
9. [Graph]
10. [Graph]

Chapter 5

1. a.
2. d.
3. a and c.
4. The bisector is perpendicular to the chord, and points toward the center of force; the arc and chord become unity as the two points of the arc merge.
5. (a) Newton used the term centripetal acceleration before he was prepared to call its cause gravity. (b) Centripetal acceleration acting on a mass requires for its existence, force, which opposed, by Newton's Third Law, is sometimes described as centrifugal force. (c) Centrifugal force is an apparent force, not a real one, that appears to arise as a result of a moving object's inertial tendency to continue in a straight line. (d) It takes a force to move a mass, by Newton's Second Law, owing to the inertial quality inherent in mass.
6. Their accelerations are proportional to their distances from the center.
7. 1/8 to 1; 1/16 to 1.
8. 628 m/s; straight-line, tangent to the radius at point of release; forever until acted upon by an external force.
9. 2.943 m/s^2; 5.86 × 10^{17} N.
10. 1.65 to 1; 0.083 to 1.

Chapter 6

1. (c).
2. 9.66 days.

Appendix: Solutions to Problems

3. 1.91 times Earth's.
4. 10.56 and 0.28, respectively.
5. 1 to 2.3.
6. 1 to 0.79.
7. 125 m or 12.5 cm.
8. 10.8 m; 0.00007 m/s squared; about the same.
9. About 1 to 2.3, or about the same as the answer to Problem 5, since the Moon's velocity around the Sun also about 30 km/s, and its distance from the Sun is about the same as Earth's. Calculating by masses, the ratio is 1 to 2.2.
10. 0.00000343 N.

Chapter 7

1. 1.234×10^{-13}.
2. 1 to 6.263.
3. 1.582.
4. Phobos: 1.847×10^5 km/day; 2.14 km/s. Deimos: 1.1675×10^5 km/day; 1.35 km/s.
5. Phobos: 0.488 m/s^2. Deimos: 0.0778 m/s^2. Ratio: 6.264 to 1.
6. Eight times longer.
7. 1/16; yes.
8. 1/2 as fast. See also the proof of Corollary VI: velocities vary as the inverse square roots of the radii.
9. 3.71 m/s^2.
10. The same. The source of the accelerations are the same.

Chapter 8

1. 18.05 km/s.
2. It is the same.
3. 4.49 years.
4. 4.49 years.
5. 7.99×10^{-4} m/s^2.
6. It is the same.
7. 4.64×10^{15} N.
8. It is the same.
9. 4.762 days.
10. 5.79×10^{18}; kg; 0.02%.

Chapter 9

1. At the equator: 1.957×10^6 kg. At the poles: 1.964×10^6 kg.
2. The equation is Kepler's Third Law.

3. 84.8 min; 7.88 km/s.
4. An increase of 0.98 m/s^2.
5. 3,343 kg/m^3.
6. 3.71 m/s^2; 3.696 m/s^2.
7. Density increases 49.3 times; g increases 13.45 times.
8. (a) 1/5; (b) 1/8; (c) square root of two times faster; (d) one over the square root of 2 less.
9. 11.274 m/s^2; 0.0543 m/s^2.
10. (a) 0.01214 m/s^2; 8.884 days; (b) 318 to 1.

Chapter 10

1. 1 to 0.1165 or 8.5837 to 1.
2. Pluto: 1.305×10^{22} kg; Charon: 1.5203×10^{21} kg.
3. Pluto: 2,042.2 km; Charon: 17,529.2 km; 0.1165 to 1.
4. a. 1.77; b. 16.97.
5. 8.5837 to 1.
6. Pluto: 23.26 km/s; Charon: 199.64 km/s. Ratio is 0.1165 to 1.
7. 222.9 km/s.
8. 1.36166×10^{21} kg.
9. Yes. Each is 3.0351×10^{23}.
10. On Pluto: 2.64892×10^{-4}; on Charon: 2.27375×10^{-3}. Their periods are 6.38522 days, by all methods.

Chapter 11

1. Each is approximately the same (about 7.5×10^{-6}), and approximately independent of the individual masses, since the masses of the planets are dwarfed by the mass of the Sun.
2. About 12% for Uranus and 3.4% for Neptune.
3. 1/22,894.
4. 1/19,411.
5. It should be close or equal to the previous answer.
6. Uranus: 8.6825×10^{25} kg; 1/22,901. Neptune: 1.0245×10^{26} kg; 1/19,414.
7. 14.5; 17.15.
8. 1.242 g/cm^3; 1.1611 g/cm^3. It would not be affected.
9. The mass should be close to or equal to the answer in Problem 6.
10. 3.506×10^{26} kg; 3.177 days.

Appendix: Solutions to Problems   423

Chapter 12

1. 5.753 km/s; 2.266 km/s.
2. 205,029 days; 561.34 years.
3. 3.61 km/s; results are the same as for Problem 1.
4. 5,753.4 m/s; 2,266.6 m/s.
5. 3,610.6 m/s; 1.28×10^{-6} m/s^2; perihelion, 3.25×10^{-6} m/s^2; aphelion, 5.05×10^{-7} m/s^2.
6. It is the same as in the previous problem.
7. Perigee: potential energy is -3.8287×10^{29} J; kinetic energy is 2.7466×10^{29} J; Apogee: potential energy is -1.5084×10^{29} J; kinetic energy is 4.263×10^{28} J. Their sum is the same due to the law of conservation of energy.
8. 5.206 km/s; 3.75 km/s; 111,867.2 days; 306.726 years.
9. 5.55 km/s; 3.72 km/s; 102,937 days; 281.826 years.
10. 4.02×10^{21} kg; perihelion, 77.2 m/s; aphelion, 69.67 m/s.

Chapter 13

1. Circular velocity: 377.587 km/s; perihelion velocity: 533.987 km/s; escape velocity: 533.989 km/s. The comet at perihelion is extremely close to solar system escape velocity.
2. 1.42571×10^{11} J; 7.1286×10^{10} J; 2 to 1; the energy of the comet at perihelion is twice the energy of its orbit in the circle.
3. Hyperbolic; p is 0.96489 AU; 41.338 km/s.
4. [graph]
5. 41.77 km/s.
6. 0.585975 AU; velocities at the respective orbits: Jupiter, 17.07 km/s; Mars, 33.43 km/s; Earth, 41.53 km/s; Venus, 49.03 km/s; Mercury, 67.34 km/s; velocity at perihelion, 54.57 km/s.
7. 7.464×10^{24} J.
8. Simplify the equation after substituting $a(e+1)$ for r at Q, and $a(e-1)$ for r at q.
9. e is 0.43287761; p is 2.9422 AU; q is 2.053 AU; Q is 5.188 AU; perihelion velocity, 18.74 km/s; aphelion velocity, 11.79 km/s.
10. It is the ratio of the reciprocal of their semi-major axes; that is, their relative total orbital energies will be $1/a$ to $1/a'$; Holmes, by 4.93 times.

Chapter 14

1. 1.035×10^6 J.
2. 1,618 m/s.
3. q, 9,376 km; Q, 23,458 km; a, 16,417 km; e, 0.4289; periapsis kick, 417.53 m/s; apoapsis kick, 330 m/s.

4. 1.3×10^6 J.
5. q, 1 AU; Q, 5.2 AU; a, 3.1 AU; e, 0.6774; periapsis kick, 8.79 km/s; apoapsis kick, 5.643 km/s; 2.73 years.
6. a, 0.9565 AU; e, 0.06135; p, 0.95287 AU; P, 341.67 days.
7. e, 0.0521; a, 1,950.2 km; p, 1,944.9 km; P, 128.8 min; periapsis velocity, 1.67 km/s; apoapsis velocity 1.51 km/s.
8. 1.257×10^6 J.
9. 4.942×10^6 J.
10. p, 3,774.5 km; apoapsis velocity, 3,400.46 m/s; apoapsis velocity, 3,336.26 m/s; 5.6724×10^6 J.

Chapter 15

1. 199.7°; 212.1°; 208.5°.
2. 310.8°; 306.3°.
3. 0.033466° per day; 12.22°.
4. 137.4°.
5. Mean daily motion, 0.0000823°; longitude of the ascending node, 144.57°; argument of the perihelion, 311.32°; longitude of the perihelion, 95.89°; mean anomaly, 357.82°; heliocentric longitude 93.71°.
6. Longitude of the ascending node, 131.81°; longitude of the perihelion 39.87°, heliocentric longitude, 330.34°.
7. 80.3°.
8. 102.9°.
9. 77.9°; yes.
10. a, 2279.9 years; P, about 108,864 years; May 1, 09:27:27.6 UT; 37.394 days.

Chapter 16

1. [Drawing].
2. 89.49°.
3. 1.61 AU.
4. 6.2582 rad; 358.6°.
5. 6.0898 rad; 348.92°.
6. 95.88°; 0.97 AU.
7. Close encounter with that asteroid.
8. 0.276 rad; 15.8°.
9. 26.7°.
10. 5.61 AU.

Appendix: Solutions to Problems

Chapter 17

1. ~5.06×10^{-7} m/s² on each side.
2. ~5.06×10^{-7} m/s².
3. New moon, 1.06×10^{-6} m/s²; full moon, 5.94×10^{-7} m/s².
4. ~2.44×10^{-5} m/s².
5. 6.15×10^{-3} m/s².
6. 252.3 times.
7. 2.6×10^{-3} m/s².
8. (a).
9. 3.34×10^{-6} m/s².
10. Perihelion, 6.6×10^{-6} m/s²; aphelion, 1.9×10^{-6} m/s²; the difference is 4.75×10^{-6} m/s².

Chapter 18

1. 3.58 Martian radii.
2. 0.455 radii, or 1,546 km.
3. 1.46 Jovian radii.
4. Roche, ~$1.12(M/\rho)^{1/3}$; meters.
5. 2.35.
6. About 1.77 million kilometers.
7. 4 Earth radii.
8. 1.8 radii.
9. Because, if we assume a coefficient of 1.8, at 1.948 radii it is just outside the Roche limit; it is non-fluid.
10. Loose particles do not have the cohesion of existing bodies.

Chapter 19

1. Solar, 5.81 mm/s²; Earth's pull (toward Sun), 0.18 mm/s²; centripetal 5.99 mm/s².
2. 29.79 km/s.
3. ~500 m.
4. 58,023 km; 326,377 km.
5. Earth's pull, 3.711 mm/s²; Moon's pull, 1.46 mm/s²; centripetal, 2.28 mm/s².
6. L1 ~ 1.082 million kilometers; L2 ~ 1.086 million kilometers.
7. L1 ~ 51.96 million kilometers; L2 ~ 54.22 million kilometers.
8. Each is 2.19×10^{-4} m/s².
9. Sun, 99.9%; Jupiter .1%.
10. Each is 5.932×10^{-3} m/s².

Index

A

Acceleration
  continuous, 12
  due to Earth's gravity, 21, 23
  lunar-induced, 359, 366, 371
  ratio of solar to lunar, 364
  solar-induced, 365
  tidal, 360–362, 364, 365, 370, 374, 375
Alpha Centauri, determination of mass of, 231–234
Amalthea, moon of Jupiter, 141, 190, 391
Angular momentum, 70–72, 239, 254, 256, 304
Angular velocity, in circular and elliptical orbits, 244–245
Aphelion, 62, 65, 66, 77, 81, 241–246, 248, 254, 256, 257, 262–265, 269, 272, 276, 279–284, 300–302, 311, 320, 334, 339, 353
Apoapsis, 65, 310, 311
Apogee, 65, 67, 68, 249–251, 254, 269, 290–295, 297–300, 302, 307, 356
Apogee, increased velocity or "kick" at, 298
Apollo Command and Service Module, 297
Apollo 11 Moon mission, 306–312
Apsidal distance, 69, 70, 246, 250, 288
Argument of the perihelion, 316, 319, 348
Aristotle, 2, 17, 391
Asteroid, 2010 TD 54, 322–327, 342, 343–347
  eccentric anomaly of, 341–343
  elements of, 324
Asteroid, 2007 WD5, 347–354
Astronomical unit, 56, 61, 73, 81, 125, 131, 140, 161, 163, 165, 187, 220, 227–231, 237, 258–260, 265, 271, 276, 279, 280, 283, 333, 364, 389, 390, 392, 399
Attraction of masses, 7–9

B

Barycenter
  of binary system, 199
  of Earth–Moon system, 195, 197–199
Binary system, 155, 193–210, 255, 370, 411, 413
Bound\ orbits, 270, 274–276, 296
Brahe, Tycho, 61, 161, 217

C

Cartesian coordinate system, 329, 330
Center of mass, 12, 64, 141, 155, 180, 193, 215, 253, 357, 411
Centrifugal force, 14, 103–106, 155–157, 165, 169–171, 173, 174, 176, 177, 357, 358, 380, 403
Centripetal force, 13, 83–85, 88–95, 97–99, 101–105, 107, 113, 116, 117, 127–137, 145, 147–151, 158, 160, 181, 209, 219, 360, 401, 406, 408
Ceres, 73–74, 258, 277, 335
Circle, 40, 59, 84, 108, 128, 149, 172, 195, 218, 243, 270, 289, 319, 338, 366
Circular orbit, total energy of, 268
Clarke, Arthur C., 289, 310
Clarke's, Interplanetary Flight, 289, 310
Clerke, Agnes, 144, 287
Comet Lovejoy C/2011 W3, 391–394
Comets
  energies of, 288
  Kreutz group of, 391, 393
Comet Shoemaker-Levy 9, 388–391, 394
Comet West C/1975 V1-a, 327–335
Conic sections, equation for, 74–78
Conservation of energy, law of, 251–257, 267, 269, 271

428   Index

Conservation of mechanical energy,
    33–35, 251
Constant acceleration, constant, 2, 20–23, 43,
    113, 116, 124
Coordinate system, 6, 313, 330
Copernican heliocentric view, 17, 62
Curvilinear motion, 13–15

D
Dark matter, 234–238
Deimos, Martian satellite, 144, 145, 310
Deriving periods with g, 46–53
Deriving velocities with g, 199–203
Deviation from Keplerian proportionality,
    162–163
Differential gravitational forces, 358–360
Differentiation, 32
Directrix, 74, 75
Distance-time-squared relationship, 27,
    33, 97
Dwarf planets, 81, 262, 264, 335

E
Earth
    as not a perfect sphere, 175–176
    orbital velocities of, 366
    translational motion of, 367
Earth–Moon system
    L4 and L5 points of, 413–418
    L1 point of, 411–412
    Roche limit of, 378–384
Earth–Sun system, L1 and L2 points of,
    399–403
Eccentric anomaly, 339–354
Eccentricity, 61, 116, 141, 190, 209, 223, 241,
    269, 290, 315, 339, 368
Ecliptic, 79, 279, 283, 299, 313–335, 348
Ecliptic coordinate system, 313
Effect of Earth's spin on g, 169, 171–175
Elements of the orbit, 305, 306,
    317–319, 324
Ellipse
    empty focus, 64, 81
    parametric equations for, 329
Elliptical orbits
    apogee velocity, 290, 293
    apsidal velocities, 244, 245,
        251–257
    compared with circular, 242–244
    gravitation, 245–246
    perigee velocity, 290, 293

total energy of, 269–270
Energy
    summary of orbital energy relationships,
        274–278
    total in a circular orbit, 268
    total in an elliptical orbit, 269–270
Energy equations
    (*vis viva* equation), 271
    summary of key, 274–278
    used to define orbit to Mars, 303–304
Epoch, 79, 303, 318, 321–325, 328, 331,
    348–350, 352–354
Equal areas in equal times, 60–62, 90, 93,
    239, 248
Equipotential surface, 369
Escape velocity, 12, 273, 276, 287, 296–298
Eugenia, asteroid, 164
Evanescent arc, 72, 93

F
Fictitious circle, 265, 319, 320, 323,
    326, 339
First point of Aries, 314, 319
Flamsteed, John, 135, 136
Focus, 13, 60–62, 64–67, 71, 74, 75, 80, 81, 84,
    87, 96, 117, 185, 194, 239, 243,
    271–273, 275, 284, 300, 315, 321,
    324, 337, 338, 341
Forces
    balance of at L1 point, 402, 407
    in combination, 3–4
    components, 5, 18, 173
Free-falling frame of reference, 6
Freely falling object, 18, 21, 23, 31

G
Galilean equations, 33, 38, 42, 43, 48, 95–98,
    116, 171
Galilei, Galileo, 2, 3, 5, 7, 17–35, 37–39, 41,
    42, 50, 55, 84, 95–97, 110, 114, 117,
    124, 128, 134–136, 290, 310
Galileo's *Dialogue concerning Two New
    Sciences*, 19
Gaussian constant, 258–264, 279, 299–301,
    311, 333
Gauss, Karl Friedrich, 258, 259, 262, 335
Gauss' *Theory of Motion of the Heavenly
    Bodies Moving about the Sun in
    Conic Sections*, 258
Geosynchronous satellite, 142–145, 184,
    279, 353

Gravitational
    acceleration, 10, 22, 41, 102, 107, 157, 169, 200, 220, 245, 358, 397
    constant, 57, 124, 162, 177, 182, 186, 187, 204, 206, 216, 224, 226, 227, 234, 249, 259, 290, 293, 305, 308, 359, 364, 368, 378, 399, 401, 404, 410, 411, 413
    forces, differential as cause for tides, 356, 358–360, 368, 378, 383
    potential energy, 251–253, 270, 369
    well, 253, 268, 299
Gravitation and the laws of motion, 11–13
Gravity, 6, 17, 37, 84, 107, 133, 148, 167–190, 195, 215, 245, 307, 355, 377, 397
Gravity and distance, 9
Great Comet of 1680, 281–288

H

Halley's comet, 78, 81, 278, 281–288, 333, 353
Heliocentric ecliptic longitude, 324
Heliocentric orbital velocities, using Gaussian constant to find, 260–261
Heliocentric periods, using Gaussian constant to find, 259–260
HM Cancri, determining the mass of, 233
Hohmann transfer, 290–293, 310
Hooke, Robert, 52
Hooke's law, 52, 56
Huygens, Christiaan, 37–57, 103, 109–111, 113, 115, 117
Huygens' The Pendulum Clock, 38, 42, 45, 56, 115
Hyperbola, 76, 77, 84, 267, 274–276

I

Impetus, 2
Inclination, 3, 19, 39, 56, 79, 108, 279, 283, 311, 313, 315–317, 319, 325, 328, 333, 342, 347, 348, 367, 368, 372
Inclined plane experiment, 17, 19
Increasing distances from Earth, 21, 384
Independence of forces, 5, 38
Inertia, 2, 3, 7, 11, 13–15, 51, 84–87, 89, 93, 99, 102, 105, 124, 150, 155, 157, 169, 357, 358
Inertial frames of reference, 5–7
Injection velocity, 290, 293, 300, 304–308
Instantaneous velocity, 27–31
Inverse square law, Jupiter's satellites as a test for, 135–138

Inverse square relationship, 9, 131, 134
Isochronism, 38, 42

J

Jones, Harold Spencer, 214, 237
Julian century, 321
Julian date, 331, 335
Julian day, 320–322, 327
Jupiter
    encounter with Comet Shoemaker-Levy 9, 388–391
    moons of, 134–137, 148

K

Keplerian mass ratio, 217, 218, 222, 237
Keplerian motion, 70
Keplerian proportionality constant, 138–140, 145, 162–163
Kepler, Johannes, 339
Kepler's Astronomia Nova, 61
Kepler's auxiliary circle, 339–341
Kepler's equation, 259, 341, 348, 351
Kepler's First Law, 62
Kepler's Harmonic Law, 229–231
Kepler's Laws of Areas, 70–72, 241, 245
Kepler's Laws of Planetary Motion, 32, 59–82, 84, 127–145
Kepler's Second Law, 70, 72, 90, 136, 245, 248, 251, 269, 281
Kepler's Third Law
    as modified by Newton, 158–161, 215
    reduction to simplest form, 228–230
    using to find mass, 228
Kinetic energy, 33, 34, 39, 52, 56, 251–255, 265, 268–270, 272, 273, 287, 304, 305, 307
Kreutz group, 391, 393
Kuiper Belt, 210, 264, 272
Kuiper Belt Objects (KBOs), 225

L

Lagrange, Joseph-Louis, 398
Lagrange's Analytical Mechanics, 398
Lagrangian points, 397–418
L1 and L2 heliocentric velocities, 408–409
Laws of conservation of energy and momentum, deriving apsidal velocities from, 251–257
Leibniz, Gottfried Wilhelm, 31

Limit, 8, 9, 18, 25, 29, 31–33, 83, 87, 89, 93, 97, 163, 239, 368, 378–388, 390–394
Line of force, 8, 9
Line of nodes, 316, 328, 333
Longitude of the ascending node, 316, 317, 319, 320, 335, 348
Longitude of the perihelion, 313, 315, 317, 319, 320, 328, 329, 335, 345
Lunar g, 179, 185–186
Lunar mission
    designing an orbit for, 293–298
    travel time for, 293

M

Major axis, 61, 64–66, 242, 280, 282, 283, 286, 296, 297, 299
Mars
    discovery of satellites of, 144
    encounter with asteroid 2007 WD5, 347–354
    energy equations to define orbit to, 303–304
    orbital velocities of, 246–247
    planning a mission to, 299–300
    velocity needed for trip to, 300–303
Mass, 1–15, 41, 63, 83, 107, 128, 147, 167, 193, 213, 239, 267, 290, 355, 377, 399
Mass of Jupiter
    modern determination of, 190
    Newton's calculation of, 220–223
    using Amalthea, 223, 225
Mass of our galaxy, 234–238
Mass of Pluto, 213–214
Mass of Quaoar, calculation of, 226
Mean anomaly, 320, 324–327, 338, 341, 342, 346–354
Mean heliocentric longitude, 319, 320, 322, 324, 325, 328
Mean planet, 320, 324
Method of differentials, 32
Milky Way galaxy, determination of mass, 234, 235
Modified Gaussian constant, 261, 263, 279
Moon
    fall distance, 112–114, 120, 171
    lifting force of, 367–375
Moons of Jupiter and Kepler's Third Law, 134–137
Moons of Saturn and Kepler's Third Law, 134–135, 137
Motion
    in elliptical orbits, 239–265
    uniform, 6, 21, 25, 26, 33, 85, 92
    uniform circular, 51, 52, 54, 55, 59, 92, 104, 116
    uniformly accelerated, 23–26, 95

N

NASA/JPL Solar System Dynamics website, 335
Near Earth Asteroids (NEAs), 278
Near Earth Objects (NEOs), 128, 157, 353
Neptune, 35, 62, 63, 106, 190, 209, 210, 213, 214, 225, 236–238, 264, 277, 321, 335, 387, 388, 394
Newcomb, Simon, 144, 237
Newton, Isaac, 1, 13, 17, 43, 83, 84, 87, 97, 104, 107, 108, 135, 161, 171, 225, 357
Newton's First Law, 2, 5, 84–87, 93, 124, 171, 173, 180, 403, 406
Newton's *Mathematical Principles of Natural Philosophy* (*Principia*), 13, 31, 83
Newton's Moon test, 107–125
Newton's Second Law, 5, 6, 11, 12, 14, 88, 100, 103, 110, 148–150, 154, 155, 159, 167, 171, 179, 203, 360, 362, 373
Newton's System of the World, 13, 87
Newton's Third Law, 11, 151, 171, 203, 245
Newton's Universal Law of Gravitation, 9–10, 152, 153

O

Oberon, satellite of Uranus, 236–237
Observed g and true g, 170
Orbit
    elements of, 305, 306, 311, 313–318, 324
    synchronous, 373, 374
Orbital energy, 253, 269–271, 288, 303–305
Orbital energy relationships, summary of, 274–278
Osculating elements, 79

P

Pan, moon of Saturn, 134–135, 384–388
Parabola, 65, 76, 77, 84, 267, 272–277, 282
Parabolic velocity (escape velocity), 276, 277
Parallax, 230
Parameter, 48, 75–77, 81, 82, 114, 184, 186–188, 203, 250, 263, 273, 275, 282–284, 287, 288, 296, 304–312, 329, 338, 387, 388

Index 431

Parking orbit, 295, 303, 304, 306–308
Pendulum, 3, 37–57, 101, 109, 129, 170
Pendulum clock, 38–40, 42, 44, 45, 56, 103, 115
Pendulum in space, 122–125
Periapsis, 64, 310, 311, 335
Perigee, increased velocity or "kick" at, 365
Perihelion, 59, 142, 241, 269, 300, 313, 337, 391
Period of a pendulum, 38, 41, 43, 46, 48, 52–54, 57, 123
Phobos, Martian satellite, 145, 394
Planetary radius and densities effect on g, 167–168
Planets
  coefficients for mean longitude, 321
  conjunctions, 334
Pluto and Charon, orbits of, 209–210
Potential energy, 33–35, 52, 251–255, 265, 268–270, 272–274, 304, 369
Principia
  Lemma VII, Book I, 93
  Phenomena I, Book III, 147
  Phenomena II, Book III, 137
  Proposition I, Book III, 137–138
  Proposition II, Book III, 138
  Proposition IV, Book I, 90, 105, 128–131, 148
  Proposition IV, Book III, 43, 108
  Proposition IV, Corollaries, Book I, 95, 128
  Proposition VIII, Book III, 148, 218, 220
  Proposition VIII, Corollary I, Book III, 218
  Proposition VIII, Corollary II, Book III, 219
  Proposition XI, Book I, 117
  Proposition XIX, Book I, 171
  Proposition XX, Book III, 171
  Proposition XXIV, Book III, 357
  Rules of Reasoning in Philosophy, Book III, 134
Principle of equivalence, 11, 41
Projectile motion, 13, 14
Proportionality constant, 21, 33, 138–143, 145, 150, 162, 163, 262

R
Radians, 47, 53, 70, 101, 103, 104, 119–121, 173, 179, 182, 220, 243, 340, 342, 344, 347, 349–354, 399, 402
Radius vector, 64, 70, 75, 77, 239, 240, 242, 246, 254–256, 271, 279, 280, 291, 304, 309, 338, 340, 353, 354
Reduced mass, 203, 211, 257
Rees, Martin, 189
Relative motion, 5–7, 85, 193

Restraining force, 51
Restricted three body problem, 397
Resultant force, 4, 5, 18
Roche coefficient, 383, 388, 390, 392, 394
Roche, Edouard, 378
Roche limit, 378–388, 390–394
  of Earth–Moon system, 378–384
  for Saturn, 385

S
Sagitta, 89, 90, 118, 119
Saturn, Roche limit of, 384–388
Seconds pendulum, 42–45, 56, 57, 113
Sedna, dwarf planet, 81, 335
Semi-latus rectum, 250, 275, 282
Semi-major axis, 61, 104, 141, 149, 199, 215, 241, 269, 293, 315, 340, 385, 418
Semi-minor axis, 61, 67, 69, 77, 81, 241, 283, 284, 288, 329, 330
Simple harmonic motion, 50–53, 55, 56
Simple harmonic oscillator, 51
Sirius, determination of mass, 230–231
Solar g, 179, 187–190
Solar mass, as unit, 163, 216, 218, 222, 225, 227–229, 231, 233–237, 258–260, 300, 409
Solar system bodies, energy relationships of, 276
Spring displacement, 52
Square of the period of an orbit, 61, 63, 81, 128, 161, 219
Straight-line motion, 2–5, 13, 14, 169, 184
Strict proportionality of Kepler's Third Law, 161–162
Summary of selected Proposition IV, Book I corollaries, 145, 218
Synchronous orbit, 290, 291, 373, 374
Syzygy, 371

T
Three-dimensional space, 5, 6
Tidal accelerations, 360–362, 364, 365, 370, 374, 375
Tidal distortion, 368, 371, 373
Tidal forces, general equation for, 78, 356, 360–365, 367–370, 372–374, 378–383, 386, 391, 393
Tides, 83, 229, 355–375, 377–394
Trans-Earth injection, 307
Trans-lunar injection (TLI), 307
Trans-Neptunian objects, 214
Triton, satellite of Neptune, 190, 236–238
True anomaly, 322, 324, 334, 337–354

U

United States Naval Observatory, 321
Uranus, 62, 63, 164, 209, 210, 236–238, 277, 289, 321, 387, 388

V

Velocity(ies)
    angular, 71, 104, 120, 121, 182, 243–245, 399, 411–413, 415
    of an artificial satellite, 249–251
    average, 22, 24, 28–30, 339
    around center of mass, 202
    change with time, 294, 311
    constant, 5, 13, 22, 90, 98, 99, 253
    of Earth and Moon around center of mass, 204–205
    heliocentric at L1 and L2 points, 408–409
    using to find mass, 209
Venus, 56, 62, 63, 116, 125, 164, 219–222, 277, 288, 289, 321, 334, 398, 417

Vernal equinox, 313, 314, 316–320, 324, 326, 328–330, 345, 350, 352, 353
Versed sine, 89, 90, 93, 94, 96, 105, 108, 172, 173
Vis viva equation, 271, 273, 274, 283, 288, 305

W

Weight, 7, 12, 13, 18, 19, 87, 154–156, 171, 174, 176, 187, 194, 196, 218, 219, 298, 360, 373
Whiston, William, 108
Work, 13, 19, 33, 35, 38, 50, 52, 61, 78, 81, 97, 108, 111, 115, 116, 129, 132, 136, 140, 143, 148, 153, 171, 173, 176, 196, 201, 208, 211, 220, 227, 234, 251–253, 258, 262, 272, 274, 281, 283, 300, 305, 339, 341, 348, 369, 386, 398, 399